TECHNOLOGY AND THE WEALTH OF NATIONS

The Dynamics of Constructed Advantage

Edited by
Dominique Foray and Christopher Freeman

Published in association with the
Organisation for Economic Co-operation and Development

Pinter Publishers
London and New York

Distributed exclusively in the United States and Canada by St. Martin's Press

25 Floral Street, Covent Garden, London, WC2E 9DS, United Kingdom

First published in 1993

Distributed exclusively in the USA and Canada by St. Martin's Press, Inc., 175 Fifth Avenue, New York, NY 10010, USA

British Library Cataloguing in Publication Data

A CIP catalogue record for this book is available from the British Library.

ISBN 1 85567 082 8

Library of Congress Cataloging-in-Publication Data

Technology and the wealth of nations: the dynamics of constructed advantage / edited by D. Foray and C. Freeman.
p. cm.
Includes bibliographical references and index.
ISBN 1-85567-082-8
1. Technological innovations – Economic aspects. 2. Technology – Economic aspects. I. Foray, Dominique. II. Freeman, Christopher.
HC79.T4T4396 1993
338′.064 – dc20 92-27953
CIP

Typeset by Best-set Typesetter Ltd, Hong Kong
Printed and bound in Great Britain by
Biddles Ltd, Guildford and King's Lynn

Contents

Preface

In editing this selection of papers we have had to make some difficult choices. There were many good papers presented to the OECD Technology/Economy Programme (TEP) Conference in Paris in June 1990 but we were constrained to select only a few of these. In making this selection our objective was to present a sequence of chapters, representing a coherent view of the new developments in the economics of technical change, as outlined in the Introduction to the book and the shorter introductions to the four Parts.

However, we did not confine ourselves strictly to the Conference Papers. In order to clarify the arguments and to strengthen the coherence of the book as a whole, we not only selected and arranged them in a definite sequence, we also invited the cooperation of the authors in revising the papers for this particular publication. We were fortunate enough to receive their full cooperation extending in some cases to the preparation of entirely new papers. We would like to thank them most warmly for their assistance, as well as Robert Boyer, Arndt Sorge and Luc Soete for their contributions. We would also like to thank the OECD and the French Ministries of Research and Technology and of Industry for their generous assistance to us in the preparation of the French and English editions of the book.

The OECD 1990 Conference in June 1990 was entitled: 'Technology and Competitiveness', but as we selected and edited the papers we became more and more aware that our theme had become much broader than this. We therefore chose a new title: 'Technology and the Wealth of Nations'. However, our attention has been drawn to a paper by Robert Ayres ('Technological Forecasting and Social Change', vol. 33, pp. 189–201, 1988) with an almost identical title. Rather than change our title we chose to keep it, with acknowledgements to Robert Ayres, as we believe this title best reflects the content of our book.

Dominique Foray
Chris Freeman

Foreword

In the course of the 1980s, growth did not attain the levels expected from the potential offered by new technologies.

Governments, acting individually, sought to attenuate problems arising from the play of market forces under conditions of rapid structural change.

This led the OECD to develop a strategy for collective action known as 'structural adjustment'.

This approach has paid off.

The basis for this success lies in the idea of a three-fold convergence:

- convergence of objectives (member countries have to solve common problems in order to unfetter their growth potential);

- convergence of interests (potential growth is more balanced at international level than real growth);

- convergence of solutions (the most effective remedies are also those that are most generally applicable).

These 'structural adjustment' policies have encouraged globalisation which is one of the most striking current economic phenomena: markets and productive sectors have become transnational; economic actors are tightly interconnected. Enterprises have found themselves exposed to the full force of international competition, while barriers at the frontiers are disappearing and direct foreign investments are increasing.

Strong differences between national systems have appeared and have led to a resurgence of the idea of the competitivity of nations over and beyond that of the international competitivity of enterprises. The idea of diverging pathways for growth is starting to resurface alongside investigation of the determinants of the potential for growth (and no longer simply the means of achieving growth).

The search for competitive advantages is replacing that of exploiting comparative advantages.

These developments fall within the domain of economic facts, their theoretical interpretation and their impact on policy.

Economic trends: growth, when taking place, is accompanied by serious imbalances which are perceived by many as being inequitable:

- At international level: the gulf between developed and developing countries is widening as is that between the latter countries and the 'dynamic Asian economies';

- At regional level: factors of production are tending to accumulate in the most favoured locations; many countries are witnessing reversal of the trend towards a reduction of regional development disparities;

- At the social level: the status of less skilled workers is declining in relative terms (greater risk and duration of unemployment, lower pay);

- At enterprise level: in many sectors, the balance of competition is threatened; the globalisation of markets and of enterprise strategies often negate national measures to curb dominant positions.

Economic theory: in the 1980s there was a rapprochement between micro- and macro-economic theory, economics and management sciences, as well as sociology and political economics. The debate over the 'productivity paradox' relaunched thinking about growth. Unlike the dominant theory of the last 30 years, the new theory stresses the endogenous nature of technical progress and emphasises the correlative importance of the phenomenon of increasing returns and cumulative advantages.

Policy concerns: although the pace of reform within the framework of structural adjustment strategies continues, a major obstacle now starting to emerge is the lack of rules of the game for the promotion of technological development.

One of the striking facts to emerge from examination of the arguments put forward to justify the correct orientation of government economic policy is that international competitiveness has become the unifying theme. The translates into programmes of action aimed at strengthening development potential by targetting the linkage between technology and human resources.

Whereas the primary objective of structural adjustment was to restore the system of incentives governing the use of productive resources, the policies arrayed under the banner of competitiveness place emphasis on the actual creation of resources.

The scope of such policies is vast. In most cases, it includes not only encouraging industrial firms to adapt to new technology, but also the consolidation of the regional bases for economic development, the strengthening of the web of small and medium-sized enterprises, and the development of activities deemed to be of strategic importance for national economic growth.

It is hard to assess their compatibility at international level. In particular, the criterion of 'three-fold convergence' does not apply to them in a simple way. Clearly, the convergence of objectives, solutions and interests does not easily take place. The diversity of national approaches does not imply their complementarity. Similarly, convergence of policies could lead to neglect of certain sources of innovation and growth.

- The perception that a country has of the risks it should avoid, and the opportunities it should grasp, is largely influenced by the verdict it reaches with regard to its 'chronic problems' and 'major strong points'. Of crucial importance in this respect is the opinion it has regarding whether or not its economic performance is commensurate with that of its scientific and technical potential.

– The 'acquired' structural features determine what a country could reasonably expect to 'construct'.

– Whereas the action taken by many of them amounts in effect to a challenge to the normative theory of comparative advantage, countries find themselves lacking an alternative conceptual framework that would allow them to adopt a common outlook towards the changes, in a context of accelerating internationalisation, in the links between technology, the competitiveness of firms and national economic performance.

Ensuring the compatibility of national policies towards competitiveness and mastering the effects of globalisation of corporate strategies within the framework of a strategy for 'harmonising growth trajectories', aimed at fostering the conditions for 'sustainable globalisation', are the challenges for the 1990s.

For the first time one volume brings together analyses and insights from senior researchers from various disciplines and countries which advance our understanding of how technology overturns traditional beliefs about the foundations of the wealth of nations. The cross-fertilisation of ideas and the coherence of approaches provide the theoretical basis and tools for understanding the 'technology dimension'.

Robert Chabbal
Jean Guinet
Paris, 1991, OECD

General introduction

Dominique Foray

> That the advance of knowledge lies at the core of modern growth process is more than an inference from the growth accounts. It is a perception enforced by well over a century of common experience. Economists have therefore applied themselves to learning more about the ways that practical knowledge is gained and exploited. A new outlook has developed and spread. It is not yet well defined. (Moses Abramovitz, *Thinking About Growth*, Cambridge University Press, 1989)

The mass of information and stylized facts available on the economics of technical change and industrial innovation has grown immense. This abundance reflects the importance of these phenomena to the understanding of economic growth. There remains nonetheless a great deal of work to be done in order to distil this unwieldy quantity of information down to a few analytical principles that would enable us to understand the factors of technological competitiveness and the links between scientific and technical dynamics and the wealth of nations.

Undeniable progress has been achieved in the analytical work carried out over the past few years, especially in the areas of the economics of innovation and the economics of research and development.[1] Certain recent syntheses have been able to discern the principal theoretical axes underlying this literature.[2] In addition, the OECD's TEP programme has played an essential catalytic role by encouraging a consideration of the *positive accumulation of knowledge* which should henceforth be realized in order to constitute the *core of an alternative theory of technical change*. In the past this accumulation has operated negatively: the successive results of historiography, the economics of innovation and the economics of R&D, have accumulated as so many discrete, external critiques of the standard theory of technical progress.

How then does one precipitate such an accumulation, so as to illuminate the links between technology and competitiveness? I believe the key to this question lies in the problem of demarcating the *avenues by which certain economic conditions may influence the future dimensions of technological knowledge and the process of innovation*. We know that two linkages are implicated here: i) technological diffusion and interdependence corresponding to the 'mediating process through which past choices among productive techniques can indirectly influence the future *global* drift of technological change' (David, 1975); and ii) research activity in basic and industrial science, as a creation procedure associated to incentive systems. These two mechanisms of technical progress have always been treated separately, whether in the descriptive analyses of the factors of technical progress[3] or in growth models.[4] This separation is basically a result of the following different levels of analysis involved.[5]

- The first level covers the dynamic properties ('localized positive feedbacks') of technological diffusion processes. Diffusion represents the locus of exploitation of the development power of a technology and of the decrease of uncertainty regarding its actual capabilities and performances. The (indirect or direct) relationships of interdependence between agents constitute here the essential structure of the process of positive feedbacks. In fact, structural interdependence imbues the decentralized technical choice, which is essentially static and microeconomic, with macroscopic and dynamic dimensions.
- The second level corresponds to the mechanisms of incentives and organization that harmonize the general characteristics of an investment activity (research) with the specific characteristics of the goods concerned (knowledge). Thus this second level illuminates the strong organizational nature of technological competitiveness.

Clearly the opposition here elaborated does not refer to any dichotomy between science and technology. Interdependence on the one hand, incentives on the other, concern both domains of science and technology.

The chapters in this book develop the argument that the articulation between these two levels represents the key to understanding the factors of technological competitiveness. The structure of the book also suggests that the tension between diversity and uniformity does not correspond to any dichotomy between technology and institution (as is argued clearly by Sorge, this volume). It stems rather from the *encounter* between the processes of technological diffusion and interdependence (which induce standardization: Antonelli, David and Callon, this volume) *with* incentives and organization structures (which induce diversity: Hollingsworth, Watanabe, Marsden, Sabel and Piore, this volume).

While these two levels may be kept strictly apart in theory, they are far from easy to separate in the real world. Two sequences are however distinguishable:

-sequence 1: Interdependence → incentive → interdependence

Faced with developmental interdependences (structural tensions), some incentives and entrepreneurial organizations (concerted activities) are necessary to generate the cumulative elements (sequence of complementarities) inherent in such 'development blocks' (Dahmen, 1989). As suggested by Mowery and Rosenberg (1989) in describing 'Taylor and White's development of high-speed steel (1898), and the subsequent development, in the 1920s, of sintered tungsten carbide', the majority of scientific discoveries and therefore of incentives and innovative organizations take form through the very process of technological diffusion and developmental interdependence.

-sequence 2: Incentive → interdependence → incentive

Conversely, in the case of incentives to create a technical project, some structural interdependences are required to diffuse the generated knowledge. As suggested by Rosenberg and Vicenti (1986), the 'Tubular Bridge' (a technical project performed by Robert Stephenson and William Fairbairn in the 1850s) 'itself had only a limited useful life in subsequent years, and in this sense merits not much more than a footnote as far as the history of bridge building is concerned'. But the project was magnified by the diffusion of experience which 'provided knowledge of great value to a wide array of productive activities' (*ibid.*).

In order to achieve a global understanding of the links between technology and economics, one must eschew a choice between these two domains. This would lead inevitably to an incomplete view of technological competitiveness. Rather, one should identify the points of contact and the principles of articulation between these two conceptual orders. The goal of this book is therefore to demonstrate how the logic of diffusion and interdependence and the systems of incentive and organization *achieve coherence* for certain fleeting moments in history, in limited geographical areas, thereby enabling technology to make a decisive, dazzling contribution to the wealth of nations.[6]

Increasing returns, interdependence and technical progress

When one considers that technologies are definitively completed and constituted as of the commencement of their diffusion, the analysis of technical choices will logically indicate the existence of diminishing returns. In other words, the outcome of a sequence of choices will be determined by an increase in exploitation costs, which in turn results from the progressive scarcity of inputs, or from negative externalities (pollution) that must be compensated for, or even from the growing difficulty of access to the sites of exploitation. Thus, in the case of competing technologies, the selection of one of two alternatives will generate 'negative feedbacks' which will, in a subsequent period, increase the probability of choosing the other technology. Under decreasing returns, a particular historical choice among techniques does not rule the future. In such cases, only the figure of the inventor will enable us to analytically account for the creative process. Faced with this conception, a distinct line of thought developed which proposed an analysis of technical choice based on the principle of increasing returns.

The basic difference between the above two representations resides in their distinct theoretical conceptions of technology. Under the standard theory, this conception is static: technology is fixed once and for all. However, according to the new conception, the diffusion and repeated use of a given technology will lead to the improvement of its design and implementation, while at the same time facilitating the lowering of production costs, so that the second artefact will in fact be better than the first, and the third will be better than the second, and so on. Thus, the diffusion process itself is fundamentally dynamic and will generate, via a series of mechanisms, the continual improvement of the given technology. The choice of a particular technology, rather than generating 'negative feedbacks', leads instead to 'positive feedbacks', so that the decision to select a technology increases the chances that it will once again be chosen in the future. Under these conditions, 'choices of technique become the link through which prevailing economic conditions may influence the future dimensions of technological knowledge' (David, 1975).

Technical choices, positive feedbacks and the historical process

A sequence of technical choices can be represented formally as a stochastic process possessing the attribute of 'path-dependency'. In a broad sense, path-dependency reveals the extreme sensitivity of the motion and evolution of a technological system upon intervening events occurring early on the path. In the domains of scientific and technical dynamics, a recognition of the path-dependent nature of the diffusion processes allows us to raise the level of our representations to meet the requirements and observations of the historians. L. Puiseux (1989), in particular, notes with respect to the succession of choices involved in French nuclear technology, that certain bifurcations provoke the definitive orientation of the trajectory into a given basin of attraction. It is therefore a case of 'essentially historic' dynamic processes, in which the randomness of a path diminishes as decisions are taken. Thus, certain historians have shown us that the first technical choices taken assume an extraordinary importance and are never forgotten in the course of the dynamics.

The combustion engine achieved precedence over the steam engine as a result of its adoption at the dawn of the automotive industry.

> Theoretically, steam could certainly have presented an attractive alternative to the combustion engine had quantities of ingenuity and capital been spent upon it comparable to those spent, from the automobile's inception, in making the combustion engine the technological marvel it is today. (CNRS, 1990)

In the domain of increasing returns, it is history (and not science or technique) that decides that one technology becomes more efficient than another. Thus, the above analytical reversal (from negative to positive feedbacks) gives the economics of technical change its necessary theoretical foundation – that of historical evolutionary economics.

The case of the steam engine, surpassed and forgotten as a result of the success of the combustion engine ('a technical aberration', *ibid.*) suggests a reconsideration of the notion of 'technological superiority'. There is always a technology which is more efficient than another at any given moment of time. But we must be aware that, in a 'counterfactual world', an alternative technology, had it been adopted, diffused and developed, might have eventually provided superior performance.

As L. Puiseux (1989) observes: 'the first steps were of a decisive and indelible importance . . . there are tracks, branchings, genealogies, that is to say, a historicity'. On this point I should like to cite the subtitle of an article in the French newspaper *Le Monde*, which appears to have been drafted with the sole objective of propounding the herein thesis: 'The steam engine could have supplanted the combustion engine, but the whims of history decided otherwise.' It is important to grasp this historicity conceptually. It obliges an evolution in our analytical representation of technical change.

Localized technical change as an 'agent of irreversibility'

By indicating that technical change does not correspond to a complete shift of the productive function but consists rather in the shifting of one point in this function, Atkinson and Stiglitz (1969) facilitate a first theoretical evolution. Learning is thus associated with a specific process (one point in the production function), so that in allocating resources a firm must not lose sight of the possibility that any subsequent technological change (to $t + 1$) will wipe out the accrued gains from localized learning associated with the technology chosen at t. The firm cannot opt for a 'transitional' change of technology in the belief that by doing so it will be able to keep up with all the successive changes in relative price structure, since localized learning makes it likely that it will have to carry on with the 'transitional' technology after the re-establishment at $t + 1$ of the price structure that characterized $t - 1$. Thus, the concept of localized technical change acknowledges the importance of history.[7]

However, the concept of localized technical change represents only one element of the path-dependent mechanism. In fact, path-dependency must be understood as an attribute of the collective process of diffusion, not of the individual choice of each agent.

Path-dependency and the collective process of diffusion

Positive externalities and indivisibilities are agents of path-dependency in that they determine the existence of localized positive feedbacks associated with an adoptive choice: the choice of one option at t_0 increases the chances that the same option will be selected at t_1 (Arthur, 1989). Under the conditions of unlimited increasing returns, 'the process eventually must lock-in to a monopoly of the market by one technology; a point is reached after which every agent regardless of inherent preferences, will select the same technology' (David and Greenstein, 1990). However, the designation of the victor (the standard) will depend totally on the first random choices.[8]

For example, Robin Cowan (1990) has identified the event which rapidly and definitively led the nuclear technology system to a 'lock-in' situation (to the so-called 'American' technology, based on pressurized-water reactors). This technology, specialists today agree, is 'inferior' to certain others. This 'historical accident' resulted from the post-war military purchase order addressed to Westinghouse and General Electric for the construction of a submarine reactor. Subsequently, when it became necessary (after the first Soviet nuclear detonation in 1949) rapidly to develop a civil nuclear programme, it was the familiar technology that was chosen, even though its suitability for military applications could not in any way guarantee a successful transfer to other areas.[9] This event is 'accidental' from the perspective of the dynamic process, in that it propelled the process definitively towards one of its basins of attraction. However, it was not accidental from a historical perspective, under which the decision was clearly coherent.

Thus, as suggested above, path-dependency is an attribute of the collective process of diffusion; not of the individual choice of each agent. Two systems which are very close to one another at t_0 (for instance, the US and Canadian nuclear systems in 1949) will diverge inexorably if they are placed in two different basins of attraction, as when an event (a seed), appearing at t_{-1}, had very early on separated the destinies of the two systems.

Decentralized choices and interdependence between agents

The model so far examined is a particular case of a more generic model of social interactions, in which processes of collective behaviour possess the attribute of path-dependency. The understanding of the emergence and diffusion of social norms, technological standards or institutional rules requires the exploration of the structure of interdependence between agents having to make sequential choices, under conditions of localized positive feedbacks.

Structures of technological interdependence are various and manifest themselves via the following different forms of externalities (Antonelli, Callon, David, this volume):

- First of all, there are those that reveal themselves in network externalities. These latter generate phenomena of propagation of behaviour and strategies, without need for 'physical materiality' for the manifestation of positive feedback effects.
- Next, there are the interdependences that manifest in the cumulative nature of technological evolution (i.e. in the existence of a technological trajectory). The sources of positive feedbacks are mainly in the area of production conditions (economies of scale) as in that of utilization modalities (learning by using).
- Finally, there are interdependences manifested in external pecuniary economies à la Scitovsky (1954), which generate phenomena of investment coordination and sequences of technological complementarity. This brings us to the approach of 'development blocks' which refers to a 'sequence of complementarities which by way of a series of structural tensions, i.e. disequilibria, may result in a balanced situation' (Dahmen, 1989).

In this volume, David analyzes the operation of such interdependence structures in the propagation of a decentralized choice. He stresses in particular the 'locally bounded' nature of the positive feedbacks which impinge upon user's choice: an agent formulates his decision with respect to the sole 'reference group' of importance to him in terms of maximizing his own utility function, from the perspective of positive externalities. However, reference groups can be disjoint or non-disjoint, such that intersecting networks of local relationships, allowing for the transmission of a decentralized choice, either exist or do not exist. In the case of non-disjoint reference groups, interdependence reminds one of the game of dominoes where the pieces, placed on end, form a closed figure: all the pieces inevitably fall if a single one of them falls. The system of potential adopters is then qualified as 'super-critical'. In such a system, a single initial impulse is sufficient to elicit the emergence of a unique standard. In other cases however the reference groups within which agents make decisions are disjoint, so that several standards can coexist. The coexistence of multiple standards is then a manifestation of the sub-criticality of the system. Thus, positive feedbacks are limited to subsets of the entire user community and a particular standard fails to percolate through the entire structure. A decentralized choice can therefore be more or less pervasive. *Whether it is or not depends on the structure of social, technical and economic networks pre-dating the dominance of a particular technology or reference. It also depends on the capacity of the given technology to modify, through its very diffusion, the system's structures of interdependence.*

The need for new analytical and political tools

We are witnessing today a sort of general rehabilitation (i.e. trans-disciplinary) of the search for causality (otherwise referred to as that of the 'genetic-causal moment') (Cowan and Rizzo, 1991).[10] In the context of this novel intellectual climate, two preliminary results may be described. First of all, in following the above perspectives economic analysis will have to deal with new exigencies.[11] Second, the scope of policy under conditions of 'positive feedbacks' will have to be updated.

Historical economics and network analysis

First, the study of 'the historical origins of' has become an indispensable prerequisite for understanding the dynamic processes of economic change. In our book, several papers (Orsenigo, Chiaramonte and Dosi, David) argue that narrative history – timings and circumstances – matters because of the durable impact of events. Decisions and individual behaviours have to be explored too, because of their possible influence on the long-run course of the technological dynamics. Thus, the approach in terms of autonomous system is significant (Foray and Garrouste, 1991). Similarly important are the analyses (cognitive models, mentality studies) which enable us to understand the visions of the future under which decisions are taken (Swann, 1992).

Second, a major research perspective now appears, dealing with the necessity of formalizing the structural interdependence from both static and dynamical points of view.

Policy scope: mastering the global dynamics of technical choices and interdependence

By investigating how choices are made, certain researchers today are seeking to formalize the optimal behaviour of agents having to choose between 'the exploitation of high-payoff actions that have been undertaken many times and are therefore well understood, and exploration of seldom-tried actions that potentially may have higher average payoff' (Arthur, 1991). As the agent learns progressively to resolve this dilemma, the context of the dilemma evolves in response to the particular decisions taken. In each period the policy-maker must choose between the adoption of a known technology, which will enable him to maximize immediate returns, and experimentation with a new technology. Here we recognize the opposition of a principle of static efficiency to one of dynamic efficiency (Klein, 1977), or as is formulated by E. Dahmen (1989), between static competitiveness and dynamical development power. Cowan (1991) shows how different rationales are possible: the short-sighted user will choose the better known technology, aiming to achieve maximal immediate returns. The selection of the lesser known technology (experimentation) would diminish incertitude and therefore clarify the choices for subsequent adopters, but this result would be of no use whatsoever for this particular user. The public policy-maker, on the other hand, will prefer that the better technology prevails and therefore will select the lesser known technology, thereby favoring experimentation, under the constraints of maximization of the expected value of the overall adoptive process.

The crucial question for the policy-maker is how to pass from the phase of diversity to the phase of selection and standardization. It is necessary to identify the moment where it is advisable to change the learning system (the regime of knowledge accumulation). There is a certain learning resulting from diversity, arising from an evaluation of the respective qualities of each alternative and leading to the selection of the 'best' solution. There is another learning from uniformity, which involves the improvement of performance of a chosen alternative, thanks to the identification of 'empirical irregularities that point to underlying structural conditions deserving further investigation' (David and Rothwell, 1991).

It must be emphasized that under conditions of 'positive feedbacks', public authorities wishing to direct the process in a given direction may only have the briefest of time-periods in which to act. These windows of opportunity will often coincide with a great uncertainty as to the respective qualities of the different alternatives. Thus, 'the study of history, along with that of underlying structural conditions that define the location of "basins of attraction" and the "watershed lines" (separatrixes) between them, would therefore be an integral part of policy analysis, inasmuch as it would help to anticipate and identify such propitious moments for action' (David, this volume).[12]

The second policy problem at the microeconomic level returns us to the need to be able to recognize *ex ante* the requirement of connectivity of an economic system, i.e. the critical value of percolation which must be attained according to the desired scope of dynamic implication for a given decentralized choice. Some *ex ante* policies of increasing interdependence will be necessary, in order to enable a unique standard to percolate through the entire structure. The creation of interdependence covers a wide range of actions, from the implementation of intersectoral systems of information and communication to the development of technological infrastructures (Tassey, 1991) which facilitate the adoption of a new technology and the propagation of its economic effects.

In summary, the following are the general features of the above perspective:

- Microeconomic technical choices in perpetual oscillation between static competitiveness and dynamical development, between learning benefits issuing from standardization and learning benefits derived from diversity.
- An imperative for the creation of structural interdependences between agents, which are necessary to ensure the pervasiveness of decentralized technical choices.

In short, two mechanisms which form a global dynamics of technical choices and interdependence, with major implications for the future global drift of technical change.

Interlude: from interdependence to incentive

What economic conception of technology lies behind the 'positive feedbacks' and interdependence approach? This conception, affirming the most recent findings of the sociology of science and techniques (Callon, this volume), maintains in essence that there is no distinction between the phases of creation and diffusion: adoption and usage determine a particular good's mode of existence. This is especially clear in the case of technologies of interconnection: a telephone cannot exist without two users. The creative process in this particular case involves the construction of the network of adopters, and not the perfection of the apparatus. But ultimately this is equally true for all other goods: usage and diffusion are always experimental phases which will generate improvements and technological ruptures. In this view, technology is mainly a matter of interdependence.

However, such an economics of technical change may seem perplexing in that it appears to do quite well without the figure of the inventor. The engine of the process is interdependence.

We find a similar absence in certain histories of techniques. For example, in his analysis of innovations the French historian Bertrand Gille begins by establishing the great importance of technical interdependence and in 1978 elaborates in this view the concept of 'technical system': technico-economic coherences cause technical progress both in terms of opportunities[13] and of favourable conditions.[14] Thus, we observe in the work of Gille 'an instinctive recoil of technical invention and the rehabilitation of technical pressure, which almost always appears determinant' (Beaune, 1984).

However, 'although the global coherence of a technical system determines, via the intermediary of tension between its elements, the place where innovation must emerge from the technical installation [. . .] the concept of technical system does not explain that 'gesture' which produces a machine' (Lazzeri, 1981). Some key element is missing, according to these various commentators, in the representation of the sources of technical progress. That element is the concept of 'inventiveness', which leads directly to the economic analysis of incentive.

Invention, incentive and technical progress

The economics of research is today the scene of an intense debate: what are the characteristics of the outputs of this activity, and what can one deduce from these in terms of a scheme of incentives and organization? The vagueness of the answers so far made to these questions is a result of the fact that here, once again, two research programmes are in confrontation with each other, so that responses that appear pertinent in the one case, in the light of the hypotheses advanced, are not pertinent in the other case.

The first programme relies for support on Arrow's work (1962) on the economics of information and therefore considers research activity in terms of perfectly codifiable and transmissible results. The second research program, especially as developed by D. Mowery and N. Rosenberg (1989) and K. Pavitt (this volume) examines research as a process with constituent elements that are strongly appropriable.

It is interesting to 'unreel' the two programmes side by side, because at each point of comparison (economic returns, articulation, science-technology, evaluation of research units, analysis of cooperation), the contradiction is maintained. Each programme leads in the end to the formulation of research policy problems for which 'the other side' cannot account.

There remains but one area of intersection, and that is with respect to the concept of incentive failure (Cohendet *et al.*, this volume). This concordance is based on the attribution to research output (result or process) the characteristic of public good externality (or as is commonly said today, the characteristic of non-rivalry).[15] This is clearly a question of major incentive failure. It is also clear that certain forms of knowledge are easily transferable (subject only to reproduction costs), which slightly magnifies the problem of incentive:[16] the choice between making and buying is therefore only a matter of costs and is not a question of quality or of the applicability of the knowledge. The level of quality, which in essence is equivalent to that of the photocopy (as of a manual), is assured whatever the option retained.

This leads us to the question of the efficiency of incentive systems. These may be based either on the establishment of property rights or on the allocation of public resources to the research activity. The patent system, for example, can never be wholly satisfactory. It cannot take into account the historical accumulation of knowledge during the exercise of research activity (Scotchmer, 1991). Moreover, it cannot truly safeguard the exercise of a property right because the exclusion of others comes at the cost of revealing the information on which the right is based.[17] In other words, this incentive system comes up against the dilemma which opposes the protection of the inventor and the social demand of knowledge diffusion. How do we choose between 'the "static" disadvantage of a patent monopoly and the "dynamic" advantage of encouraging invention' (Hirshleifer, 1971)?

While the above question practically exhausts the scope of the first research programme, there remain to be studied, within the framework of the second programme, the forms of non-transferable technical knowledge, which represent the essential part of knowledge issuing from research processes. Here we encounter the concept of 'tacit knowledge', that is to say, knowledge which is inseparable from the collective work practices from which it derives. Tacit knowledge nonetheless remains (I insist on this because of the confusion that reigns on this issue) a public good, in Samuelson's sense of the term. But it is a public good which is transferred only with difficulty. That is, it is a public good, not a free good (Pavitt, this volume). This very quality of transferability is fundamental to the analysis of research organizations, which is at the heart of the second programme.

Research as result and research as process

The opposition between the two programmes proceeds from the fact that, in the one programme the analytical propositions generally correspond to the standard results of the problems of incentives and allocation of resources in the economics of information, whereas in the other programme the propositions issue from an economics that recognises the specificity of the process of production of scientific and technical knowledge, as distinct from the process of production of information.

Economic returns from research

From the first point of view (research as a result), economic returns are reducible to the traditional framework of cost-benefit analysis. This analysis has nonetheless been greatly refined in work which develops a probability approach to economic returns (Hirshleifer, 1971): the decision to commit oneself to a research programme depends on a sequence of probabilities. A first probability concerns the state of nature (it may or may not in fact be possible to create a particular alloy). The second probability concerns the effective discovery of such an alloy. By informing us on the state of nature, 'basic research information may contribute economic benefits by improving the allocative efficiency of research resources among competing areas of applied research, even if it does not result in discoveries that open up entirely new areas for applied research' (David, Mowery and Steinmueller, 1992).

When research is examined in terms of process, the output is no longer reducible to a given quantity of information, even if a significant part of the output indeed consists of information. The economic returns should also take into account the improvement of research technologies (e.g. instrumentation), training capacities and the opportunities for connection and networking offered by the pursuit of a given research activity (Pavitt, this volume).

The science-technology distinction

What is the economic content of the distinction between scientific and technological research? Dasgupta and David (1986) provide an intellectually stimulating response within the framework of the 'results-oriented' programme; the two domains are solely distinguished by their adoption of different rules for the disclosure of information. In the case of technology, disclosure is discouraged by the possibility of preserving private exploitation of information; in the case of science, disclosure is required in the context of publicly financed research. Thus science and technology are only distinguished by norms of social organization, and especially by those governing the disclosure of knowledge.[18]

In the framework of the 'process-oriented' programme, the distinction between science and technology involves the analysis of complexity. Two complexities are at work, and are in conflict with each other: the abstract complexity of science (the world of *ceteris paribus*) and the concrete complexity of technology (Pavitt, this volume).

The social utility of the research unit: duplication versus redundancy

From the viewpoint of the results-oriented programme, an obvious property of research is the indivisibility of its use: 'there is no value-added when a discovery is made a second or a third time' (Dasgupta, 1987). Thus, we do not need to re-invent the wheel. In consequence, observes Dasgupta, 'the winning research unit is the sole contributor to social surplus' (*ibid.*).

When research is looked at from the process point of view, the property of indivisibility in usage disappears and the argument linking social utility to the research unit becomes completely erroneous. Numerous research tasks do not fit into the category of fixed costs. Their reproduction in thousands of laboratories is therefore necessary, even if the results obtained are similar. Rosenberg's historical analysis (1991a) is edifying on this point: at the beginning of the twentieth-century science entered industry via elementary work, and not at all through great inventions. It was primarily a question of the perfecting of tests, measurement operations and quality controls.

Should we seek to eliminate duplications, understanding them as a factor of inefficiency (Dasgupta, 1987), or should we instead seek to organize the redundancy, considering it an essential element in the design of organizations (Nonaka, 1990)? Clearly, the response to this question depends upon whether research activity is conceived in terms of result or of process.

The analysis of R&D cooperation

From the point of view of the results-orientated research programme, cooperation is basically considered as an incentive mechanism enabling a resolution of the dilemma between social diffusion of knowledge and protection of the inventor. As a result, the stability of the agreement only depends upon the appropriation and equitable sharing of results: 'it must be assured that the sharing of joint benefits gives to each as much as he could have obtained in a non-collusive context' (Jacquemin, 1987).

For the process-oriented research programme, cooperation seeks the implementation of collective processes capable of generating know-how, research technologies and standards, different from what each of the entity-members could have produced alone. Other factors of stability also appear, particularly with respect to the generation of collective learning, which in turn enhances the stability of the agreement because any firm abandoning the agreement would forgo the benefits of such learning (Cohendet and Gaffard, 1991). The production of infratechnologies and common instrumentation during the course of cooperation is also a factor of its stability.

Problems of research policy

The two research programmes described above lead us in the end to consider quite distinct problems of research policy.

- In the results-oriented framework, science, as an institution founded on specific rules of incentives and disclosure, appears to be threatened. Science and technology utilize the same human resources and are therefore competitors for these resources. Thus, the phenomenal increase in the economic returns from technology considerably increases the maintenance cost of scientific resources and definitively disturbs the equilibrium of the allocation of resources: the rate of entry into scientific institutions decreases and the average quality of researchers declines proportionately. The combined effect of a decrease in the quality and quantity of the flow of new knowledge clearly has an impact on the efficiency of technological activities themselves, which are strictly linked to scientific activity: 'Along that dismal path both science and technology appear less and less economically attractive in comparison with other fields of human endeavour until, presumably, enterprises engaged in technology find it worth while to finance research committed to public disclosure and stabilize the system at some much lower level' (Dasgupta and David, 1986).

  The most obvious expression of this problem of resource allocation lies in the application of market principles to scientific institutions: the rule of priority, as a system of specific incentives, is progressively replaced by other forms of property rights, incompatible with the norm of complete disclosure. Thus, it is no longer rare to see researchers at scientific conferences refuse to disclose certain results. In addition, we can observe a significant drop in the publication of scientific articles, a result of information-retention strategies. In short, collegiality itself is threatened.

The classical factors of attraction of resources towards science (style of life, teaching, honour) are no longer sufficient in the context of the gigantic increases in economic returns from technology. One consequence is the appearance of new institutional forms, 'networks of flexible cooperation' created around interests common to science and technology. As M. Callon (1989) suggests, these networks go to the heart of both academia and industry: 'In the morning the mathematician drafts an article on equations of partial derivatives, and in the afternoon he sells an algorithm to Dassault or the Aerospatiale.' These cooperative networks do more than render obsolete the opposition between public science and private technology. They even shake the foundations of scientific altruism.

- From the point of view of the process-oriented framework, the principal problem is that of transformation of knowledge. Two kinds of transformation are important: (i) from basic to applied knowledge, and (ii) from codified (generic) to tacit (specific) knowledge.
- The first kind of transformation deals with the organization of the science-technology interface, i.e. the passage from one complexity to another, in which success alone enables scientific research to constitute a factor of technological competitiveness. A Nobel Prize in and of itself is not an indicator of technological competitiveness: 'the correlation is inverse. That is to say, countries that were ranked high in terms of Nobel Prizes in science tended to be ranked low in economic performance and vice versa' (Rosenberg, 1991b).
- The second kind of transformation deals with the organization of the interface between the firm and its environment. The generic knowledge is only rarely directly usable by firms.[19] Some organizational interfaces are thus required to transform the available (but not usable) knowledge into a specific (usable) one.

Thus, we are faced here with problems of organization (and not of allocation) of resources. As a result the framework of analysis is quite different, even if the solutions presented are both based on the same type of institutional creation: networks of cooperation.

The organization of interfaces

Concerning the science-technology interface, there is no unique model of linkages between scientific and technological progress. Science is either leader or follower. The temporal gaps between a scientific discovery and the technological innovations linked to it are highly variable according to the industries and the historical epochs. Thus, no unique organizational model can cover the science-technology interface. 'Indeed, the diversity of institutional mechanisms that emerged in the American context is most striking, and it is fair to say that the underlying rationale for the particular matter that emerged and succeeded is not very well understood' (Rosenberg, 1991a).

Regarding the other kind of interface (between the firm and its environment), organizational conditions may be more or less favourable to the transformation of generic knowledge. Firms should therefore invest in R&D capacity so as to confer a characteristic of applicability to general, codified knowledge, in the light of specific sites, practices and usages. Only the articulation of the R&D function with the other functions in the firm enables it to transform a generic knowledge into a specific one.

Here, the cost of adoption is no longer a natural attribute (a parameter) of knowledge, it is instead a function of the research effort undertaken by the firm. In order to sustain this argument, one must agree that the goal of R&D is not limited to innovation. R&D also plays an important role in learning and expertise, directed on the one hand towards the 'processing' of available external knowledge and on the other hand to the 'processing' of research requests emanating from within the organization.[20]

The need for new analytical and policy instruments

Yet another change of intellectual regime is now perceptible: 'there does exist a research domain, basic or applied, subject to the title "Organisational economic theory," which covers, not a homogenous theoretical position, but rather a sort of continuum of theoretical positions, from the closest to the most removed from the paradigm of the "standard economic theory"' (Favereau, 1989). This theoretical evolution entails, as above, new analytical exigencies, and reveals new policy domains (Hollingsworth, Piore, Lundvall, this volume).

The requirements of the organizational economic calculus

Without a doubt, it is Williamson's transactional school that goes the furthest in developing a certain type of organizational economic theory, at least in the area of industrial economics and therefore also with respect to R&D. As was noted above however, the double-function of internal R&D implies that the integration of research cannot simply be expressed in terms of transaction cost economics, which only offers a negative explanation of this phenomenon (Foray, 1991). Rather, integration should first be explained in positive terms, since by entering into the firm R&D acquires supplementary attributes as it becomes absorbed by the organization. By being integrated, R&D as an innovative capacity also acquires a learning (or absorptive) capacity; when it is separated it loses this latter quality.[21] This integration property provides a positive explanation for the emergence of an organization.

In the light of this analytical representation, the concepts of 'composite quasi-rent and irreversibility costs' form the analytical framework for an organizational economic theory.

This change in the nature of activities with the development of the organization can be interpreted with the aid of the concept of 'composite quasi-rent'. This is a result of the fact that 'the value of a collection of resources dependent on continued association for their maximum product exceeds their summed market values' (Alchian and Woodward, 1988). The notion of composite quasi-rent thus clarifies the organizational underpinnings of integration.

But as J.L. Gaffard establishes (this volume), the integration of resources also creates irreversibilities, because the process of integration also renders resources more specific. With integration, the range of choices available to the firm decreases, to the point where the specificity of resources becomes an obstacle to change.

There is therefore an 'organizational equilibrium' that must be determined between the need to integrate resources, as a condition of the enrichment of the organization, and the need to leave these resources on the market, as a requirement of reversibility. The determination of this equilibrium is at the heart of the work of Aoki (1984), Amendola and Gaffard (1988). Departing from this theoretical basis, it is undoubtedly possible to continue by defining different historical models of organizational balances. For example, in the United States most research work was developed within the firm. In contrast, in England and France external research organizations were much more important (Rosenberg, 1991a). In Japan, finally, one may observe the continuous policy effort for maintaining an organizational balance between internal and external capabilities (Freeman, 1987).

Policy scope

Institutional diversity, which should be encouraged in order to enable the exploitation of bodies of knowhow and knowledge, nonetheless conceals certain operational constants. The preservation of such constants must be a fundamental goal of any policy of interface between science and technology (Carlsson *et al.*; Lundvall, this volume).

First of all, a firm's possession of internal research capacities is irreplaceable. There is no possible substitutability between internal and external research (Foray and Mowery, 1989). One should therefore focus on aspects of complementarity, which are linked to the irreducibility of the learning function provided by internal research.

Second, as Rosenberg suggests (1991a), 'the achievement of commercial success, as opposed to purely technical success, usually required feedback loops that provided valuable information, drawn from experience that flowed "upstream" to those involved in product design and modification'. The creation of strict articulations between the different resources mobilized in the innovation process is therefore central to the science-technology interface: 'Achieving the right combination of information flows, on the one hand, and incentives on the other, seem to have been vital ingredients of success' (*ibid.*). But the manner in which these combinations are elaborated depends also on models of innovation that are almost cultural (Watanabe, Hollingsworth, this volume). As Aoki suggests (1984), certain combinations (manifested in the Japanese firm) seem to express a model of innovation wherein production processes play the role of driving force ('pull by techne'). Other combinations (manifested in the American firm) accord with a model of innovation attributing great importance to the dynamic of scientific knowledge ('episteme pushing').

Third, the definition of the efficiency of public research is evolving. For example, how do we evaluate a research institute? What is the theoretical space of its efficiency? Should we not in particular describe the results which have enabled an effective exploitation in industry or in other research milieux? These questions imply that the efficiency, or better yet, the pertinence of the research centre should be fundamentally subordinated to the quality of the relation which links it to other sites of technological accumulation.

Fourth, a fundamental contribution of the theory of cooperation consists in bringing up to date the essential role of the length of the cooperative agreement as a condition of success (Sabel, this volume): the predetermined duration creates a consciousness of the future, which alone allows for the avoidance of opportunistic behaviour (Axelrod, 1984). Public policies can play a decisive role in this regard if they assign themselves the task of 'enlarging the shadow of the future', so as to augment the durability and frequency of interaction. In pursuing this logic, it is certain that 'a good way to increase the frequency of interactions between two given individuals is to keep others away' (*ibid.*). It is therefore advisable to form clubs (Antonelli and Foray, 1992), which is a strategy that Japanese electronic firms have admirably carried out (Wakazugi, 1990), and which has today become a principal objective of American and European policy.

Fifth and finally, there is an effect of territorialization of organization and cooperation systems. In this case, the external economies of organization will be added to the benefits of collective learning, so as to reinforce the stability of cooperation. The district structure is a perfect illustration of this phenomenon (Piore, this volume; Freeman, 1991).

But whether or not cooperation becomes part of a local history, what determines in the final analysis the duration of the organization is whether the game is a positive-sum one: 'There will be winners and losers during every round of the game, but it is a positive-sum game in the sense that, owing to the increase in demand, the gains will exceed the losses' (Klein, 1988).

Technology and the wealth of nations

At certain historical moments, in certain geographical spaces, the entering into coherence of the two mechanisms of technical progress (interdependence and diffusion mechanisms, and systems of incentive and organization) enables technology to contribute decisively to the wealth of nations. In conclusion, I will try to describe a few institutional elements which, under certain conditions, can create coherence and, under other conditions, can destroy it.[22] Two categories of elements of coherence will be distinguished.

First, let us consider the institutional elements which simultaneously produce technical choices and interdependence on one hand and incentive systems on the other. That is to say, let us look at the very intersection of the two mechanisms of technical progress.

- The role of *major military research programmes* is quite clear in this regard. From 1950 to 1970, these programmes constituted a decisive factor in the creation of coherence. On the one hand, they defined technical choices capable of energizing technological knowledge for society as a whole (Rosenberg, 1977). On the other hand they played a significant incentive role by permitting contractors to maintain a critical mass of researchers.

  In a subsequent time period, this coherence disappears. Technical choices become so specific that their influence on the dynamic of technological knowledge is substantially diminished. As for incentives, they become problematic in that military R&D teams become indifferent to costs. The critical mass of researchers, maintained owing to military R&D contracts, is not actually transferable to civil production programmes.[23]
- The *policies of support for basic research* (Pavitt, this volume) constitute a second

important element of coherence. They elicit technological choices in the areas of instrumentation and research technologies, which have a major impact upon the future dimensions of technical knowledge. Moreover, they permit the improvement of the quality of applied research, thereby playing a significant incentive role.

- The *policies of support for cooperative research and standardization*, which Mowery and Rosenberg (1989) group together in terms of 'adoption-focused policy', are determinative of the coherence of diffusion mechanisms and incentive systems (Carlsson *et al.*, this volume). These policies enable an increase in the degree of interdependence between agents, so that the implication of decentralized technical choices becomes more and more important. They also improve the quality of research projects and are therefore incitative in this respect (Dufourt, 1991). However, these policies may also, in certain circumstances, be a factor of incoherence in that they endanger 'the appropriate balance between protection of intellectual property rights and support for technology adoption' (*ibid.*).
- *Education and training policy* – 'the most important long-term investment for the future' (Freeman and Soete, 1987) – is obviously a critical element of coherence, in that it forms by itself interdependence structures which strongly facilitate the diffusion of knowledge. On the other hand it makes the incentive systems more efficient (Marsden, Watanabe, this volume).

Other elements fall more clearly onto one or the other level (either that of diffusion and interdependence or that of creation and incentives), but their relative 'plasticity' allows them to play a determinant role in the entering into coherence of the two mechanisms of technical progress.

- The element of *property rights* perfectly illustrates the case of the adaptability of incentive mechanisms to determined technological situations. Ordover (1991) shows how the Japanese patent system permits a reconciling of incentives and diffusion of knowledge. Two characteristics are crucial here: (1) the patent system is designed to induce innovators to rapidly disclose strategic information, and (2) the narrow scope of patent coverage and very relaxed novelty requirement severely reduce the exclusionary value of the patent right. Ordover concludes therefore that the patent system 'rewards those who reverse engineer and modify, often in minor ways, the existing inventions and penalizes those who wish to protect their major technological breakthroughs'. There is therefore a coherence between a specific incentive scheme and a mode of accumulating knowledge that favours minor innovations and imitations. As a result the mechanisms of incentives adapt themselves to a specific logic of technological interdependence and diffusion. This represents a truly exceptional effort in favour of coherence. In most countries, one can observe on the contrary a great rigidity in the system of property rights, in disregard of the profound transformations that have occurred in the modes of accumulation of technological knowledge.
- Another case of the plasticity of the incentives mechanisms is provided by Dahmén (1989) who explains, with the historical cases of Sweden and Finland, how *fiscal and credit policies* may play a key role in some historical periods, by supporting entrepreneurial activities in promoting development blocks. Uncompleted blocks 'are usually indicated by price and cost signals on various markets. This could mean low current profits, or even losses, in some areas but also promising prospects, if the steps necessary to complete the blocks are within reach. [. . .] Entrepreneurs have to respond to signals by active search' (*ibid.*). Some incentives policies like R&D tax credit, may help firms to meet the requirements of developmental interdependence in order to exploit cumulative elements inherent in the completed development block.

- The creation of coherence can also result from the plasticity in *the modes of technological interdependence and diffusion – standardization or diversity*. There are, in effect, possible adaptations with respect to an ensemble of incentives and organization characteristics. The typical example is that of the 'lack of standardization in the U.S. nuclear industry in contrast to the French nuclear industry' (David and Rothwell, 1991). Which choice is more coherent, one in favour of a dynamics of learning through diversity, or one in favour of a dynamics of learning through standardization? The answer will depend on the organizational structure (either centralized or decentralized) of the system, which detemines its capacity to most fully profit from the learning issuing from one or the other of these two modalities of knowledge accumulation.

It is thus on the basis of this articulation between the dynamics of interdependence and diffusion and the systems of incentives and creation, that our book is constructed. Part I proposes different important analytical views and empirical studies of this articulation. Part II provides several brilliant illustrations of the so-called 'model revolution', which addresses the necessity of integrating the different features of interdependence and incentives presented above. Part III focuses on the problem of interdependence in order to highlight the driving forces of technological uniformization and convergence. The last Part returns to the diversity stemming from incentives and organizational systems.

It is now clear that the tension between diversity and standardization, 'which touches on the welfare of people and the wealth of nations' (David, 1988), is primarily an analytical one. It does not correspond to any dichotomy between technology and institutions (Sorge, this volume). It stems rather from the *encounter* between deterministic and random processes of diffusion and interdependence (which induce standardization because of the uniformization power of choices with positive externalities) *with* incentives and organization systems (which induce diversity because of their necessary adaptation to the great variety of problems of creation) (Rosenberg and Birdzell, 1986). In certain historical and geographical conditions, this encounter generates a coherent dynamics of interdependence and incentives (Boyer, 1992). Technology thus becomes a source for the wealth of Nations.

Notes

1. Cf. clearly the work of reference, edited by Dosi, Freeman, Nelson, Silverberg and Soete (1988).
2. C. Freeman (1990), see in particular the general introduction as well as the J.E.L. survey of G. Dosi, also included in the book edited by C. Freeman.
3. Cf. Introduction to Part I, by C. Freeman, this volume.
4. Cf. Introduction to Part II, by R. Boyer, this volume.
5. Great exceptions however are provided in this book by Orsenigo, Chiaramonte and Dosi, Carlsson and Jacobsson, and Lundvall. All these chapters try clearly to integrate both levels (interdependence and incentives) either in empirical studies or in more theoretical approaches.
6. This General Introduction is not intended to specifically introduce all the chapters in this volume; their scope and importance is presented in the Introductions to each of the four Parts, drafted by C. Freeman, R. Boyer, D. Foray and A. Sorge.

7. At the macroeconomic level, 'economies with different past experiences may have divergent futures' (Atkinson and Stiglitz, 1969). A bad start perpetuates its effects in that learning goes on being learned (Stiglitz, 1987). To a certain extent this implication punctures Veblen's hypothesis that the accumulation of experience can under certain conditions constitute an obstacle to technological change. Cf. M. Abramovitz (1989) for a discussion of this hypothesis.
8. From the point of view of the formal theory of stochastic processes, the sequence of choices has the properties of a generalized Polya urn process (cf. Arthur, Ermoliev and Kaniovski, 1987).
9. Canada is the only country to have followed this trajectory to its logical conclusion. The technology (natural uranium and heavy water) was initially chosen independently of military constraints and is indeed more pertinent for civil applications.
10. According to Paulré (1985), 'perhaps the main obstacle to the emergence of causal thinking in economics was the growing gap between the complex causal structures of reality and the simplicity of the type of relation according to which the philosophers and logicians believed the causality of a phenomenon should be studied. Today, in contrast, the concepts of feedback, causal structure and the analytical tools of models intended to identify possible regimes of evolution, the basins of attraction allow for the conceptualization and modelling of interpretive schemes that take into account dynamic interdependences internal to economic systems and which connect the decisional level to the level of physical causality.'
11. See Amable, Aghion and Howitt, Chiaramonte and Dosi and Gaffard (this volume) for the integration of some path-dependent features into macro models.
12. See Foray and Grübler (1991) for some tools (morphological analysis and evolutionary tree) in this perspective.
13. 'Once the first discoveries were made everything, or almost everything, took care of itself. The distortions between the different phases of production themselves compelled complementary inventions: there follows a sort of sequence which culminates in the re-establishment of the disturbed equilibrium' (Gille, 1978).
14. For example, 'It would have been impossible to develop nuclear power or many new chemical processes and materials in the last fifty years, without new scientific instruments' (Freeman, 1982). To borrow an expression from Maunoury (1968), scientific progress widens the 'horizon of creativity' of chemical and nuclear engineers.
15. This is a purely technical characteristic, first defined by Samuelson: the utilization of a good by an agent does not exclude its utilization by others.
16. Or as Romer (1986) would put it, the possibility of transmission 'de-idealizes' the property of non-rivalry.
17. Even if the establishment of a patent only requires the disclosure of a part of the information necessary for the implementation of the invention, it nonetheless furnishes a primary indication with respect to the possible success or 'practicability' of a research orientation.
18. Dasgupta and David (1986) show that two sides exist with respect to property rights: the system of patents for technology and the rule of priority for science. The latter constitutes a remarkable social innovation, in that it distinguishes between the phenomena of intellectual property and that of exclusive possession. This is an invaluable system of incentive because it is compatible with the need for disclosure, which is the rule so far as science is concerned.
19. One can nonetheless hypothesize that great progress has been made today with respect to the technologies of codification of tacit knowledge (e.g. expert-systems software). The respective roles of tacit and codifiable knowledge would then be reversed in the research process.
20. 'The ability of the firm to exploit even the limited services that were available via

contract was affected critically by the presence of an in-house laboratory' (Rosenberg, 1991a).

21. Cf. D. Mowery's work on AT&T's divestitures and their effect upon the structure of Bell Laboratories (1988).
22. This is the same goal pursued by R. Nelson (1990) in his most recent work on the United States.
23. 'Military R&D programs may also encourage "learning" of expensive and ultimately inefficient habits [. . .]. Such learning "spillovers" may reduce the effectiveness of R&D personnel in competitive commercial markets where close attention to cost considerations can be a matter of commercial success or failure' (Mowery and Rosenberg, 1989).

References

Abramovitz, M. (1989), *Thinking About Growth*, Cambridge University Press.

Alchian, A. and Woodward, S. (1988), 'The firm is dead; long live the firm', *Journal of Economic Literature*, XXVI.

Amendola, M. and Gaffard, J.L. (1988), *The Innovative Choice: An Economic Analysis of the Dynamics of Technology*, Basil Blackwell.

Antonelli, C. and Foray, D. (1992), 'The economics of technological clubs', *Economics of Innovation and New Technology*, forthcoming.

Aoki, M. (1984), *The Cooperative Game Theory of the Firm*, Oxford University Press.

Arrow, K. (1962), 'Economic welfare and the allocation of resources to invention', in Nelson (ed.), *The Rate and Direction of Inventive Activity*, National Bureau of Economic Research, Princeton.

Arthur, B. (1989), 'Competing technologies, increasing returns, and lock-in by historical events', *The Economic Journal*, 99.

Arthur, B. (1991), 'Designing economic agents that act like human agents: a behavioral approach to bounded rationality', *American Economic Review Paper and Proceedings*.

Arthur, B., Ermoliev, Y. and Kaniovski, Y. (1987), 'Path-dependent processes and the emergence of macro-structure', *European Journal of Operational Research*, 30.

Atkinson, B. and Stiglitz, J. (1969), 'A new view of technical change', *The Economic Journal*, 79.

Axelrod, R. (1984), *The evolution of cooperation*, Basic Books.

Beaune, J.C. (1984), 'Formes philosophiques de l'invention technique' *Milieux*, 18.

Boyer, R. (1992), 'New directions in management practices and work organization: general principles and national trajectories', OECD.

Callon, M. (1989), *La science et ses réseaux*, Ed. La Découverte.

CNRS (1990), *L'aventure scientifique et technique de la machine à vapeur*, ouvrage collectif, ed. du CNRS.

Cohendet, P. and Gaffard, J.L. (1991), 'Innovation et entreprise', in Greffe, Mairesse and Reiffers (eds), *Encyclopédie Economique*, Economica.

Cowan, R. (1990), 'Nuclear power reactors: a study in technological lock-in', *The Journal of Economic History*, 3.

Cowan, R. (1991), 'Tortoises and hares: choice among technologies of unknown merit', *The Economic Journal*, 101.

Cowan, R. and Rizzo, M. (1991), *The Genetic-causal Moment in Economic Theory*, STARR Center for Applied Economics, New York University.

Dahmen, E. (1989), 'Development blocks in industrial economics', in Carlsson (ed.), *Industrial Dynamics*, Kluwer Academic Publishers.

Dasgupta, P. (1987), 'The economic theory of technology policy', in Dasgupta and Stoneman (eds), *Economic Policy and Technology Performance*, Cambridge University Press.

Dasgupta, P. and David, P. (1986), 'Information disclosure and the economics of science and technology', in Feiwel (ed.), *Arrow and the Accent of Modern Economic Theory*, New York University Press.

David, P. (1975), *Technical Choice, Innovation and Economic Growth*, Cambridge University Press.

David, P. (1988), *Information Technology, Social Communications, and the Wealth and Diversity of Nations*, CEPR, Stanford University.

David, P. and Greenstein, S. (1989), 'The economics of compatibility standards: an introduction to recent research', *Economics of Innovation and New Technology*, 1.

David, P. and Rothwell, G. (1991), *Performance-based Measures of Nuclear Reactor Standardization*, CEPR, Stanford University.

David, P., Mowery, D. and Steinmueller, E. (1992), 'Analysing the economic payoffs from basic research', *Economics of Innovation and New Technology*, forthcoming.

Dosi, G., Freeman, C., Nelson, R., Silverberg, G. and Soete, L. (1988), *Technical Change and Economic Theory*, Frances Pinter.

Dufourt, D. (1991), 'Les politiques technologiques: une nouvelle rationalité de l'intervention publique dans le système productif?', in De Bandt and Foray (eds), *L'évaluation économique de la recherche et du changement technique*, ed. CNRS.

Favereau, O. (1989), 'Vers un calcul économique organisationnel?', *Revue d'Economie Politique*, 2.

Foray, D. (1991), '"The secrets of industry are in the air": industrial cooperation and the organizational dynamics of the innovative firm', *Research Policy*, vol. 20, 5.

Foray, D. and Mowery, D. (1989), 'L'intégration de la recherche industrielle: nouvelles perspectives d'analyse', *Revue Economique*, 3.

Foray, D. and Grübler, A. (1991), 'Morphological analysis, diffusion and patterns of technological evolution', in Nakinenovic and Grübler (eds), *Diffusion of Technologies and Social Behavior*, IIASA, Springer Verlag.

Foray, D. and Garrouste, P. (1991), 'The pertinent levels of analysis in industrial dynamics', in Hodgson and Screpanti (eds), *Rethinking Economics: Market, Technology and Economic Evolution*, Edward Elgar.

Freeman, C. (1982), *The Economics of Industrial Innovation*, Frances Pinter.

Freeman, C. (1987), *Technology, and Economic Performance: Lessons from Japan*, Pinter Publishers.

Freeman, C. (1990), 'Introduction', in Freeman (ed.), *The Economics of Innovation*, Edward Elgar.

Freeman, C. (1991), 'Networks of innovators: a synthesis of research issues', *Research Policy*, 20, no. 5.

Freeman, C. and Soete, L. (1987), *Technical Change and Full Employment*, Basil Blackwell.

Gille, B. (1978), *Histoire des Techniques*, Gallimard, Encyclopédie de la Pléïade, (English trans.), (1986) *History of Techniques*, 2 vols, Gordon and Breach.

Hirshleifer, J. (1971), 'The private and social value of information and the reward to inventive activity', *American Economic Review*.

Jacquemin, A. (1987), 'Comportements collusifs et accords en recherche-développement', *Revue d'Economie Politique*, 1.

Klein, B. (1977), *Dynamic Economics*, Harvard University Press.

Klein, B. (1988), 'Luck, necessity, and dynamic flexibility', in Hanusch (ed.), *Evolutionary Economics*, Cambridge University Press.

Lazzeri, C. (1981), 'Rationalité et irrationalité de la technique', *Dialectiques*, 32.

Maunoury, J.L. (1968), *La genèse des innovations*, PUF.

Mowery, D. (1988), 'Assessing the effects of divestiture on Bell Telephone Laboratories', *Technovation*.

Mowery, D. and Rosenberg, N. (1989), *Technology and the Pursuit of Economic Growth*, Cambridge University Press.

Nelson, R. (1990), 'U.S. technological leadership: where did it come from and where did it go, *Research Policy*, Summer., pp. 117–132.

Nonaka, I. (1990), 'Redundant, overlapping organization: a Japanese approach to managing the innovative process', *Californian Management Review*.

Ordover, J. (1991), 'A patent system for both diffusion and exclusion', *Journal of Economic Perspectives*, vol. 5, 1.

Paulré, B. (1985), *La causalité en économie*, PUL.

Puiseux, L. (1989), *Bifurcations et irréversibilités dans l'histoire économique de l'énergie depuis 1945*, CIRED-EHESS.

Romer, P. (1986), 'Increasing returns and long-run growth', *Journal of Political Economy*, October.

Rosenberg, N. (1977), 'Civilian spillovers from military R&D spending: the American experience since World War II', in Lakoff and Willoughby (eds), *Strategic Defense and the Western Alliance*, Lexington Books.

Rosenberg, N. (1991a), *The Role of Science in Industrial Development*, CEPR, Stanford University.

Rosenberg, N. (1991b), *Scientific Research and Economic Performance*, CEPR, Stanford University.

Rosenberg, N. and Birdzell, L. (1986), *How the West Grew Rich*, Basic Books.

Rosenberg, N. and Vicenti, W. (1986), *The Britannia Bridge: The Generation and Diffusion of Technological Knowledge*, MIT Press.

Scitovski, T. (1954), 'Two concepts of external economies', *Journal of Political Economy*.

Scotchmer, S. (1991), 'Standing on the shoulders of giants: cumulative research and the patent law', *Journal of Economic Perspectives*, vol. 5, 1.

Stiglitz, J. (1987), 'Learning to learn: localized learning and technological progress', in Dasgupta and Stoneman (eds), *Economic Policy and Technology Performance*, Cambridge University Press.

Swann, P. (1992), 'Rapid technology change "visions of the future", corporate structure and market structure', *Economics of Innovation and New Technology*, forthcoming.

Tassey, G. (1991), 'The functions of technology infrastructure in a competitive economy', *Research Policy*, 20.

Wakazugi, R. (1988), 'A consideration of innovative organization: joint R&D of Japanese firms', in Heertje and Perlman (eds), *Evolving Technology and Market Structure*, The University of Michigan Press.

Part I
The sources of localized learning and the science–technology interface

Introduction to Part I

Chris Freeman

Economists have discussed the problems of financing fundamental research for many decades now. Ever since the seminal papers of Nelson (1959) and Arrow (1962) there has been a widespread consensus that this was a clear case of market failure. It was generally believed that industry would not invest a great deal in basic research because of the well-recognized problems of appropriability, the high degree of uncertainty and the very long-term nature of much of this research. But whilst a low *private* rate of return would deter all but the largest and most profitable firms from investing in basic research, a relatively high *social* rate of return could be expected from public investment, both in universities and in other publicly funded research institutions. Economic theory thus provided relatively strong justification for the increasing research expenditures of many governments through such institutions as the National Science Foundation, the CNRS and various similar Research Councils. The relationships between universities, concentrating on basic research, industry concentrating on applied research and development and public institutions with hybrid functions became a major concern of policies for science and technology and a focus of much policy research.

Several research projects in the 1960s and 70s demonstrated unequivocally that the role of the science-technology infrastructure was crucial for competitive success in industrial innovation. Project TRACES and Project HINDSIGHT in the United States, Project SAPPHO in the United Kingdom and many similar research studies all showed the vital importance of external networks of scientific and technical advice and information for efficient innovation performance (Freeman, 1982). However, although this evidence was conclusive in the case of radical innovations, such as the electron microscope or gas chromatography, other empirical studies of innovation showed that numerous minor innovations and improvements to products owed little or nothing to university research.

In this type of incremental innovation the crucial factors affecting innovation success were user–producer relationships and the in-house accumulation of technological capability within the innovating organization. External networks with outside sources of scientific and technical advice and information could still be important even for incremental innovation, just as user–producer relationships were also important for radical innovations but the balance differed considerably. Science and technology 'push' tended to predominate in the more radical type of innovation while market demand 'pull' was much stronger with incremental innovations.

In the 1980s several new developments have combined to change both the conceptualization of these problems and their policy implications. Whilst it was always recognized that the science-technology relationship was important, this had become strikingly evident with the generation and diffusion of three generic technologies deeply affecting most industries – biotechnology, advanced materials technology and computer technology. In all three cases the relationship between 'science' and 'technology' was so close that it was sometimes virtually impossible to distinguish between them. In the case of biotechnology, not only the fundamental theory but also the critical techniques of gene-splicing, cloning, etc., originated almost entirely in university research. Whereas university physicists, chemists and engineers had been accustomed for a long time to consultancy and collaboration with industry in a variety of different forms, for many university biologists it was a novel experience to be courted, consulted and recruited by industrial R&D departments. In the case of materials technology, computer technology (and semiconductor technology) the institutional relationships were less one-sided but here too the interface between basic science and industrial technology was clearly of great importance. This would be seen alike from 'scientometric' analysis of publication patterns, as from the patterns of funding and from the mobility of qualified engineers and scientists across the institutional divides.

Second, the evident intimacy of the science-technology relationship led policy-makers and policy analysts to raise questions about the effects of *national* basic research performance on *international* competitiveness. The 'free-rider' debate became not merely a debate in theoretical economics but a policy issue affecting especially US–Japanese relationships. Exacerbated by the huge US trade deficit and the enormous success of Japanese firms in world competition, especially in electronics, some US observers concluded that Japanese industry was following a 'free-rider' strategy in relation to the US leadership in basic science and this became a source of some nationalistic controversies and tensions.

Finally, although economists were generally not successful in their attempts to measure the rate of return to investment in basic research either on the national or the international level, scientometric techniques did advance rather rapidly making it possible to deploy 'output' indicators of relative performance by individuals, laboratories, universities, companies and countries in diverse fields of research. The use of these techniques and especially of citation analysis reinforced our understanding of the importance of networks in science and technology but also tended to emphasize the role of 'information' inputs to innovation and of publications relative to the less obvious and less measurable, but no less important work of engineers and technicians in developing unpublished (and often unpublishable) tacit knowledge about how to make new products and operate new processes. The knowledge accumulation process within firms and other institutions was thus understated and underestimated both in economic theory and in public policy, relative to the more obvious and measurable flows of scientific information.

The renewed recognition of the importance of basic science for technological competitiveness led to a greater willingness to provide resources for its expansion both in industry and in government in Japan, in the United States and in most European countries. But this was accompanied by frequent misunderstanding of the true role of basic science in competitiveness and a considerable exaggeration of the 'free-rider' problem in nationalistic terms.

The four chapters in Part I all contribute original insights into the role of basic research in national competitiveness as well as in the general advance of world-wide technology. Whilst confirming the intimacy and importance of the science-technology interrelationship, especially in biotechnology, they also throw new light on the contribution of basic research to industrial competitiveness.

The first chapter by Pavitt offers a new justification for public endowment of fundamental research in terms not of specific information outputs (as measured by scientometric techniques), but in terms of training and learning. The research skills and techniques acquired in university research, Pavitt argues, are valuable in many later applications in the careers of engineers and scientists, and their value is not confined to industrial R&D departments. They enable firms to accumulate and absorb knowledge and to relate to external networks in ways which would not be possible in the absence of these skills and techniques. This argument was implicit in much previous discussion of the complementarities in research, teaching and learning and in the debate on the results of the US Department of Defense Project HINDSIGHT. This showed that the main contribution of US universities to US defence technology came not from specific discoveries or pieces of information but from the recruitment of graduates with good general knowledge of both recent and earlier scientific techniques and advances. But this type of interpretation was controversial at the time and tended to be obscured and forgotten with the advance of scientometrics and especially with the rise of generic technologies with their rather obvious connections between scientific advance and industrial innovation.

Orsenigo's chapter illustrates in considerable depth just how close this relationship has become in the case of biotechnology. He shows not only that companies were compelled to gain access to an entirely new body of scientific knowledge located originally almost entirely in universities, but also that they were obliged to coordinate and integrate inputs from a variety of subdisciplines crossing both academic boundaries and those of company organizations. Nor was the emerging industry exempt from the accumulation of tacit knowledge associated with the scaling up of industrial processes. On the contrary this proved to be extremely important, so that on the one hand there were acute problems of appropriability associated with the redefinition of patent law to cover new products in this field, but on the other hand new entry barriers grew up associated with knowledge accumulation in process technology.

These new problems of appropriability are taken up in the chapter by Cohendet *et al.* Confirming the results of earlier research at Cambridge and Yale they show that in most industries patent protection is of secondary or negligible importance. Once knowledge is recognized as localized, specific and cumulative then it is clear that protection depends far more on the dynamics of the knowledge base than on formal systems of patent protection. Only where imitation is relatively easy and information easily transferable as in some new products for the drug and chemical industries, does patent protection assume great importance. The validity of the conclusions of Cohendet *et al.* on the protection afforded by cumulative learning and 'running fast' is shown by the experience of such industries as aero-engines and oil refinery construction. In both cases the importance of tacit knowledge, person or group-embodied competence and skills, and organizational learning involving complex coordination, combine to create such formidable entry barriers that a few companies have been able to sustain world-wide leadership over many decades.

But neither these nor any other companies can be entirely self-sufficient in their technological learning and innovation. They too are obliged to assimilate many new advances in information technology, in advanced materials technology and in other areas which they cannot generate autonomously even if they wished to do so. The generation and maintenance of competitive advantages therefore depends on a creative combination of in-house learning and technology accumulation with fruitful participation in a variety of external networks and recruitment of new engineering, scientific and other skills embodying new knowledge from external sources. The final chapters in Part I by Carlsson and Jacobsson confirm this proposition from an entirely different angle. Carlsson's chapter shows that the national science and technology infrastructure was essential for the success of the Swedish engineering industries in the assimilation and development of manufacturing automation.

The chapters in Part I of this book thus provide the background both for the discussion of modelling technical change in Part II and for the more extended discussion of national systems of innovation in Parts III and IV.

References

Arrow, K. (1962), 'Economic welfare and the allocation of resources for invention', in NBER, *The Rate and Direction of Inventive Activity*, Princeton University Press.

Freeman, C. (1982), *The Economics of Industrial Innovation*, Pinter, London.

Gibbons, M. and Johnston, R. (1974), 'The roles of science in technological innovation', *Research Policy*, vol. 3, no. 3: 220–42.

Nelson, R. (1959), 'The simple economics of basic scientific research', *Journal of Political Economy*, vol. 67: 297–306.

1 What do firms learn from basic research?*

Keith Pavitt

1. Introduction

In this chapter, I shall argue that academic science (defined as research and related training activities, performed in universities mainly with public money, for the provision of public knowledge) is an essential input into industrial technology (defined as development and related research and learning activities, performed in firms with private money for private profit). However, the representation of its output solely in terms of codified, useful and published information is unduly restrictive, and potentially misleading as a guide to both company and public policies.

In section 2, I argue that basic science should be considered primarily as an input into economic activity. In section 3, I show that the nature of this input is not restricted to codified and immediately useful information, but also comprises person- and institution-embodied research skills, instrumentation and networks. In section 4, I show how this broader view is necessary for any useful analysis of the benefits accruing to countries that fund basic science in an open, multi-country world. On these bases, I argue in section 5 that many policies designed to achieve greater practical relevance in the funding of basic research are misguided; that the economic justification for the public subsidy of basic science needs to be revised, and that the activities of basic science are better conceptualized as the creation of a capacity to solve problems, than as the production of codified and immediately useful information.

2. The economics of basic science

Science as an economic activity

The importance of science as an economic activity was recognized right at the beginning of the process of industrialization. In Chapter One of *The Wealth of Nations* (1776), Adam Smith observed that technical advances were made not only at the point of production, but also by suppliers of capital goods, and by

* This chapter draws heavily on my earlier paper, 'What Makes Basic Research Economically Useful' (*Research Policy*, 20, 1991), and more generally on the research programme of the Centre for Science, Technology and Energy Policy, financed by the Economic and Social Research Council (ESRC) at the Science Policy Research Unit.

'philosophers and men of speculation', which is what scientists were then called.

Perhaps less well known are the predictions of Alexis de Tocqueville in *Democracy in America* (1830). He observed the down-to-earth and problem-solving nature of US society, and predicted the rapid expansion of science, for three reasons: first, as a form of conspicuous intellectual consumption, funded from the great accumulations of private wealth that de Tocqueville rightly predicted the US system would produce; second, as a foundation for the education of the large number of applied scientists that de Tocqueville predicted (again rightly) that modernizing society would require; and third, as a source of fundamental knowledge needed to facilitate and guide the solving of practical problems.

In other words, the rapid growth of modern science must be seen as part of a more general process of the specialization and professionalization of productive activities in modernizing societies. To this, we must add Marx's important insights into the major influence of economic and social demands on the rate and direction of scientific advance, through the problems that they pose, the empirical data and techniques of measurement that they generate, and the financial resources that they make available (see Rosenberg, 1976). For all these reasons, economists are right to argue that large expenditures on science can be neither understood nor justified solely on cultural and aesthetic grounds; they inevitably have important economic and social dimensions. However, we shall see in section 3 below that the links today between science and technological practice are far from straightforward.

The public subsidy of science

In the meantime, progress has been made in the analytical justification for the regular and large-scale government funding of basic research practised in all modernizing countries. As is often the case, principle followed practice rather than led to it, since governments in some countries (and most notably Germany) had already been funding basic research for a very long time. After the Second World War, the United States followed suit, and in the early 1950s established the National Science Foundation. In 1959, R. Nelson published his pioneering paper entitled 'The Simple Economics of Basic Research', in which he argued that – left to itself – a competitive market will invest less than the optimum in basic research. This is because a profit-seeking firm cannot be sure of capturing all the benefits of the basic science that it sponsors, given major uncertainties about the benefits for the sponsoring firm, and the difficulties it faces in extracting compensation from subsequent imitators. At the same time, a policy of secrecy aimed at stopping such imitation would be sub-optimal, since it would restrict applications with small marginal cost. If, in addition, profit-seeking firms are risk-averse, or have short-term horizons in their decisions to allocate resources, private expenditures on basic research will be even more sub-optimal.

Nelson's insights have been developed and modified over the past 30 years, notably by Arrow (1962) and Averch (1985) in the United States, and by Kay and Llewellyn Smith (1985), Dasgupta (1987), Stoneman (1987) and – most recently – Stoneman and Vickers (1988) in the UK. Risk aversion, low or zero marginal cost of application, and difficulties in appropriating benefits, have become standard explanations for the public subsidy of science.

At the same time there has been a subtle shift in emphasis. Nelson's original paper was grounded in research on the development of the transistor (1962), and his paper is spliced with examples of the development and application of science. Over time progressively fewer references have been made to the empirical evidence, and more to the standard theorems of welfare economics. Whilst it might be advantageous in the economics classroom to assume that basic science is instantly applicable and easily transferable, we shall now see that such assumptions are empirically invalid, and have effectively restricted debate on two of the major issues of contemporary science policy: the nature and extent of the contributions of science to technology, and the impact of national science on national technology (Mowery, 1983).

The real world is more complicated. There is a common confusion between the reasonable assumption that the results of science are a 'public good' (i.e. codified, published, easily reproduced and therefore deserving of public subsidy), and the unreasonable assumption that they are a 'free good' (i.e. costless to apply as technology, once read). In a paper entitled 'Why do Firms do Basic Research (with Their Own Money)', N. Rosenberg (1991) argues that basic research financed and performed in (mainly large) firms often grows out of practical problem-solving, and that the two are highly interactive. He also argues that in-house basic research is essential in order to monitor and evaluate research being conducted elsewhere:

> This point is important in identifying a serious limitation in the way economists reason about scientific knowledge and research in general. Such knowledge is regarded by economists as being 'on the shelf' and costlessly available to all comers once it has been produced. But this model is seriously flawed because it frequently requires a substantial research capability to understand, interpret and to appraise knowledge that has been placed on the shelf – whether basic or applied. The cost of maintaining this capability is high, because it is likely to require a cadre of in-house scientists who can do these things. And, in order to maintain such a cadre, the firm must be willing to let them perform basic research. The most effective way to remain plugged in to the scientific network is to be a participant in the research process.

3. The impact of science on technology

Technology as science?

It is comfortable as well as convenient to treat science and technology as the same thing, given the similarities in their inputs (scientists, engineers, laboratories) and their outputs (knowledge), and given the well-known examples of outstanding basic science performed in corporate laboratories. However, this neglects the very different nature and purpose of the central activities of university and business laboratories. In universities, basic research seeks generalizations based on a restricted number of variables, and results in publications and reproducible experiments. In business, a combination of research, and (more important) development, testing, production engineering and operating experience accumulates knowledge on the many critical operating variables of an artefact, and result in knowledge that is not only specific, but partly tacit (uncodifiable) and therefore difficult and costly to reproduce (Kline, 1991).

Given these differences, basic research as a public good is more likely to meet the conditions for private underinvestment, as defined by Nelson and others, which explains the higher proportion of public funding in basic research than in development in all OECD countries. Analysts conscious of the distinction between science and technology have made a major contribution to the policy debate by stressing the complementary nature of private and public investments in science and technology, with the former concentrating on the short term and specific, and the latter on the long term and the general. They have also warned of the dangers and inefficiencies of heavy public funding of commercial development activities (Eads and Nelson, 1971; Jewkes, 1972). However, insufficient analytical attention has been directed to the interface between science and technology.

Calculating the economic returns to basic research

Resource-starved academic scientists no doubt welcome studies demonstrating a high economic return to basic research. One such study has just been completed in the United States, by a distinguished economic expert on R&D – E. Mansfield (1989). It is one of the most ingenious and persuasive of its kind but, as pointed out by P. David and his colleagues (1988), calculations of this kind do not satisfactorily reflect the nature of the impact of science and technology:

> The outputs of basic research rarely possess intrinsic economic value. Instead, they are critically important inputs to other investment processes that yield further research findings, and sometimes yield innovations, . . . Policies that focus exclusively on the support of basic research with an eye to its economic payoffs will be ineffective unless they are also concerned with these complementary factors.
>
> The alternative conceptualization . . . that we have developed focuses on basic research as a process of learning about the physical world that can better inform the processes of applied research and development. Rather than yielding outputs that are marketed commercially, basic research interacts with applied research in a complex and iterative manner to increase the productivity of both basic and applied research. The development of links between the basic and applied research enterprises are critical to the productivity and economic payoffs of both activities. (pp. 68–9)

The variability of science's impact on technology

We know from the results of past research that these links between basic science and technology are in fact complex along at least four dimensions (Williams, 1986).

(i) The *intensity of direct transfers of knowledge* from basic science to application varies widely amongst sectors of economic activity, and amongst scientific fields. The most systematic analyses have been made in the United States, on the basis of patent citations to journals (Carpenter, 1983; Narin and Noma, 1985; Narin and Olivastro, 1992), and of a survey of industrial R&D directors (Nelson and Levin, 1986). They both confirm strong links in chemicals and drugs firms to basic research in biology, whilst the links of electronics firms are also intense but to more applied research activities in physics. In mechanical and transport technologies, on the other hand, the links to the published results of basic science are weak.

(ii) The *level of the impact of basic research on technology* also varies widely from the generation of epoch-making new technologies (e.g. electricity, synthetic materials, semiconductors; see Freeman *et al.*, 1982), through accumulated improvements in continuous flow industries resulting from routine chemical analysis (Rosenberg, 1985), to insights and methods for dealing with applied problems. In all cases, operationally viable technology requires combinations with knowledge from other sources, including design and production engineering (Rosenberg, 1991).

(iii) The *nature of the impact* of basic science on technology is not restricted to direct transfers of codified knowledge, but also includes *access to research methods, instruments and uncodified knowledge* (Faulkner and Senker, 1991).

(iv) In this context, knowledge transfers are mainly person-embodied, involving *personal contacts, movements, and participation in national and international networks* (Gibbons and Johnston, 1974).

The broad demands for research skills

The importance of these last two impacts must not be underestimated. One important function of academic research is the provision of trained research personnel, who go on to work in applied activities and take with them not just the knowledge resulting from their research, but also skills, methods, techniques, and a web of professional contacts that will help them tackle the technological problems that they later face.

In one of their less well-known studies, Irvine and Martin (1980) have shown that Masters and Doctoral graduates from British radio-astronomy benefited in subsequent non-academic careers from the research skills – rather than the research knowledge – that they obtained during their post-graduate training. A more comprehensive survey undertaken at Yale University suggests that this is the rule rather than the exception.

The relevant results are summarized in Table 1.1. They show the responses of 650 US industrial research executives, spread across 130 industries, who were asked to rank the relevance to their technology of a number of fields of pure and applied science. Table 1.1 lists the number of industries in which each scientific field was given high ranking according to two criteria: first, the relevance of the knowledge base in the science to the technology; second, the relevance of the academic research in the science to the technology. As the authors point out:

> Industrial scientists and engineers almost always need training in the basic scientific principles and research techniques of their field, and providing this training is a central function of universities. Current academic research in a field, however, may or may not be relevant to technical advance in industry, even if academic training is important. (Nelson and Levin, 1986)

Table 1.1 The relevance of scientific fields to technology (USA)

Field of science	Number of industries (out of 130) ranking scientific field at 5 or above (out of 7) in relevance to its technology of:	
	Science (i.e. skills)	Academic research (i.e. knowledge)
Biology	14	12
Chemistry	74	19
Geology	4	0
Mathematics	30	5
Physics	44	4
Agricultural science	16	17
Applied maths & OR	32	16
Computer science	79	34
Materials science	99	29
Medical science	8	7
Metallurgy	60	21

Source: Nelson and Levin (1986); Nelson (1987)

Table 1.1 shows that, in most scientific fields, whether pure or applied, academic training and skills are relevant over a far larger number of industrial technologies than is academic research. The exceptions are the pure and applied biological sciences where we know from other studies that academic research is at present very close to technology (Narin and Noma, 1985; and see also Chapter 2). These results show clearly that most scientific fields are much more strategically important to technology than data on direct transfers of codified knowledge would lead us to believe.

Unplanned applications

A final major influence of science on technology is through unplanned applications, where useful knowledge emerges from research undertaken purely out of curiosity, without any strategic mission or expectation of application. Two US studies – one undertaken in the late 1960s and the other some 20 years later – both show the importance of such research for achieving relatively short-term technological objectives (Illinois Institute of Technology, 1969; Mansfield, 1991). In both cases, important innovations would have been substantially delayed without contributions from unprogrammed research performed in the ten years preceding the commercial launch of the innovations. Furthermore, in both studies, unprogrammed research contributed about 10% of the important knowledge inputs.

One implication of these findings is that programmed R&D should be built on a wider spread of non-programmed research. Analysts like Nelson (1959), and Kay and Llewellyn Smith (1985), have gone further and used various examples to suggest that more useful knowledge is produced in the long term by allowing basic scientists to pursue their own interests, than by fixing practical objectives for their work. It is a view that needs to be considered seriously by analysts in future (see, for example, Council for Science and Society, 1989).

4. Are the applications of science (at last) being internationalized?

The localized benefits of basic research

The analytical apparatus developed by economists to justify public subsidy to basic research has in general assumed a closed economy. This is paradoxical given that the main stimulus for public policies for science and technology has not come from any notions of (national) market failure, but from what is perceived as best practice in a world system of international competition where technological leads and lags are of central importance. It is also perhaps fortunate that the subject has not been pursued too often within this mainstream analytical framework: if we assume that basic research is a 'free good', an open international world would in principle permit any one country to live off the rest of the world's basic research (Kay and Llewellyn Smith, 1985).

As we have already argued, a country's ability to assimilate the results of other countries' basic research depends in part on the performance of basic research itself. An active national competence in basic research is therefore a necessary condition for benefiting from research undertaken elsewhere in the world; indeed it can be viewed as a national scientific intelligence system. And since most transfers of knowledge and skills between science and technology are person-embodied, the constraints of distance and language have meant that nation-based transfers between science and technology have been the exception rather than the rule.

Are science's benefits now being 'Globalized'?

Now, it is argued, conditions are changing. The barriers of distance and language are lower than they used to be. The training and employment of scientists and engineers is becoming more international in scope. And firms are internationalizing their R&D activities, which enable them more easily to benefit from academic science in a number of countries, through personal contacts and the hiring of scientists and engineers with research experience from local systems of higher education.

Does this mean that linkages between science and technology will become internationalized? Does it mean that an increasing proportion of the benefits of national governments' investments in basic research will 'leak away' through the migration of research scientists and engineers, and through the operation of foreign-controlled firms, to other countries?

This is a subject that deserves further research. Suffice to suggest at this stage that the degree of 'leakage' depends in part on the proportion of a country's corporate technological activity that is controlled by foreign firms, and which reflects their capacity to monitor and absorb local basic research

Table 1.2 Foreign controlled domestic technology compared to nationally controlled foreign technology (based on US patenting, 1981–6)

Home country	US patenting from inside country by foreign firms (as % of country's total US patenting)	US patenting by national firms from outside home country (as % of country's total US patenting)
Belgium	39.7	14.7
France	10.0	3.4
FR Germany	10.5	6.9
Italy	11.6	2.2
Netherlands	8.7	82.0
Sweden	3.9	11.3
Switzerland	6.0	28.0
UK	19.1	16.7
W. Europe	6.2	8.1
Canada	16.9	8.0
Japan	1.2	0.6
USA	3.1	3.2

Source: Patel and Pavitt, (1991)

skills and knowledge. Similarly, the importance of the foreign technological activities of nationally owned firms will reflect a country's capacity to benefit from basic research undertaken in other countries.

Table 1.2 is a first attempt to measure and compare these variables across countries. The first column gives the proportion of each country's US patenting originating from foreign-controlled firms. It shows that, in most countries, large foreign firms still play a relatively small role in national technological activities; only in Belgium, Canada and the UK do they account for more than 15% of the total. The second column gives the US patenting of nationally controlled firms from outside their home country, again as a proportion of total national patenting in the United States. It shows, as might be expected, that large firms based in small countries undertake a higher proportion of their technological activities outside their home countries. For the Netherlands, they amount to more than 80% of the national total, and more than 15% for Switzerland and the United Kingdom.

Taken together, the two measures show that most national technological systems are relatively self-contained. Both measures of internationalization are less than 20% of total technological activities in 8 out of the 11 countries. In Belgium and Canada, foreign-controlled domestic technological activities are much greater than domestically controlled foreign technological activities, whereas for the Netherlands, Sweden and Switzerland the opposite is the case. When Western Europe is considered as a whole, the degree of internationalization is much less than for most of the European countries taken individually, but still greater than that of either Japan or the United States.

These results show that complete internationalization of links between science and technology is not at all likely in the immediate future. In most countries, national basic science will still be feeding into largely nationally controlled technology, and close links with foreign science through personal contacts and recruitment will in most cases be small compared to national links. Contrary to conventional wisdom, Japan is not well positioned to benefit from foreign countries' basic science, since their firms undertake such a small proportion of their technological activities outside Japan. Our data suggest that the Dutch are much better at it.

5. Analytical and policy implications

Misguided policies seeking practical 'relevance'

Our analysis suggests that the objectives of many policies seeking to make basic research more directly useful may turn out to have been badly misconceived. Policies to concentrate on basic research that is directly applicable ignore the considerable indirect benefits across a broad range of scientific fields resulting from training, from instrumentation and methods, and from unplanned discoveries. They also neglect the all-important fact that the application of basic research depends overwhelmingly on the size and persistence in investment in downstream activities by business firms. Dealing with deficiencies in business R&D by making basic research more 'relevant' is like pushing a piece of string.

A revised case for public subsidy for basic science

Our analysis further suggests that the justification for public subsidy, in terms of complete inappropriability of immediately applicable knowledge, is a weak one. In fact, the results of basic research are rarely immediately applicable, and making them so also increases their appropriability, since – in seeking potential applications – firms learn how to combine the results of basic research with other firm-specific assets, and this cannot be imitated overnight. In three other dimensions, the case for public subsidy is stronger.

The first justification was originally stressed strongly by Nelson (1959), but has been neglected since then: namely, the considerable uncertainties before the event in knowing if, when and where the results of basic research might be applied. We now know from transaction cost theory that high uncertainty is one reason why markets are not necessarily efficient (Williamson, 1975). The probabilities of application will be greater with an open and flexible interface between basic research and application, which implies public subsidy for the former. The case for such a subsidy is strongest for 'unstrategic' fields of curiosity-driven research, the application of which cannot be foreseen.

A second, and potentially new, justification grows out of internationalization of the technological activities of large firms, discussed in section 4. Facilities for basic research and training can be considered as an increasingly important part of the infrastructure for downstream technological and production activities. Countries may therefore decide to subsidize them, in order to attract foreign firms or even to retain national ones. Recent interest in so-called 'science parks' might sometimes be one manifestation of this trend. Clearly there are dangers of competitive subsidy, the implications of which should keep game and trade theorists busy for some time.

The final and most important justification for public subsidy is training in research skills, since private firms cannot fully benefit from providing it when researchers, once trained, can and do move elsewhere. There is, in addition, the important insight of P. Dasgupta (1987) that, since the results of basic research are public and those of applied research and development often are not, training through basic research enables more informed choices and recruitment into the technological research community.

Better conceptualizations of science and technology

This last justification illustrates a broader conclusion emerging from this chapter (see also Mowery and Rosenberg, 1989): economists and other social scientists will benefit enormously in both the accuracy and impact of their analyses, if they discard their conceptualizations of science and technology as activities producing easily transmissable and applicable 'information', and recognize them instead as search processes and skills embodied in individuals and institutions. In this context, they would more easily appreciate the importance of basic research as both training and a cumulative body of knowledge. As we have seen, this was clear to de Tocqueville a long time ago. It was also clear to one of the major figures in the development of modern policies for basic science, Vannevar Bush, who wrote in 1945:

> [t]he responsibility for the creation of new scientific knowledge – and for most of its application – rests on that small body of men and women who understand the fundamental laws of nature and are skilled in the techniques of scientific research. (p. 7)

References

Arrow, K. (1962), 'Economic welfare and the allocation of resources for invention', in R. Nelson (ed.), *The Rate and Direction of Inventive Activity*, Princeton UP, New Jersey.

Averch, H. (1985), *A Strategic Analysis of Science and Technology Policy*, Johns Hopkins Press, Baltimore.

Carpenter, M. (1983), 'Patent citations as indicators of scientific and technological linkages', Computer Horizons Inc., New Jersey.

Council for Science and Society (1989), *The Value of 'Useless' Research: Supporting Science and Scholarship for the Long Run*, London.

Dasgupta, P. and David, P. (1988), 'Priority, secrecy, patents and the socio-economics of science and technology', *Center for Economic Policy Research*, Publication No. 127, Stanford University.

David, P., Mowery, D. and Steinmueller, W. (1988), 'The economic analysis of payoffs from basic research – an examination of the case of particle physics research', *Center for Economic Policy Research*, Publication No. 122, Stanford University.

Eads, G. and Nelson, R. (1971), 'Governmental support of advanced civilian technology: power reactors and the supersonic transport', *Public Policy*, 19, pp. 405–28.

Freeman, C., Clark, J. and Soete, L. (1982), *Unemployment and Technical Innovation*, Pinter, London.

Gibbons, M. and Johnston, R. (1974), 'The roles of science in technological innovation', *Research Policy*, 3, pp. 220–42.

Illinois Institute of Technology (1969), *Report on Project TRACES*, National Science Foundation, Washington.

Irvine, J. and Martin, B. (1980), 'The economic effects of big science: the case of radio-astronomy', in *Proceedings of the International Colloquium on Economic Effects of Space and Other Advanced Technologies*, Ref. ESA SP 151, European Space Agency, Paris.

Kay, J. and Llewellyn Smith, C. (1985), 'Science policy and public spending', *Fiscal Studies*, 6. pp. 14–23.

Kline, S. (1991), 'Models of innovation and their policy consequences', in Inose, H., Kawasaki, M. and Kodama, F. (eds), *Science and Technology Policy Research. What should be done? What can be done?*, Mita Press, Tokyo.

Mansfield, E. (1991), 'The social rate of return from academic research', *Research Policy*, 20, pp. 1–12.

Mowery, D. (1983), 'Economic theory and government technology policy', *Policy Sciences*, 16, pp. 27–43.

Mowery, D. and Rosenberg, N. (1989), *Technology and the Pursuit of Economic Growth*, Cambridge UP, Cambridge.

Narin, F. and Noma, F. (1985), 'Is technology becoming science?', *Scientometrics*, 7, pp. 369–81.

Narin, F. and Olivastro, D. (1992), 'Status report – linkage between technology and science', *Research Policy*, 21, pp. 237–49.

Nelson, R. (1959), 'The simple economics of basic scientific research', *Journal of Political Economy*, 67, pp. 297–306.

Nelson, R. (1962), 'The link between science and invention: the case of the transistor', in Nelson, R. (ed.), *The Rate and Direction of Inventive Activity*, Princeton UP, New Jersey.

Nelson, R. (1987), *Understanding Technical Change as an Evolutionary Process*, North-Holland, Amsterdam.

Nelson, R. and Levin, R. (1986), 'The influence of science, university research and technical societies on industrial R&D and technical advance', Policy Discussion Paper Series Number 3, *Research Program on Technological Change*, Yale University, Newhaven, Connecticut.

Patel, P. and Pavitt, K. (1991), 'Large firms in the production of the world's technology: an important case of non-globalisation', *Journal of International Business Studies*, 22, pp. 1–21.

Pavitt, K. (1987), 'The objectives of technology policy', *Science and Public Policy*, 14, pp. 182–8.

Rosenberg, N. (1976), 'Karl Marx on the economic role of science', in *Perspectives on Technology*, pp. 126–38, Cambridge UP, Cambridge.

Rosenberg, N. (1985), 'The commercial exploitation of science by American industry', in Clark, K., Hayes, A. and Lorenz, C. (eds), *The Uneasy Alliance: Managing the Productivity-Technology Dilemma*, Harvard Business School Press, Boston.

Rosenberg, N. (1990), 'Why do firms do basic Research (with their own money)?', *Research Policy*, 19, pp. 165–74.

Senker, J. and Faulkner, W. (1991), 'External acquisition of technological knowledge', SPRU (mimeo).

Smith, A. *The Wealth of Nations*, Routledge (1985 edition).

Stoneman, P. (1987), *The Economic Analysis of Technology Policy*, Oxford, UP, Oxford.
Stoneman, P. and Vickers, J. (1988), 'The assessment: the economics of technology policy', *Oxford Review of Economic Policy*, 4, pp. i–xvi.
de Tocqueville, A. (1830), *De la Democratie en Amerique*, Union Generale d'Editions, Paris (1963, abridged edition).
Williams, B. (1986), 'The direct and indirect role of higher education in industrial innovation: what should we expect?', *Minerva*, 2–3, pp. 145–71.
Williamson, O. (1975), *Markets and Hierarchies: Analysis and Antitrust Implications*, Free Press, New York.

2 The dynamics of competition in a science-based technology: the case of biotechnology

Luigi Orsenigo

1. Introduction

The analysis of the process of emergence and development of biotechnology provides an important example of how technological change proceeds in modern industrial economies and suggests relevant insights into the forms of competition and the origins of competitiveness in science-based industries. Biotechnology is certainly, in many respects, an extreme case. Thus, some of the indications stemming from its observation cannot be generalized directly to other technologies and industries. Yet, precisely because of its extreme features, this case illustrates more clearly some issues which have both a practical and a theoretical relevance:

(a) The role of science and scientific institutions as a major source of technological opportunities and competitiveness and hence the problems connected with the transfer and transformation of scientific knowledge into technological innovations.
(b) The patterns of competition among business firms and their forms of organization in environments characterized by fast technological change.
(c) The role of institutions and of the national systems of innovation in affecting the patterns of competition and the competitiveness of individual companies and countries.

The practical significance of these issues is quite evident to any businessman or woman and policy-maker. From a theoretical perspective, the case of biotechnology offers interesting evidence and insights into some very general questions that are at the heart of the contemporary debate on the determinants of competitiveness and the patterns of competition at the firm and at the country level.

Indeed, the analysis of biotechnology requires an investigation into the properties of environments characterized by (a) ample opportunities for innovation, the exploitation of which is bounded only by the ability and the willingness to become involved in costly and uncertain learning processes; (b) high degrees of uncertainty about the future evolution of technology and high degrees of variety in the technological capabilities, expectations and strategies of a large number of agents: companies active in different industries, universities and other research centres, financial institutions, etc.; (c) irreversibilities and dynamic increasing returns arising from learning activities which continuously recreate opportunities for innovation along particular and specific trajectories.

Against this general background, this chapter argues that the emergence and evolution of biotechnology can be fruitfully analyzed as a process of collective learning among differentiated agents, involving both competition and cooperation and the construction of adequate organizational solutions and institutions for the coordination of such learning processes. The specific features of this process are determined by the properties of the technology, i.e. by the nature of the technological regime, by the level and distribution of scientific and technological capabilities across companies and countries and by the institutional set-up in which innovative activities occur, i.e. by the structure and behaviour of the national systems of innovation.

The chapter deals mainly with the process of emergence of biotechnology and is organized as follows. Section 2 examines the basic properties of biotechnology as a technological regime, the resulting features of the patterns of innovative activities and the conceptual background of the analysis. Section 3 analyzes the role of science and scientific institutions in the development of biotechnology. Section 4 examines the patterns of evolution of innovative activities and of their forms of organization as well as the prevailing forms of competition. Finally, in section 5 the main conclusions are summarized and some policy implications are briefly discussed.

2. The nature of the technological regime and the resulting pattern of competition

Recent developments in the economics of technological change have suggested that the sectoral patterns of innovative activities are determined to a substantial extent by the very nature of the relevant technology itself, i.e. by the specific properties of the relevant technological regime (Nelson and Winter, 1982; Pavitt, 1984; Malerba and Orsenigo, 1990).

A technological regime is defined as a combination of particular knowledge bases, sources and degrees of technological opportunities, conditions of appropriability, forms and degrees of cumulativeness of technological advances. Indeed, in the case of biotechnology, the specific nature of the emerging technology identifies the problems and the requirements that each individual company and country had to face when undertaking innovative activities in this area.

The nature of the knowledge base

First, biotechnology is a strongly science-based technology. It emerged from the tremendous growth of molecular biology in the post-war years and its development took place in the laboratories of universities and other research centres. Such research was spurred on by purely scientific motivations and was funded essentially by public sources, primarily the NIH in the United States.[1]

The subsequent development of industrial innovations rests on the prior understanding of the scientific principles underlying them and, especially in the early stages, industrial and scientific research have substantially overlapped.

Contemporary biotechnology represents also a radical change of previous technological trends. The basic discoveries of recombinant DNA and hybridoma technology have opened new research directions and have revitalized more traditional technologies (e.g. 'second generation' biotechnology). The knowledge base underlying technological innovations in this area draws now on a wide spectrum of scientific disciplines and technical know-how ranging from biochemistry, microbiology, cell culture and fermentations to molecular biology, genetics, immunology, virology, cell biology and tissue culture.

Relatedly, the search procedures and heuristics as well as the main trajectories of potential technological advances are radically different from the past. They involve the *ex-ante* programming and design of molecules having specific properties, rather than the reliance on the slow and 'random' process of natural mutation. Thus, genetic engineering might be considered to have generated a new technological paradigm (Dosi, 1982), in that it has created a new body of knowledge characterized by particular heuristics, rules, directions of research, products and processes.

However, contemporary biotechnology emerged also as the result of the integration of various scientific disciplines and technologies and it is inherently characterized by a high degree of multidisciplinarity. The source of such knowledge is differentiated and located in a multiplicity of agents and institutions. The need to organize specialised, multidisciplinary training programmes for biotechnologists is a direct consequence and manifestation of this feature. Even more important, the translation of scientific knowledge into new products and processes requires the fusion of the new techniques with the older technologies, especially where the processes for large-scale production are concerned.

Opportunities, pervasiveness and uncertainty

Biotechnology offers in principle enormous technological opportunities for industrial innovation and is characterized by high degrees of pervasiveness, since its potential applications concern a large variety of products and processes in a wide number of industries and markets, ranging from pharmaceuticals and agriculture to chemicals, food, energy, etc. Thus, technological opportunities relate not only to the potential direct economic benefits that can be expected from any single innovation, but above all they are generated from the ample scope for diversification in a wide spectrum of related technological areas and markets which is offered by biotechnology.

However, the technical feasibility of the potential applications of biotechnology, the time required to develop them and the economic revenues which can be generated by innovations in this area are extremely uncertain. Thus, expectations on the future developments of the technology differed drastically even among professional consultants. World market estimates for biotechnology products ranged from less than $10 billion to over $60 billion.

Appropriability and cumulativeness

Uncertainty is exacerbated by the confusion surrounding the extent and the means through which the revenues stemming from innovations can be privately appropriated. Especially in the early stages of the development of biotechnology, the relevant knowledge base had to a large extent a generic nature and it was available and accessible through public sources. At least in principle, the relevant knowledge can be codified and can be acquired by everybody having sufficiently high scientific and technological capabilities. As the basic technological and scientific capabilities started to diffuse, purely technical lead-times have become very short, making imitation relatively easy (Harrigan, 1981; OTA, 1984). Thus, patent protection represents a crucial requisite for the private appropriation of the profits generated by innovations. The confusion over the patenting conditions of biotechnology innovations and the rapid developments in patent legislation have been therefore a major factor affecting innovative activities. Although in recent years important progress has been made in the definition and harmonization of patent legislation, several problems remain open (OECD, 1985), particularly as it concerns the conditions under which patents can be obtained on products or only on processes. Litigation and secrecy have thus become a prominent feature of the biotechnology industry.

Moreover, the control of what David Teece (Teece, 1986) has defined as co-specialized assets (e.g. the experience in and the access to the procedures for approval and registration of products, as well as the availability and control of marketing networks) has become a further fundamental condition for capturing the profits from innovation (Teece, Pisano and Shan, 1987).

As Cohendet *et al.* argue in the next chapter, as a technology develops, other means of appropriability become more important. Gradually, specific skills having a more tacit and idiosyncratic nature are accumulated, especially where large-scale production is concerned. The variability and the biochemical complexity of bioprocesses makes reverse engineering practically impossible and gives rise to steep learning curves (Harrigan, 1981; OTA, 1984; Hacking, 1986).

As a consequence, moreover, as the technology develops, the degree of cumulativeness of technological advances tends to increase. Whilst in the very early stages the emergence of a novel knowledge base had meant that past accumulated capabilities had little relevance (or even a negative influence) for current innovative activities, as time went by the situation rapidly reversed. The growing importance of past accumulated capabilities stems from different factors. First, cumulativeness has a purely cognitive dimension. It relates to the fact that learning necessarily builds up on past experiences and capabilities and to the limited abilities of any one organization in exploring successfully areas of knowledge which are too far from its current core capabilities. Second, it is apparent at the organizational level. The experience of conducting highly complex research at the frontier of technology in a wide range of areas, managing complex and sophisticated organizational structures is likely to represent an important capability in its own right which is itself learnt over time (Chesnais, 1986; Stiglitz, 1987). More generally, learning entails exclusion effects, specialization and irreversible phenomena which are the sources of non-convexities and increasing returns.[2]

These properties of the technological regime identify some major factors of competitiveness for individual companies and countries as well as some general features of the patterns of competition.

First, the science-based nature and the novelty of the knowledge base implies that the industrial applications of biotechnology impinge on the formation of strong scientific capabilities as a necessary prerequisite for the accumulation of technological capabilities. Would-be innovators must acquire knowledge which is to a large extent completely new to them, develop new competences and research procedures and fuse them with other capabilities and competences. Since such esoteric skills were located in the laboratories of some selected academic institutions and were often embodied in a few individuals, they must get access to these external sources and develop organizational mechanisms to coordinate the process of transfer and acquisition of knowledge.

Second, the complexity and pervasiveness of the knowledge base requires that the development of technological capabilities rests on the accumulation of high competences in a wide range of scientific disciplines, old and new technologies and, above all, on their integration. Thus, would-be innovators have to develop both strong internal capabilities as well as contacts with a relatively large number of agents in order to acquire the necessary pieces of knowledge and to establish a network of linkages with different companies and institutions which individually have accumulated only fragments of the relevant knowledge. The integration and the development of the various knowledge bases is therefore a crucial step in the evolution of biotechnology and the ability to do so is an essential skill defining competitiveness.

Third, the exploitation of the opportunities offered by biotechnology requires not only high R&D expenditures, but also a reorientation and a redefinition of the existing boundaries across academic departments in universities, across divisions at the company level and across industries at the aggregate level. Diversification is therefore a major trajectory of technological advance in this area and a major organizational necessity.

From this point of view, the evolution of biotechnology can be analyzed primarily as a process of collective learning by differentiated agents, involving the coordination of various learning sources and procedures, the construction of adequate organizational solutions and the progressive emergence of established research directions.

Such learning processes, however, involve difficult organizational problems, first of all, the conventional ones implicit in the production, exchange and transfer of knowledge (Arrow, 1962; Teece, 1988), related to the appropriability of the results of the research on the one hand, and to the need to develop long-term, stable relationships in order to make possible the exchange of specific and often tacit knowledge. More generally, when it is recognized that individuals and organizations have limited capabilities, that learning activities are local and interactive (Atkinson and Stiglitz, 1969; Kline and Rosenberg, 1986), imply cumulativeness and exclusion effects and proceed through the formation of robust routines which store existing capabilities and guide future research, organizational devices for the coordination of individual learning activities are necessary and inherently complex (Cohen, 1987; Nelson and Winter, 1982; Dosi and Egidi, 1990). In particular, such organizational solutions have to satisfy the trade-off existing between the need to coordinate complementary learning activities along particular trajectories by developing robust codes and mechanisms of communications and the need to explore new directions which are external to the organization and require radical changes in the current routines. Thus, in any one organization trade-offs permanently exist between coordination and decentralization and between allocative and learning functions (Nelson and Winter, 1982; Pavitt, 1989; Dosi and Aoki, 1989; Orsenigo, 1989).

The nature of the technological regime can account to some extent for the specific organizational forms that are more likely to prevail. For instance, high degrees of cumulativeness and of complexity of the knowledge base, low degrees of appropriability are factors pushing towards higher centralization of innovative activities. High opportunity conditions imply at the same time high expected profits from the direct control and use of knowledge (and therefore incentive towards centralization) but also the need to acquire and experiment with new external sources of knowledge.[3]

Fourth, innovative activities and decisions in this area take place in conditions of high uncertainty. Given the multidisciplinary and pervasive nature of the knowledge base, entry might occur from existing companies active in different sectors and from new companies. A large variety of strategies is in principle possible, ranging from specialization in particular technological niches to the accumulation of capabilities over a broad spectrum of areas. Moreover, in these contexts, decisions to invest in the new technology cannot be myopic, but must take into account the long-run benefits of investment deriving from the acquisition of further capabilities to innovate in the future (Cohen and Levinthal, 1988; Stiglitz, 1987; Pavitt, 1989). However, given the cumulative nature of technological progress, such decisions are likely to embody strong features of irreversibility. Accumulated capabilities influence the possibilities to innovate in the future and the research strategies of the agents. Moreover, whilst early entry may provide a substantial 'first mover' advantage *vis-à-vis* competitors, companies might later find themselves locked-in in the 'wrong' trajectories. The success of any one strategy or trajectory of research is likely to be determined in this kind of environment by the past history of accumulation of technological capabilities among the relevant agents and by 'small events' (Arthur, 1989) and cannot be predicted at the outset. Finally, if appropriability conditions are such that late entrants can to some extent benefit from the experience accumulated by early entrants through imitation or simply because part of the knowledge which is accumulated becomes at least partly public, a 'wait and see' strategy may turn out to be more profitable. However, such conditions are hardly observable and predictable and given the high degree of uncertainty, agents are forced to take decisions, the rationality of which will be revealed only considerably later (Silverberg, Dosi and Orsenigo, 1988; Silverberg, 1988).

Variety in the capabilities, expectations and strategies of the agents is likely therefore to be an unavoidable feature of competition but also a necessary condition for technological advance. At the same time, the existence of a strong and differentiated industrial base, capable and willing to undertake such risky learning activities over very long time-horizons is therefore a further fundamental factor of competitiveness. Thus, accumulated capabilities have become increasingly important in defining the rates and the directions of innovation. The possession of capabilities in 'second generation' biotechnology, and more generally in life sciences, as well as the existence of large and diversified R&D laboratories have progressively emerged as crucial variables influencing innovative activities and competitiveness.

Finally, the learning activities of individual agents, their strategies, the modes of competition and/or cooperation among them and, more generally the patterns of innovative activities are strongly influenced by the institutional set-up, primarily as it concerns the academic system, the financial system, patent laws and research regulations and obviously government intervention. Moreover, the attitudes of the public towards genetic engineering change considerably across countries and influence heavily research strategies and government legislation.

The different conditions of these variables across companies and countries defined the boundaries within which learning and competition took place. The economic and institutional conditions of any one country defined the particular patterns of innovative activities. Let us now examine these factors in some more detail.

3. The role of science and scientific institutions

The level and distribution of scientific capabilities in biotechnology, as well as the nature of the scientific system, have been important variables affecting the rates and patterns of technical change in any one country. In particular, the American scientific superiority provided the United States with a substantial technological lead, especially in the early stages (Tables 2.1 and 2.2).

Indeed science is a crucial cognitive component of the knowledge base underlying biotechnology. Its importance goes far beyond the simple utilization of scientific discoveries for creating new products and processes. Perhaps more important, biotechnology research implies to a large extent the use of scientific procedures and heuristics and it is now becoming clear that genetic engineering is primarily a research tool rather than a direct source of techno-

Table 2.1 Patent activity by country (%)

	Biotechnology		Genetic engineering	
	1980–4	1963–84	1980–4	1963–84
USA	55.4	54.8	64.7	59.4
Japan	42.3	45.7	39.2	36.8
FRG	17.4	14.6	9.8	11.2
UK	7.5	8.0	9.8	7.9
France	6.4	7.2	8.8	9.9
Italy	4.7	3.6	1.0	1.3

The US share is calculated over the total number of US patents. The share of the other countries is calculated over the total number of foreign patents registered in the United States.

Source: Orsenigo, 1989 (OTAF/SPRU Data Base)

Table 2.2 Scientific output by country: scientific publications in genetic engineering and biotechnology (%)

	Genetic engineering (1) (1978–86)	Biotechnology (2) (1982–3)
USA	49.9	32.2
Japan	7.3	13.9
FRG	6.9	4.8
UK	7.7	9.9
France	6.0	4.1
Italy	n.a.	1.2

(1) Genetic engineering papers in chemical abstracts and biological abstracts.
(2) Biotechnology papers in current biotechnology abstracts.

Sources: Adapted by Stankiewicz, 1986; Federchimica, 1986

logical innovations. Thus the behaviour and interests of scientists had a crucial role in transferring generic scientific knowledge into feasible technological applications and in selecting the possible avenues of technological research. For instance, the capabilities and interests of scientists (which had primarily a medical origin) may account for the emphasis laid in the early stages of industrial biotechnology on the development of particular products (e.g. pharmaceuticals, growth hormones, interferon, etc.), on the use of specific microorganisms and on the relative lack of attention given to the problems connected with scaling-up and production. In Europe and Japan, scientists had been very important in eliciting and in shaping public policies.

On the other hand, industrial companies have become an increasingly important source of scientific research, as their large and growing share in the genetic engineering literature shows (Table 2.3). In other words, a process of interpenetration of the traditional functions of industry and universities has marked the development of biotechnology. Universities have become increasingly involved in technological, for-profit activities, but industry has assumed a prominent role as producer of scientific knowledge. As such, this direct involvement of industry in scientific research represents by itself a striking phenomenon. As is well known, companies invest usually very little in basic research, with the notable exceptions of pharmaceutical and chemical firms. Given the public, uncertain and generic nature of such knowledge, companies do not usually find sufficient economic incentives to invest in its production (Arrow, 1962; Rosenberg, 1988), unless it is directly relevant for technological innovation and/or it provides substantial 'first mover' advantages. This is certainly the case in biotechnology. A further and possibly more important motive can account however for such a heavy involvement of business firms in basic research, namely the fact that such research represents an important intermediate input in the inventive, rather than in the productive process (Rosenberg, 1988; Cohen and Levinthal, 1988). In other words, the acquisition of basic, generic capabilities increases in the long run the capability of innovating in the future, although in unexpected and unpredictable directions,

Table 2.3 Institutional origin of genetic engineering literature (%)

	Universities		Other research institutes		Companies	
	80–1	84–5	80–1	84–5	80–1	84–5
USA	80	52	15	17	5	30
Japan	45	59	35	16	8	25
Europe	44	45	52	35	4	20

Source: Adapted by Stankiewicz, 1986

directly and through an increased ability to monitor and to evaluate research conducted elsewhere (Rosenberg, 1988).

However, despite the proximity, and in many cases, the overlapping of scientific and technological research, these two forms of knowledge remain different, both at the cognitive and at the institutional level. The development of technological innovations requires, as was emphasized above, the fusion of different disciplines and capabilities. Since scientific knowledge was mainly located in the university laboratories, companies had to develop linkages with the scientific community in order to have access to it. Thus, scientific knowledge has to be transferred from universities to industry and has to be transformed into technology.

4. University–industry relationships

The role of scientific institutions in the production of technically relevant knowledge is quite evident (Table 2.4) and the intensity of industry–university collaboration is certainly a distinctive feature of biotechnology. In the late 1970s and early 80s industry–university relationships have experienced an explosive rate of growth and have assumed a large variety of institutional forms, especially in the United States. In Europe and in Japan, the strengthening of industry–university ties has been a major focus of public policies.

Besides the substantial involvement of industry in basic research and the exceptionally high intensity of industry–university ties, two more major features characterize these relations in the case of biotechnology: (a) the direct involvement of scientific institutions and individual scientists in the economic exploitation of knowledge; and (b) the targeted and long-term nature of the relationships.

Whilst a large fraction of industry–university ties developed following traditional routes such as contract research arranged for relatively short periods of time and finalized to solve specific problems, long-term multimillion dollar research agreements between companies and academic centres played a prominent role, resulting in more than $150 million being channelled into scientific institutions between 1980 and 1983 (Kenney, 1986). These contracts usually involve the commitment of the sponsoring corporation to fund university basic research in return for some proprietary advantage on the result of

Table 2.4(a) Institutional origin of US biotechnology patents by country: 1980–4

	Companies	Private indiv.	Non-profit inst.*
USA	67.9	13.2	18.9
Japan	93.0	2.5	4.5
FRG	84.8	9.1	6.1
UK	71.9	7.0	21.1
France	42.6	14.9	42.5
Italy	95.4	2.3	2.3

* Universities, public research centres, hospitals, government agencies

Table 2.4(b) Patenting activities of public agencies and scientific institutions by country* (% of total country patents): 1980–84

	Gvt. agencies	Univ. and hosp.	Other res.centres
USA	4.8	15.1	2.4
Japan	1.9	—	3.3
FRG	—	—	7.3
UK	27.3	—	—
France	10.6	—	31.9
Italy	2.3	—	—

* Patents held by private individuals are not included

Source: Orsenigo, 1989 (OTAF/SPRU Data Base)

such research, typically the right for exclusive licensing of all discoveries made in connection with the research agreement. Research concerns usually a relatively broad but specifically targeted area which has some definite commercial interest to the company.

This type of contract allows the funding company to acquire not only a window on a new, fast developing technology, but also the skills of the principal investigator and of the entire laboratory and research team. Thus the company gets access to a highly qualified work force and to unique training services whenever (as is often the case) the contract considers the exchange of personnel.

In other cases, universities have attempted to exploit directly their research skills through the creation of controlled companies and/or systematic patenting policies.[4]

The most striking example of the direct role of scientists in the economic exploitation of scientific knowledge was however the proliferation of university spin-offs, whereby individual scientists started their own company with the backing of venture capital. We shall discuss the role of the new biotechnology firms (NBTFs) in more detail in the next section. Suffice it to say for the time being that the variety and the nature of these institutional forms of transfer of academic research to industry witness not only the relevance of science in a science-based technology in its early stages, but also the problems involved in this task.

Indeed, the exploitation for technological purposes of scientific capabilities is not a simple and straightforward process. On the contrary, it has raised difficult problems and required considerable organizational and institutional innovations.

First, it must be emphasized that there was little experience of collaboration between industry and biologists. To be sure, pharmaceutical and chemical companies had a long tradition of collaboration with scientists and scientific institutions. However, the biological research in the disciplines which constitute the basis of contemporary biotechnology was by and large separated from any industrial application. Genetics, molecular biology and cell biology had received very little support from industrial sources and biologists were usually very little involved in industrial activities. Some collaboration had been developing before the advent of genetic engineering in the field of biomedical research in the medical and agricultural schools, especially in the United States. Linkages existed in bioprocess engineering in most countries and were particularly strong in Japan, but they were rarely either on a large scale or systematic and concerned essentially in the training of personnel and technical consultation.

The separation between biology and industry exacerbated the traditional differences between academic and industrial research, related to their different social organization (e.g. public disclosure vs. private appropriation of the results, mechanisms of compensation based on fame and prestige vs. monetary rewards) and cognitive orientation (e.g. nature vs artefacts) (Mayr, 1982; Dasgupta and David, 1987).

Finally, industry–university relationships involve the organizational problem discussed earlier. The variety and the nature of these ties reflect the technical and economic problems involved in the transfer process. A key feature of the system of industry–university collaboration in biotechnology is certainly the increasingly proprietary nature of scientific knowledge. Scientists and universities found themselves to be endowed with the monopoly of a strategic asset, i.e. knowledge, and they set out to exploit it commercially, by restricting its circulation and accessibility.

A considerable and often bitter discussion was raised on the ethical and economic appropriateness of these trends. In any case, secrecy about the directions and the results of research, the communication of information to the corporate sponsors before its publication, the need for protecting valuable information before and after patenting in order to avoid the loss of patent rights are claimed to have become permanent features of academic biology research (Kenney, 1986). This debate is certainly not new. On the contrary, it has repeatedly emerged whenever a certain science has become particularly relevant for industrial activities. Yet, the case of biotechnology contains some elements of novelty due to the scale of the phenomenon and the speed with which the process has occurred. In any case, the debate on industry–university relationships highlights the clash of 'rationalities', interests, motivations of the scientific community and the industrial world implied by the transfer of science to technology.

In the case of biotechnology this clash was partially solved through the emergence of a strongly profit-oriented 'rationality' within the scientific community.

As far as companies are concerned, access to science was essential for building up autonomous capabilities and it represented a valuable asset which might have guaranteed a substantial lead *vis-à-vis* competitors. Close contacts with universities coupled with the availability of proprietary rights on the results of the research provided therefore an important competitive advantage.

5. Diversity of company strategies and national systems

However, in conditions of high uncertainty and in the absence of well-defined trajectories of technological advance, companies had to experiment with alternative options and possibilities. Therefore, they established differentiated contacts with a number of different institutions (universities, NBTFs, other companies) in order to accumulate basic generic capabilities in a wide range of areas. Indeed linkages with universities represented one aspect of a wider strategy of collaboration and exploration.

Yet, this strategy was viable and satisfactory to the extent that the knowledge to be transferred was of a generic nature and still not immediately relevant for production and commercialization. Moreover, such knowledge was still embodied in a relatively small number of people and institutions, at least where frontier research was concerned, and had to be merged with other more specific and idiosyncratic competences within individual companies. Thus, its transfer and application still required stable and close contacts between the partners. Appropriability considerations, finally, represented a further motive for establishing intimate relationships.

Finally, it must be noted that despite their vast diffusion and heterogeneity, the linkages concentrated heavily in a relatively small number of universities and companies, especially where the most important and systematic ties were concerned. Establishing costly agreements with scientific institutions was worthwhile insofar as the quality of the know-how was high. Moreover, it is worth stressing that the same universities had a long history of collaboration with industry in the field of high technology, medical and agricultural research. Indeed, in most of the long-term research contracts the principal investigator had previous experience of consultancy with the sponsoring company and was very well known by them.[5]

The same is true also as far as companies are concerned. Those firms which have been more consistently involved in linking with universities had systematic experience of collaboration and were also among the first to develop strong in-house research facilities and to engage systematically and ambitiously in biotechnology.

These results confirm the hypotheses that: (a) linking with academia is not a substitute but rather highly complementary to the development of strong in-house capabilities; (b) the ease, propensity and efficacy of these forms of interaction is a function of the absolute and relative capabilities of the agents involved and of the existence of previous experience and communication channels between them.[6] Indeed, in countries other than the United States, where scientific capabilities could not match the American levels and/or were concentrated in few selected research centres, industry–university linkages were more difficult to create, with the possible exception of the United Kingdom.

To be sure, in these countries too, industry–university linkages experienced a remarkable growth (Faulkner, 1986; Orsenigo, 1989). However, whilst in the United States the development of industry–university ties was an extremely rapid, spontaneous and decentralized process, in Europe and Japan considerable efforts were devoted to strengthen the scientific base and the institutional mechanisms governing industry–university collaboration. The involvement of the scientific community was strongly influenced by public action and was exerted mainly through its direct participation in the process of formation of public policies. Such policies were markedly different across countries, reflecting the specific organization and levels of the national research systems, but in general they relied to a large extent on non-academic public research institutions and sometimes creating new ones *ad hoc* in order to organize the dispersed scientific capabilities in poles of excellence.[7]

Undoubtedly, the much less flexible and commercially orientated nature of academic systems in Europe and in Japan made it more difficult for individual scientists and universities to become so deeply involved in the exploitation of their knowledge as in the United States, the only partial exception being the United Kingdom, where the organization of academic science is much more similar to the American one. Industry–university collaboration however takes place through different, usually more informal, channels.

However, no simple and direct relationship can be found, in my view, between the detailed structural features of the scientific system and the performance of any one country. The absolute level of scientific capabilities remains the single most important variable explaining the intensity and the efficiency of industry–university ties, whilst the structure of the academic system accounts essentially for the specific institutional form of such relationships. In particular, a transfer system heavily oriented towards applied research and informal collaboration nevertheless allowed Japan to progress rapidly.

The relative scientific backwardness of Europe and Japan produced two contradictory phenomena. On the one hand, the US scientific system increasingly attracted foreign researchers and resources, generating a brain drain which might substantially weaken the national research system. On the other hand, both Japanese and European companies relied extensively on the American scientific system to get access to the relevant knowledge. As Pavitt has argued in Chapter 1, it was not just a question of reading scientific papers but of much more intimate forms of access and collaboration. The research contract signed by Hoechst and the Massachusetts General Hospital in 1981 was the first major move in this direction and since then practically all major European and Japanese companies have established some form of collaboration with American institutions and new specialized biotechnology firms.

Whilst these trends strengthened the American scientific leadership, they also allowed import and appropriation of the basic scientific know-how, providing the basis for the domestic development of biotechnology. Thus, the American scientific system performed partially as an externality for foreign companies, but only to the extent that they were themselves capable of assimilating this knowledge.

6. The patterns of innovative activities and the nature of competition

The establishment of contacts with universities and other scientific institutions was the first step and represented one component of a more complex strategy towards entry into the new technology by industry.

Indeed, two main features characterize the patterns of innovative activities and the modes of competition in biotechnology: (a) the variety of agents involved and of their strategies concerning the size, directions and procedures of research, commercial perspectives, etc.; (b) the coexistence of fierce competition and strong collaboration between firms of different age and size.

Given the uncertainty and the high degree of indeterminacy surrounding biotechnology in its early stages, a high variety of strategies concerning the timing of entry, the size of the effort and the directions of research characterize the patterns of competition in this area. The availability of strong scientific capabilities was certainly a major factor affecting the incentives and the possibility to engage in biotechnology. However, other variables played an important role.

First, the availability of a strong and differentiated industrial base which was capable and willing to undertake costly and risky research activities defined the seedbed of firms potentially interested in the new opportunities and capable of exploiting them. As mentioned earlier, the very nature of the technology implied in the long run costly and complex research efforts over long time-horizons[8] and therefore adequate organizational experience in this field. Moreover, given the multidisciplinary and pervasive nature of biotechnology, the existence of a large number of such companies in a wide range of industries increased the potential for innovation not only directly, but also through the possibility of exploiting the opportunities for diversification.[9] Finally, although genetic engineering represented a radical change of research procedures and perspectives which might have rendered previously accumulated capabilities obsolete, nevertheless the possession of capabilities in second generation biotechnology made it easier at least to realize the opportunities for innovation.

Thus, the availability of these conditions raised both the size of the population of potential entrants and the probability that, *ceteris paribus*, companies might decide to enter the new technology. Finally, in the longer run, as technology comes closer to production and its development becomes more cumulative, the possession of technological capabilities in production technology and the control of co-specialized assets become increasingly important conditions of competitiveness.

The level and distribution of both scientific and technological capabilities determined therefore the patterns of evolution of innovative activities across countries and over time.

The entry of industry into biotechnology was very rapid, beginning soon after the key scientific discoveries. The rates of technological natality and mortality measured in terms of patents (see Table 2.5)[10] reached significant values in all countries already in the late 1970s, witnessing a high degree of dynamism and the substitution of old innovators with new ones. These new innovators, however, are not necessarily newly founded firms and their contribution to total patents is probably a better indicator of their technological

Table 2.5 Biotechnology: rates of technological natality and mortality in terms of patents (1980–4)

	Rate of natality	% of patents	Rate of mortality
USA	26.2	72.2	59.2
Japan	46.1	23.1	23.1
FRG	117.6	22.5	46.4
UK	47.1	27.3	41.2
France	76.5	55.3	64.7
Italy	144.4	86.4	64.7

Rate of natality: Number of firms patenting for the first time/total number of firms patenting in the previous period (1975–9). *% of patents*: Patents held by the firms patenting for the first time/total patents in the same period.
Rate of mortality: Number of firms which stop patenting/total number of firms in the previous period (1975–9).

Source: Orsenigo, 1989 (OTAF/SPRU Data Base)

relevance than the rate of natality measured in terms of the number of firms. Moreover, these data reflect very different initial conditions. Thus, the high rates observed in Italy, France and Germany reflect primarily the strong degree of concentration of technological capabilities related to biotechnology in a small group of companies, rather than the entry of a high number of new innovators.

In general, three types of companies contributed to the take-off of biotechnology:

1. Companies which had already substantial capabilities in second generation biotechnology, i.e. in fermentation and in products like antibiotics, vaccines, enzymes, etc. and/or in related technologies.
2. Companies which had no previous tradition in these activities but perceived the potentialities/threats provided by genetic engineering and were willing to diversify in new product markets on these grounds.
3. New companies specialized in genetic engineering which were explicitly created to exploit their specific capabilities.

All these companies played an important and complementary part in the development of biotechnology, although their relative role, their behaviour and the relationships among them varies markedly over time and across countries. Entrants came primarily from the pharmaceutical and chemical industry, i.e. the industries more directly related to biotechnology, but also from other sectors reflecting the existence of strong technological skills and of numerous large technologically dynamic companies across countries (Orsenigo, 1989).

In the United States innovative activities resulted from the initiative of a variety of firms of different size and industries (Table 2.6); given the high degree of variety, the relatively low degree of cumulativeness and appro-

Table 2.6 Size distribution of the firms holding US patents in biotechnology (1980–4)

Employees	0–500	501–1,000	1,001–10,000	>than 10,000
USA	6.4	3.2	14.5	75.9
Japan	5.5	5.5	20.4	68.5
FRG	2.9	6.4	9.3	81.4
UK	7.3	4.9	—	87.8
France	—	14.8	29.6	55.6
Italy	2.4	16.6	26.2	54.8

Sample: of total company patents: USA 71.7%; Japan 84.2%; FRG 98.3%; UK 88.6%; France 94.3%; Italy 100%

Source: Orsenigo, 1989 (OTAF/SPRU Data Base)

Table 2.7 Patent concentration ratios (C4). Sum of the shares of total country biotechnology patents registered in the USA held by the four largest organizations in terms of patents

	Biotechnology 1980–4	Biotechnology 1969–84	Genetic engineering 1971–84
USA	25.8	19.7	25.9
Japan	29.5	40.4	55.6
FRG	64.2	64.9	59.9
UK	52.6	44.8	53.3
France	46.8	55.1	69.3
Italy*	66.6	59.5	66.6**

* C3 ** C2

Source: Orsenigo, 1989 (OTAF/SPRU Data Base)

Table 2.8 Estimated number of firms active in biotechnology by country

	(a) 1985	(b) 1969–84*	(c) Patents per firm**
USA	495	349	8.2
Japan	304	112	10.1
FRG	69	51	7.1
UK	148	34	5.8
France	78	40	4.2
Italy	31	23	2.5

* Companies holding US patents
** Total number of patents in the period 1969–84/total number of firms holding US patents

Source: Orsenigo, 1989

priability in the early stages, the degree of concentration of innovative activities is as a consequence relatively low and the hierarchy of the main innovators changed considerably over time. In Japan and in the European countries innovative activities remained concentrated in a smaller group of large, highly diversified corporations, the excellence of which partly compensated for the lower degree of variety, especially in Japan and in Germany (see Tables 2.7 and 2.8).

7. New biotechnology firms (NBTFs)

The proliferation of new, specialized biotechnology companies was essentially an American phenomenon and was made possible, first of all, by the nature of the technological regime. High opportunity conditions and a new knowledge base weaken the entry barriers represented by the cumulative nature of technical change (Winter, 1984; Malerba and Orsenigo, 1990; Orsenigo, 1989). Their diffusion and relevance, however, was determined basically by the capabilities and structure of the scientific system (and of the financial system), insofar as the incentives to start a new company were primarily a function of the level of the scientific skills available. More generally, the proliferation of new biotechnology firms represented a further institutional solution to the problem of the transfer of science to industry linked to the excellence of the American scientific base and facilitated by the structure of the academic system and by the availability of a vigorous venture capital industry. In other countries, where these conditions did not exist (again with the partial exception of the United Kingdom), their role was partially taken by the public research system, although, as in the case of universities, European and Japanese companies linked extensively with the American NBTFs.

These companies were essentially university spin-offs usually formed through the collaboration of a scientist and a professional manager backed by venture capital in order to apply the new scientific discoveries to commercial product development. Their origin and nature identify at the same time their capabilities, functions and limitations. Their specific skills concerned research in areas directly related to the experiences and field of interest and specialization of the scientists involved. However, they lacked the experience and organization necessary to support some crucial aspects of the innovative process such as manufacturing, the testing and the administrative procedure for the approval of new products and marketing, i.e. those activities which mark the difference between a research company and a fully fledged independent producer.

These companies played a fundamental part in the development of biotechnology in two ways: directly, since they were responsible for most of the research in genetic engineering in the early years; indirectly, because they were a major source of knowledge for established corporations.

Although the aim of the large majority of the NBTFs was that of becoming fully integrated pharmaceutical companies, their strategies varied considerably both in terms of the research directions they pursued and of their attitudes towards vertical integration. As far as the directions of research are concerned, at one extreme one finds companies which tried to develop as soon as possible products that could be sold on the marketplace, attaining an early lead through specialization in a specific niche. This strategy implied concentrating on the most immediate applications of biotechnology and on products which did not require large-scale production processes or substantial marketing efforts (e.g. diagnostics based on monoclonal antibodies). At the other extreme, one finds companies which tried to develop core technological capabilities in a wide range of areas, exploiting the technological complementarities made possible by genetic engineering and pursuing on these grounds diversification. Clearly, this strategy implied longer time-horizons.

As far as vertical integration is concerned, the NBTFs faced the alternative of trying to establish very early production and marketing facilities or to rely on external partners, subcontracting the manufacturing of their products and/or licensing the commercialization rights for different geographical areas to other companies.

The interaction of these strategies can generate a large variety of behaviour and one actually finds different combinations of them. However, the dominant strategies that gradually emerged were relatively similar. Most of these companies became specialized suppliers of monoclonal antibodies. By the beginning of 1984, around 100 firms were involved in this field; more than 600 monoclonal antibodies were available and about 60 in vitro diagnostics had been approved. Even in this case, however, manufacturing and – more often – commercialization were contracted out to other companies.

Other companies concentrated their efforts on the development of pharmaceutical and agricultural products: in 1985 it was estimated that 36% of public genetic engineering companies were mainly active in the field of diagnostics, 34% in the development of therapeutical proteins, 18% in the supply of instrumentation and 6% in agriculture or chemicals (Galante, 1987; Orsenigo, 1989).

Within the field of therapeutics, companies concentrated their research efforts in a small range of areas which reflected the core experiences of the leading scientists and the more immediate technological opportunities, showing strong degrees of duplication and fierce competition. Moreover, this research was done under contract to the large corporations and the confusion concerning patents had generated intense litigation and further uncertainty.

By 1985, very few products were actually commercialized and their production and marketing had also been licensed in most cases to other large corporations.[11] Thus the product of the NBTFs was, especially in the early stages, almost exclusively research for and in collaboration with large pharmaceutical and chemical companies.

8. Interaction between NBTFs and other corporations

The interaction between established corporations and NBTFs was highly complex, involving cooperation and hierarchical relationships as well as competition and reliance on market transactions. This complexity arises from the need to integrate different capabilities and to coordinate differentiated learning processes. The exchange of technical knowledge implies substantial transaction costs (Williamson, 1985; Teece, 1981; Teece, Pisano and Shan, 1987) and therefore reliance on arms-length, market relationships is likely to be inefficient as compared to hierarchical forms of coordination of exchanges. In the case of biotechnology, appropriability conditions are the most important incentive to vertical integration, because arms-length transactions might imply the appropriation of technical know-how by the licensee. On the other hand secrecy is not an efficient means of protection of product innovations and lead times are irrelevant when the innovator does not commercialize the product, as in the case of NBTFs. Moreover, market transactions in biotechnology involved inefficiencies linked to the tacit and specific part of relevant know-how and to the need to coordinate R&D with production and marketing. Insofar as the exchange of knowledge requires close communication between the partners, the development of adequate information structures and the deployment of complementary, specialized assets, vertical integration reduces the costs of such transfer. More generally, to the extent that learning processes involve the development of particular cognitive frames and routines and proceeds cumulatively along irreversible trajectories, the integration of different skills necessarily requires forms of organization coordinating individual learning (Cohen, 1987; Dosi and Egidi, 1990; Orsenigo, 1989).

However, since the capabilities and the knowledge embodied in the NBTFs and in the established corporations were strongly heterogeneous and since technical progress in genetic engineering was very fast and uncertain, integration could not be achieved rapidly and costlessly. On the contrary, it required a gradual process of acquisition of external knowledge and experimentation with alternative sources in order both to get access to differentiated capabilities and avoid the dangers of 'locking-in' to particular, possibly inefficient, trajectories.

The evolution of the forms of interaction between companies in biotechnology can indeed be analysed as the outcome of a process of collective learning, characterized again by a large variety of strategies. Where NBTFs are concerned, at one extreme some companies linked with a high number of contractors, supplying them with specific products that were difficult to reproduce and imitate. These companies behaved in other words as specialized suppliers of high technology intermediate products, without establishing too close a relationship with specific partners. At the other extreme, linkages were restricted to a very small number of contractors with whom long-term stable relationships were built.

In general, collaboration concentrated in those areas like pharmaceuticals where NBTFs had higher competences in genetic engineering and technical change was faster but the barriers to forward integration were stronger. Conversely, collaboration is lower in the areas where NBTFs have lower competences and the relevant knowledge is more tacit (chemicals). Collaboration concerned essentially R&D, marketing and technology transfer. Notably, the relevance of R&D as a motive for collaboration is particularly low in diagnostics, where NBTFs face lower barriers in product development and production and tend therefore to integrate vertically, linking with established companies only for marketing (Teece, Pisano and Shan, 1987; Orsenigo, 1989).

Where established companies are concerned, linking with NBTFs was not very different from linking with universities. On the contrary, as already mentioned, these forms of collaboration were strongly complementary.

However, the relationship with the NBTFs was complicated by the existence of elements of potential competition and appropriability conditions acquired a crucial relevance here. In order to secure higher degrees of control and closer knowledge of the NBTFs activities, contract research was frequently supplemented with the purchase of equity and sometimes with outright acquisition, especially in those areas where R&D involves longer time-horizons (plant agriculture and chemicals) and where the tacit and cumulative nature of R&D requires more intimate relationships (Teece, Pisano and Shan, 1987).

The advantages stemming from these interactions could be fully exploited, however, only through the contextual development of in-house capabilities, which made it possible to absorb and complement the knowledge supplied by external sources. In-house research had a more applied nature, a stronger focus on scaling-up problems and longer time-horizons than that carried out in the NBTFs. The companies known to have undertaken significant in-house activities largely coincide with the companies apparently more involved in linkages with NBTFs and they also emerge as the major innovators in terms of patents.[12]

When the technology started to mature, acquiring stronger features of specificity and cumulativeness and getting closer to the stage of production, the role of in-house R&D became increasingly important and the relationship with the NBTFs became closer, evolving towards higher degrees of integration, including in many cases acquisition. Conversely, although many NBTFs tried to integrate forward, the majority did not succeed in this strategy and gradually strengthened their features of research companies and specialized suppliers.

9. Conclusions

The case of biotechnology illustrates some important features of the dynamics of competition in science-based technologies.

First, science is an increasingly important component of the innovative process and is a major source of competitiveness. This relevance goes beyond the provision of technological opportunities to affect the procedures, heuristics and methodology of research. As a consequence, the academic system becomes an increasingly important actor in the national systems of innovation. However, the growing involvement of scientists and scientific institutions entails the development of adequate forms of organization and coordination of scientific and technological research. The potential for and the ease of constructing such linkages depends to a large extent on the absolute and relative capabilities, experiences and commitment of the agents involved. From this perspective, the case of biotechnology confirms that the main role of scientific institutions in a modern system of innovation is to continue to do their primary job, i.e. producing first-rate research and education, the ability of transferring it being correlated to and a consequence of that. At the same time, however, industrial firms become relevant sources of scientific knowledge. Thus, the institutional set-up supporting innovative activities is likely to evolve towards increasing degrees of complexity.

However, technological performance does not depend exclusively on the availability of strong scientific capabilities. Even in an extreme case such as biotechnology, science remains one component of the innovative process and a form of knowledge profoundly different from technology. The main difference relates to the modes in which science and technology are articulated, coordinated and applied. Moreover, science is intrinsically more accessible and difficult to appropriate than technology. Thus the availability and the control of scientific knowledge does not guarantee high technological performances, especially in the long run, both at the company and at the country level.

On the contrary, the availability of high and differentiated technological competences and of the high related organizational capabilities is a necessary precondition for competitiveness. The innovative process requires the integration and coordination of different learning processes, knowledge bases and assets. It is intrinsically cumulative and uncertain. In these conditions, variety is a fundamental requisite for technological progress, both because it allows various alternatives to be searched and because it provides the different capabilities of which innovation is made. At the same time, the development of complex and differentiated organizations for the coordination of learning activities is a central aspect of the innovative process and a major source of competitiveness. In particular, the integration of R&D, production and marketing bears a great relevance, because they are crucial components of the innovative process and their absence greatly diminishes the ability to identify and understand the technical problems related to design and production as well as user needs.

Finally, competitiveness depends crucially on the institutional set-up in which innovative activities take place. The case of biotechnology illustrates clearly the role of institutions like universities, the financial system and governments in determining the rates and patterns of innovation in any one country. Similarly, patent legislation, the regulation of research, public attitudes towards the new technology are essential aspects defining the conditions of competition and competitiveness. More generally, the institutional set-up determines at a more micro level of analysis what sort of strategies are feasible, acceptable or successful (see the chapters by Carlsson, by Lundvall and by Hollingsworth).

This implies an increasing role for government intervention, although along lines which differ profoundly from the past. The experience of biotechnology shows that governments had a fundamental role in defining competitiveness, mainly through the support they provided to basic research in universities. The American scientific advantage was built indeed on a large flow of resource channelled to scientific research (Sharp, 1985; Orsenigo, 1989). However, governments face tremendous difficulties and problems in trying to devise adequate policies for the support and regulation of technological change. First, in conditions of uncertainty, early decisions to give priority to particular trajectories of advance are likely to be wrong and may have negative effects in the long run, insofar as technical progress entails irreversibilities (David, 1986; Arthur, 1989; Cowan, 1988). On the other hand, late intervention and/or generalized support to generic, precompetitive research implies the danger of falling behind competitors and it is likely to be ineffective. Second, however, since new technologies affect wider social issues, governments are inevitably involved. In particular, legislation, regulation and the support to industry–

university ties become crucial aspects of government intervention, which requires the accumulation of experience in those fields, the coordination of the activities of the different agencies involved and high degrees of collaboration with other national governments.[13]

The take-off and subsequent development of a new technology require therefore and cannot be separated from contextual institutional adaptations and change. More generally, in this perspective, technological change which involves different agents and which necessarily entails the contextual construction of forms of coordination of individual learning processes and capabilities at different levels of analysis. At the cognitive level, it involves the establishment of specific trajectories of research, research procedures and heuristics. At the firm and sectoral level, this process involves at the same time specialization, diversification and vertical integration, competition and cooperation. At the level of the national system of innovation, it involves the adaptation of existing institutions and the creation of new ones. In other words, technological change coincides to a large extent with a process of formation of institutions.

Whilst structural conditions like the nature of the technological regime define some general requisites for efficiency of alternative forms of coordination at different levels of analysis, nevertheless in changing innovative environments it is usually very difficult and often impossible to identify optimal and stable strategies and solutions. This is particularly the case in the analysis of national systems of innovation, given their high degrees of complexity. Moreover, their structure is rarely entirely coherent, but it results from the gradual addition and integration of agencies and bodies which often have different and multiple functions.[14]

Finally, what is efficient and/or feasible depends crucially and in an irreversible way on the past history of evolution of technological capabilities, forms of organization, etc. Thus, whilst over time particular trajectories of technological progress, companies and forms of organizations gradually establish themselves, variety in the patterns of innovation and of organization of innovative activities is likely to be a permanent feature and a necessary outcome of technological change.

Notes

1. The bulk of the molecular biology research which eventually led to the fundamental discoveries in genetic engineering was funded by subunits of the NIH such as the National Cancer Institute and its viral oncology unit. In 1982 the NIH funding of research directly linked to biotechnology was estimated to amount to $380 million (OTA, 1984). Other important sources of funds for research were the US Department of Agriculture ($35 million in 1982) and the Department of Energy ($35 million in 1982). Private funding of research in universities and medical schools by philanthropists or charities, however, had also an important role. See Sharp (1985), Kenney (1986) and Orsenigo (1989).
2. For a more detailed discussion of these issues, see: Nelson and Winter, 1982; Stiglitz, 1987; Dosi and Egidi, 1987; Malerba, 1987; Orsenigo, 1989.
3. A complete discussion is not possible here. See Orsenigo, 1989.
4. Examples are Neogen, a private for-profit company created by Michigan State

University, and Eugenics which was established by a pool of venture capitalists to commercialize the research done at Stanford and Berkeley. See Kenney, 1986.

5. Kenney, 1986. See Faulkner, 1986 for similar results.
6. Teece, Pisano and Shan (1987) and Arora and Gambardella (1990) provide further systematic support to these conclusions.
7. For detailed analyses of these issues, see: OTA, 1984; Sharp, 1985; Saxonhouse, 1985; Faulkner and Orsenigo, 1991; Orsenigo, 1989.
8. As such biotechnology R&D activities did not require in the early stages exceedingly expensive and sophisticated equipment and laboratories. Moreover, collaborative research with universities provided access to the more specialized laboratory equipment. The main costs stemmed primarily from the high salaries that had to be paid to researchers, from the length of the time-horizons involved and from the need for experimenting on a wide range of alternative areas of research. As research became more systematic and ambitious, costs started to rise rapidly. Equipment itself has evolved quickly and has now become quite expensive. The costs of setting up a small biotechnology start-up were estimated in 1980 to be around $12 million for the first three years of activity; a hybridoma company needed 'only' $4–5 million. The in-house investments of large existing companies were however on a different magnitude: Monsanto's investment in its biological sciences research centre amounted to $185 million and DuPont's to almost $150 million. The time required to develop a viable product was estimated at the time around five years, abstracting however from commercialization which would add more time and money (Kenney, 1986; Schneider, 1980; Orsenigo, 1989).
9. In this perspective, the behaviour of the chemical industry played a crucial role in the development of biotechnology.
10. Patents suffer serious limitations as indicators of innovative activities in biotechnology. For a discussion of these drawbacks see Orsenigo, 1989. However, in the absence of other reliable indicators and if carefully interpreted, patents provide at least a broad quantitative indication about the patterns of innovation.
11. In addition to the diagnostic kits, only human insulin produced by Ely Lilly under Genentech's license, the human growth hormone produced by Genentech, a pig scours vaccine produced by Molecular Genetics and two amino acids manufactured by Genex were already commercialized.
12. For similar results, see Arora and Gambardella, 1990.
13. For a more complete discussion of the role and problems of public policies in biotechnology see Orsenigo, 1989 and Faulkner and Orsenigo, 1991.
14. Important attempts in this direction can be found in: Nelson, 1988; Freeman, 1987; Dosi and Aoki, 1989.

References

Aoki, M. and Dosi, G. (1989), 'Corporate organization and finance', paper presented at the Conference on 'Finance and the Enterprise in a Historical Perspective', Terni, Italy, September 29–October 1.

Arora, A. and Gambardella, A. (1990), 'University–industry interactions and the strategies of large firms in biotechnology', Stanford, Stanford University, mimeo.

Arrow, K.J. (1962), 'Economic welfare and the allocation of resources for invention', in NBER, *The Rates and Direction of Inventive Activities*, Princeton, Princeton University Press.

Arthur, W.B. (1989), 'The dynamics of allocation under increasing returns', *Economic Journal*.

Atkinson, A. and Stiglitz, J. (1969), 'A new view of technological change', *Economic Journal*.

Chesnais, F. (1986), 'Some notes on technological cumulativeness, the appropriation of technology and technological progressiveness in concentrated markets', paper presented at the Conference on Innovation Diffusion, Venice, 17–21 March.

Cohen, M. (1987), 'Adaptation of organizational routines', Ann Arbor, University of Michigan, mimeo.

Cohen, W.M. and Levinthal, D.A. (1988), 'Innovation and learning: the two faces of R&D. Implications for the analysis of R&D investment', Pittsburgh, Carnegie Mellon University, mimeo.

Cowan, R. (1989), 'Technological Variety and Competition. Issues of diffusion and intervention', paper presented at the International Seminar on Science, Technology and Economic Growth, OECD, Paris, 5–8 June.

Dasgupta, P. and David, P. (1986), 'Information disclosure and the economics of science and technology', in Feiwel, G. (ed.), *Essays in Honour of K. Arrow*, London, Macmillan.

David, P. (1986), 'Narrow windows, blind giants and angry orphans: the dynamics of system rivalries and dilemmas of technology policy', Working paper no. 10, Stanford, Stanford University, CEPR.

Dosi, G. (1982), 'Technological paradigms and technological trajectories: a suggested interpretation of the determinants and directions of technological change', *Research Policy*.

Dosi, G. and Egidi, M. (1990), 'Substantive and procedural rationality: an exploration of economic behaviour in complex and changing environments', *Revue d'Economie Industrielle*.

Dosi, G., Freeman, C., Nelson, R.R., Silverberg, G. and Soete, L. (eds) (1988), *Technical Change and Economic Theory*, London, Pinter Publishers.

Faulkner, W. (1986), 'Linkage between industrial and academic research: the case of biotechnological research in the pharmaceutical industry', D.Phil Dissertation, Brighton, SPRU, University of Sussex.

Faulkner, W. and Orsenigo, L. (1991), 'State policies on industrial innovation in weaker economies: the case of biotechnology in Britain and in Italy', in U. Hilpert (ed.), *State Policies and Techno-Industrial Innovation*, London, Routledge.

Freeman, C. (1987), *Technology Policy and Economic Performance: Lessons from Japan*, London, Pinter Publishers.

Galante, Y. (1987), 'University–industry relationships in biotechnology in the USA and in Europe: the case of Italy', paper presented at the Conference Biobusiness '87, Brussels, 16–18 March.

Hacking, A.J. (1986), *Economic Aspects of Biotechnology*, Cambridge, Cambridge University Press.

Harrigan, K. (1981), *Strategies for Vertical Integration*, Lexington, Lexington Books.

Kenney, M. (1986), *Biotechnology: The University–industrial Complex*, New Haven, Yale University Press.

Kline, S.J. and Rosenberg, N. (1986), 'An overview of Innovation', in Landau, R. and Rosenberg, N. (eds), *The Positive Sum Strategy*, Washington, DC, National Academy Press.

Malerba, F. (1988), 'Learning, incremental innovation and cumulative technical change: a cross-industry analysis', Brescia, Università di Brescia, mimeo.

Malerba, F. and Orsenigo, L. (1991), 'Technological regimes and patterns of innovation: a theoretical and empirical investigation of the Italian case', in Heertje, A. and Perlman, M. (eds), *Evolving Technology and Market Structure*, Michigan University Press.

Mayr, O. (1982), 'The science-technology relationship', in Barnes, B. and Edge, D. (eds), *Science in Context: Readings in the Sociology of Science*, Open University Press.

Nelson, R.R. (1988), 'Institutions supporting technical change in the USA', in Dosi *et al.*

Nelson, R.R. and Winter, S. (1982), *An Evolutionary Theory of Economic Change*, Cambridge, Mass., The Belknap Press of Harvard University Press.

OECD (1985), *Biotechnology and Patent Protection: An International Review*, Paris.

Office of Technology Assessment (OTA) (1984), *Commercial Biotechnology: An International Assessment*, Washington, DC, US Government Printing Office.

Orsenigo, L. (1989), *The Emergence of Biotechnology, Institutions and Markets in Industrial Innovation*, London, Pinter Publishers,

Pavitt, K. (1984), 'Sectoral patterns of technological change: towards a taxonomy and a theory', *Research Policy*, vol. 13, no. 6, pp. 343–75.

Pavitt, K. (1987), "On the nature of technology", Inaugural Lecture, given at the University of Sussex, 23 June, Brighton, SPRU, University of Sussex.

Pavitt, K. (1989), 'Towards a theory of the large innovating firm', Brighton, SPRU, University of Sussex, mimeo.

Rosenberg, N. (1988), 'Why do companies do basic research (with their own money)?' Stanford, Stanford University, mimeo.

Saxonhouse, G. (1985), 'Biotechnology in Japan: industrial policy and factor market distortions', *Prometheus*.

Schneider, N.M. (1980), 'Prepared statement at a hearing before the Subcommittee on Science, Technology and Space of Committee on Commerce, Science and Transportation', US Senate, 20 May, in *Industrial Applications of Recombinant DNA Techniques*, US Government Printing Office, Washington, DC.

Stankiewicz, R. (1986), 'Genetic engineering: international R&D trends', paper presented at the International Symposium on Japanese Technology, Urawa Saitama, 17–19 September.

Stiglitz, J. (1987), 'Learning to learn, localised learning and technological progress', in Dasgupta, P. and Stoneman, P. (eds), *Economic Policy and Technological Performance*, Cambridge, Cambridge University Press.

Sharp, M. (1985), *The New Biotechnology: European Governments in Search of a Strategy*, Sussex European Papers, Brighton, University of Sussex.

Silverberg, G. (1988), 'Modelling economic dynamics and technical change: mathematical approaches to self-organization and evolution', in Dosi *et al.*

Silverberg, G., Dosi, G. and Orsenigo, L. (1988), 'Innovation, diversity and diffusion; a self-organisation model', *Economic Journal*.

Teece, D. (1981), 'The market for know-how and the efficient international transfer of technology', *The Annals of the Academy of Political Social Science*.

Teece, D. (1986), 'Profiting from technological innovation', *Research Policy*.

Teece, D., Pisano, G. and Shan, W. (1988), 'Joint ventures and collaboration in biotechnology' in Mowery, D. (ed.), *International Collaborative Ventures in US Manufacturing*, Ballinger Press.

Williamson, O. (1985), *The Economic Institutions of Capitalism*, New York, Free Press.

Winter, S. (1984), 'Schumpeterian competition in alternative technological regimes', *Journal of Economic Behaviour and Organization*.

3 Technological learning, economic networks and innovation appropriability

Patrick Cohendet, Jean-Alain Héraud and Ehud Zuscovitch★

Property rights are a key instrument in the individual-driven functioning of capitalistic society. Deficient appropriation mechanisms and insufficient legal instruments of appropriation often lead to troubles on the path of economic development and growth such as those experienced now by ex-socialist countries. The impact of the appropriability issue is even greater in the area of innovation because this is where the engine of the capitalistic machine lies, at least according to the Schumpeterian view of economic development which we share. Because of the transitory nature of innovation profits (Schumpeter, 1934), any change in the ability to appropriate the fruits of innovation will be of critical importance to the rate and direction of technological change and beyond to economic development and welfare. The economic understanding of innovation appropriation is still dominated by the seminal contributions of Richard Nelson (1959) and Kenneth Arrow (1962) according to whom imperfect appropriability of innovation creates an important distortion between the private and social returns on innovative activities. It is hence always possible for firms that did not contribute directly to the social R&D effort to gain access to the newly generated knowledge and use, with relatively small information or imitation costs, the innovation's economic potential. Those who really innovate cannot avoid (involuntarily) sharing profits with others. R&D results are viewed as easily (cheaply) diffused and an investment in innovative activities is carried out at a level which is lower than the socially desirable one.

This received view is now questioned by many researchers who try to get a deeper understanding of the features of these innovative activities in general and of new technologies in particular. Although there is certainly an appropriability tension of the innovators/imitators type (otherwise game theorists would remain idle), it is important to recognize in the first place, that the vanishing profits of innovation often contribute to intensify research activities, not to reduce them. Secondly and more importantly the problem should not be overemphasized. It is by no means easy to have access to research results. The

★Bureau d'Économie Théorique et Appliquée (BETA), Université Louis Pasteur de Strasbourg. We gratefully acknowledge suggestions made by Dominique Foray and Christopher Freeman on a previous draft of this chapter.

received view explicitly contains the assumption that the innovation is basically an informational phenomenon and analyses its implications in a static resource allocation framework. In our restatement of the appropriability problem we should recognize that the development of new technologies is fundamentally a localized and path-dependent learning process to which the access is quite limited even when the general characteristics are publicly known. Economic competition bears on operational and production knowledge and on technological know-how, not on the last theoretical findings cited in the Physical Abstract as Schmookler would have put it. The public good nature of the innovation was a concomitant view of the 'general purpose knowledge base'. Such a view suited well the economic approach that prevailed in the 1960s and early 1970s and was a necessary complementary piece for the theories of homothetic growth. The differentiated view of technological learning we have in mind 15 years later is radically different and we should make the necessary adaptations to the appropriability analysis. The need for and nature of such a restatement is the scope of this chapter.

The limits of the received view on innovation appropriability

The received view on innovation appropriability is to be located within the theoretical context of the 'Market Failures' approach where, together with uncertainty and indivisibility, one has to overcome market coordination troubles that prevent the otherwise competitive structure from reaching efficient allocation of resources. In this framework, the innovation itself is treated like an informational commodity, only capable of imperfect appropriation. Therefore the incentive to innovate can only be partial, compared with other economic activities and typically fails to attract enough investment. This characterization calls for two important corollaries.

- On the one hand, we find a theoretical justification to public policy as the government is required to correct the underinvestment in R&D activities. Together with the other above-mentioned sources for market failure, inappropriability is often used to justify big public research projects such as those in the aerospace and military areas.
- On the other hand, imperfect appropriability of innovation calls for the elaboration of legal devices to secure property rights as much as possible. Patents are of course the key element in such a legal construction, giving innovators monopolistic power over a well-defined limited period, and making the innovation subject to competition afterwards. We should recall as well that patents are also registered in order to make the information about the novelty publicly available and to stimulate competition and innovation in related fields (and to accelerate further developments and spillovers).

We can see how central and powerful the hypothesis of imperfect appropriation is. Indeed, together with the above corollaries (or procedures to amend market failures) it set the very foundation for the conceptual edifice of innovation theory and policy for many years. Nevertheless, the hypothesis itself and/or its corollaries were gradually questioned by a growing number of economists, sociologists, and historians of technology and other scholars who studied technological change closely and it does seem today much more restricted in its scope. The appropriability issue is presently tackled both from an internal and an external perspective. By external perspective we mean that innovation remains amorphous (its structural properties are not taken into consideration) and that the analysis focuses on the competitive structure only. By internal perspective we mean that in challenging the prevailing theoretical view one uses factual knowledge about how the innovative activities are structured and actually carried out in the different sectoral and technological areas (a traditional economist would typically invert the use of the terms internal and external).

From a purely external point of view Dasgupta and Stiglitz (1980) showed, in a game-theoretical perspective, that in a competitive market there is also a risk of overinvestment in R&D. This is because parallel (double) research is done by competing firms and since only one typically wins the jackpot there is necessarily a waste of resources. This claim should also be qualified, by the way. Innovating firms usually race within a multi-product and process portfolio setting. Furthermore it is essentially a repeated game. You win some and you lose some. No real waste of resources takes place, unless a systematic failure occurs over a longer time because one uses the same experiences for different products, sometimes around the patented one and sometimes for further developments, etc. This bring us closer to the internal critique on which we wish to put our emphasis.

Technological innovation: a 'commodity' or a 'specialty'?

The postulated availability of the information contained in the innovation ignores the very nature of technological progress. Innovation is basically a learning process. It is neither an exogenous promethean gift nor a multi-purpose knowledge base that can be oriented according to relative price changes. This neo-classical view of the 1960s accommodated well the models of homothetic growth. With the crisis of industrial restructuring and the emergence of new technologies that came later we were incited, if not obliged, to look closer at the inner characteristics of technology and they by no means matched such a view. Technological innovation is a process which occurs differently across industries and over time (Pavitt, 1984); it is at the same time localized, partly tacit and to a large extent history-rooted and with strong irreversibility character that makes it strongly path-dependent and of very limited transmissibility. Let us say a few words about each of these characteristics.

The 'local' nature actually encompasses two interrelated dimensions. The first is the recognition that technological innovation takes place within a particular structure, a specific context of industrial products and production processes. More fundamentally the local character expresses a sort of 'fixed point' that results from a cumulative process of successive adjustments of the technology to its context. In such a process a specific configuration of technological, sectoral and scale-related arguments gradually match each other to create a specific 'personality' or pattern. The consequences of this local character are far from trivial and in 1969 Atkinson and Stiglitz presented it as a selective displacement of a point or a segment rather than a shift of the overall production function. And as Sahal (1981) put it '. . . insofar as the learning tends to be localized, the very concept of a production function representing a near-infinity array of techniques sharing a common state of knowledge may have to be abandoned.'

The tacit aspect is linked to the fact that production technology is about doing things, not only about knowing things in the form of an abstract (scientific) principle. Operating know-how is something very different and is much less transmissible (Dosi, 1988). It is often dependent upon a division of labor which very often contains implicit components that make the production setting unique. It incorporates working routines and craftsmanship components and sometimes, even with the voluntary transfer of the original blueprints, it is not possible to imitate the ensemble of gesture and implicit working practice in such a way that the product quality is satisfactory.[1]

The third and most powerful argument of the technological learning process is its historical nature. The view of technological development as an evolutionary process is now well established due to the joint effort of Nathan Rosenberg's (1976, 1982) historical research and the implementation of the evolutionary approach by Nelson and Winter's continuing effort (1982, for example). Irreversibility of the technological learning process was further documented and analyzed in the work of Paul David (1975, 1986) and of B. Arthur on path-dependency. In addition to the localized aspect there is a clearly time-dependent behavior in learning. If for some random reason a given technological trajectory is selected, it will undergo successive improvements that will make it perform better and it will, by this very fact be selected for future applications. If the current choice had had to be made previously (before the first choice was made) an altogether different technique could have been chosen. This is of course a major cause for a restricted transmission of knowledge. Since we can only keep alive a relatively small number of options – our 'practice memory' capacity being very limited – purely theoretical options have a very small chance of being reactivated (Zuscovitch, 1986). After the oil shocks, no one really considered simply going back to coal-based organic chemistry. This would have created new hydrocarbons from liquified coal and then used petroleum-based techniques that have been constantly improved.

All three major characteristics of the learning process mentioned above imply, in one way or another, severe restrictions on the ability to transmit the information contained in the novelty. D. Foray went a step further in the argument (1989). He claims that the standard theory actually mixes up two types of diffusion effects. The first consists of the dissemination of the information about the outcome of an R&D activity. Such dissemination occurs without serious adjustment costs while the actual implementation of the innovation usually requires considerable adjustment costs. If the former diffusion costs are low it is because the firm invests a lot in maintaining a considerable R&D capacity that can in turn quickly absorb such information. To put it differently, the adoption costs which are also a measure of appropriability become an inverse function of the investment in R&D. As Pavitt has argued in Chapter 1 this view is particularly valid for basic research, where, although the results are publicly available it takes constant investment effort in theoretical research to be able to interpret and use abstract results. Even this really public good can only be appropriated with significant costs which are required to maintain the learning process alive.

Partnerships and controlled diffusion of spillovers

A strong feature related to the evolutionary nature of innovation as a learning process is the ability to stimulate change within and across the frontiers of the firm. This time the appropriation matter relies not only on the same innovation, but rather on the chain of induced effects. Spillover effects attracted considerable analytical attention over the last decade and attempts were made to relate their magnitude to technological infrastructure (Jaffe, 1986), or to the capital and productive structure as in Bernstein and Nadiri (1989). One usually finds the same claim that the social rate of return exceeds the private one, but in no way can we realistically conclude that the spillover effects escape altogether the innovating firms and are to be considered as being a negative incentive for R&D investment. A recent increase in the rate of technological partnership formation especially in advanced technologies (see Hagedoorn and Schakenraad, 1990) does seem to indicate that spillovers are mastered to a large extent by reciprocity agreements and that partners do find that the benefits of sharing and sometimes even diffusing knowledge overcome the welfare or profit losses. Actually many economists now believe that these R&D externalities should be taken into consideration in the new formalization of real growth. Although their macro magnitude is hard to assess (see Griliches, 1991) – hence the difficulty to say to what extent they offer an escape from decreasing returns – it is quite possible that the return to the originating firms becomes higher than their losses even though they may capture only a small part of the social overall return. We will come back to that later. To better grasp the spillover effects one usually distinguishes between productivity spillovers and inter- or intra-industry spillovers (see Griliches, 1979 or Mohnen, 1989).

Productivity spillovers express the fact that, in addition to the surplus gains for the innovating industry, these (typically materials or process) innovations tend to reduce input prices in downstream industries and therefore increase their own surplus. These however are not always considered as pure externalities since they are transmitted through the market mechanism.

The industry spillovers stem, according to the received view, from the informational nature of the innovation, considered as a commodity that can be easily appropriated. The innovating firms can expect other firms, either from the same industry or from altogether different sectors, to appropriate relevant information about the innovation properties with relatively negligible costs. These kinds of spillover are made easier with procedures and practices such as skilled labor mobility, systematic screening of data bases of relevant professional literature, reverse engineering and sometimes industrial spying. However, although we should recognize these 'negative' spillovers, we should notice at the same time that a rapidly increasing number of firms expect globally positive feedback effects[2] from the different modes of technological transfer. It is certainly the case for technological partnership agreements, for the 'Technopole' – that is, new technologies industrial breeding areas – and for the voluntary dissemination of R&D results to potential competitors in order to sustain strategic markets.[3] Inter-industry transfers are even less problematic and very often, as in the case of advanced materials, one cannot really implement innovation unless a significant scaling-up allows for price reduction that makes the whole diffusion possible. Indeed diffusion is often a controlled process where the reciprocity rule prevails in a long repeated game. In addition, very often the development of an innovative 'milieu' proves to be essential and information exchanges are the very way of participation. In all these instances spillovers are of a 'positive externality' type and there is absolutely no reason to conclude that under-investment in innovative activity would prevail.

The analytical and empirical criticism of what we have termed the received view on appropriation did not focus on its theoretical core only. Its two major corollaries were equally attacked, i.e. on the one hand, the need for governments to use public policy instruments such as big research projects in order to substitute for or palliate the lacking private initiative; and on the other hand, many authors questioned the primary role and the efficiency of the legal protection device – the patent system.

(1) The questioning of the well-founded need for big public R&D and technology programs, set in order to substitute for insufficient private incentive has never left the theoretical and political literature. It was never as fierce as after the failure of the federal program 'Cogents-centers of generic technology' (see Mowery, 1983 and Foray and Lebas, 1988). The purpose of the program launched in 1980 and stopped in 1983 was to create big research centers on generic technologies. Apparently the new knowledge generated in the project did not feed enough commercial exploitation in the industry and probably, as Foray and Lebas suggested:

> There can be no pure substitution between intra-firm research and contractual research without damaging the overall innovative performance of the system. The firms must maintain or develop an acceptable level of research capabilities in order to be able to better exploit the results of the contractual research.

(2) The questioning of the efficiency of patent legislation is equally frequent but the empirical evidence reported by Levin, Klevorick, Nelson and Winter (1987) is particularly convincing. They showed with the Yale inquiry on appropriation mechanisms, that patents were by far not the best or the preferred instrument to secure innovation property rights. Having sufficient lead time over competitors, a better mastering of production and technology thanks to experience effects (learning curve advantages) or simply trying to keep secret the new results, were all instruments that were considered as at least as effective as patents and often much preferred. Moreover, strong protection regimes for the individual firm may damage the innovative performance of the system. Levin *et al.* (1987) express it beautifully and we would like to quote them fully at this point:

> The premise that stronger protection will always improve the incentives to innovate is also open to challenge. Unimpeded diffusion of existing technology is immediately beneficial not only to customers but also for those who would improve that technology. Because technological advance is often an interactive, cumulative process, strong

protection of individual achievements may slow the general advance. This would not occur in a hypothetical world without transaction costs . . . in reality however, markets for rights to information are subject to major transactional hazards, and strong protection may preclude competitors from making socially beneficial innovations. The semiconductor industry of the 1950s and the 1960s provides an excellent example of rapid progress in a cumulative technology that might have been impossible under a regime that strongly protected intellectual property.

The interactive argument should be extended also to include inter-industry spillovers. In many instances the indirect system (shadow) benefit from releasing an important constraint exceeds by far the direct benefit to both users and producers.

The new stylized facts on appropriability

We can summarize all these findings in the following manner. There is certainly a risk of market failure in the inducement of innovative activities due to informational leakages. Yet the scope of the problem is much limited because on the one hand technological knowledge is not easily transferable. It is to a large extent naturally protected by the fact that it is partly tacit and something which is not expressed cannot be transferred. The local, context-dependent and path-dependent technological learning makes transferability further limited either by the division of labor in which the technological body of knowledge of the firm is embedded, or due to the fact that the specification of the technical solution hardly matches other problems without expensive translation costs.

On the other hand, we have the arguments related to market structure. We would not like too close a protection that will prevent induced innovation effects both at the industry level (making cumulative progress less effective) and across industries (discouraging complementary innovations). However we should not worry too much there about excessive uncontrolled spillovers because of the transaction costs involved. These very transaction costs enforce explicit cooperation, but in the case of such agreements reciprocity principles are established to ensure incentive compatibility among the agents. The rapid development of cooperative forms gives birth to a network organization that superimposes itself on the traditional industrial structure. This is due to the fact that in an information-intensive production system with a multiplicity of micro markets with particular specifications for smaller production runs, there is a growing number of bodies of knowledge that one has to master and the only possible access to such a diversity is through cooperation with other producers and users of the products. There is no doubt that the ability of the new information technologies together with others, such as biotechnologies and advanced materials, to tailor applications and hence to combine diversity and efficiency makes the 'local' argument much stronger because niche production is local by definition.[4]

So the new model of innovation comes equipped with a built-in and rather strong appropriability mechanism. The latter is due both to the 'local' context-dependent self protection at the intra-firm level and the 'controlled' diffusion of spillovers at the inter-firm level. This spontaneously stronger appropriability accelerates the rate of technological progress as a whole. Not only the industrial structure tends towards network organization, the firm itself becomes a network of scientific and technical skills capable of adopting new technologies and of selecting proper development paths among relevant alternatives. In this view the ability to have access to the know-how of firms in connected areas increases the value of the firm's own human capital and therefore the incentive to cooperate is obviously strong. This new technological system with its appropriation principles is not yet fully known empirically, let alone analytically, because hybrid entities are hard to manage from both points of view; and what we have is practically hybrids everywhere: competition/cooperation, social(network)/private goods and dynamic (innovative)/static (management) efficiency.[5] Yet we should be able to analyze or at least to comment on some of the implications for the two main corollaries, namely technology policy and patent legislation.

Policy implications

As far as technology policy is concerned the new innovation 'model' implies an altogether different type of intervention for the government when it considers stimulating technology. According to the logic underlying this model the old linear view of a clearly distinct phase of research followed by production and commercialization is basically less and less valid and the stages become much more integrated. The system evolves towards generalized forms of learning by doing. Since the research phase is not clearly distinguishable anymore, no external research can really compensate private insufficiencies. Public intervention should therefore take forms which differ from the one derived from the 'former' innovation model. Instead of big public research programs aimed at substituting for private research the new model asks for a more catalytic type of intervention, i.e. facilitating the conditions of the research learning process. The basic idea is to help the technological body of knowledge of the firm reach a critical mass and thus allow for significant research activity. At the inter-firm level the basic idea is to stimulate technological partnerships in order to obtain more extensive cross-fertilization among disciplines, technologies and research programs so that each individual trajectory of specialization becomes less risky to follow (inasmuch as a better adjustment to diversified needs is ensured beforehand and a broader application potential is available). In a competitive scheme of globalization, that is larger planet integration of the market, the risk of over-specialization can be shared in two ways. One is to institutionalize risk-sharing devices among firms and helping to organize common market segmentation and some generic research facilities; the other is by each using special segments of the fruits of the specialization of the others. In other words some sort of technological integration is needed. It may be more spatial in nature as in chemical industries sites that share common infrastructure devices such as steam power, autonomous electrical power supply and a cheaper access to joint products. It may also be more abstract in nature – sharing knowledge complementarities, or it may include both, as in Silicon Valleys and the like.

In a way, network stimulation policies are necessary not only in order to facilitate appropriability, as we argued before. In the new technological context they are indispensable to allow specialization itself. Technological learning has a strong collective dimension but it is not exported outside private initiative as in big public programmes. It is incremented from the beginning in private incentive. Cooperation practices create 'private' public goods and the real public intervention should have a catalytic function allowing this technological/social division of labor to reach the often required 'critical mass'. The analytical tools that would enable us to measure such effects are not available since catalytic contributions require another type of formal tool that can function in a typically hybrid competitive/cooperative structure. We will probably have to borrow tools from other disciplines as is often the case in periods of major change. As this last paragraph shows, classical problems of uncertainty and indivisibility do not disappear; they appear differently in the new technologies context. In a way, this network effect existed quite well before, but it did not attract theoretical attention. In our own experience in measuring the economic effects of the European space programme (BETA, 1988) perhaps the most important finding was that the technological learning was made possible through the gradual construction and the upkeep of a critical mass of skills within and among the high-tech firms of the space program. It was not only technological in nature. Organizational and quality tightening-up effects allowed the firms to develop skills with which they could face, and sometimes even develop, markets in which they could not have been present before. There was certainly a highly beneficial learning process that the space project firms appropriated, but it was to a large extent the outcome of a skill network consolidation. The cooperation, started under artificial conditions of the programme often became autonomous and the technological partnerships were activated jointly to face international markets. This ambition for the state to catalyze innovation networks' ('Milieu Innovateur') is at the center of some recent theoretical preoccupations and policy recommendations (see Callon, 1990). However, once established, cooperative agreements are not always stable, because clear-cut sharing rules are often not well defined (Jacquemin, 1987) and even when they are, they do not necessarily resist strong appropriation incentives. What we have been able to show also in the case of the space programme is that the network organization is not neutral *vis-à-vis* the nature of technological learning, and so the type of role one plays in the network (main contractor, equipment, system contractor, etc.) tends to shape the learning trajectory and related appropriability practices (Shachar and Zuscovitch, 1990).

As far as property legal instruments are concerned it is useful to distinguish as well between the individual firm practice and the new industrial structure implications. At the firm level, patents will continue to be employed along with other traditional strategies of appropriation such as lead time, trademarks, etc. Proportions may vary according to the nature of the product and of the market involved. Where new jurisdictional initiative and imagination is most required is in managing cooperation. It is quite probable that new legal forms of association should be defined in order to bring more stability to partnerships and other cooperative agreements. In France for instance, such initiatives appeared on the public/private interface in the form of 'Groupement d'intérêt économique'. Yet, if the network form of technological innovation grows to become a dominant industrial structure, new legal forms of economic association will have to emerge in order to offer better incentives in terms of appropriation of innovative activities.

Notes

1. David Landes (1990) at the inaugural conference of the last EARIE (European Association for Research in Industrial Economics) conference in Ferrara brought precisely such an example where French gun blueprints were sent to the United States to be manufactured safely (in times of war) and even so, production was not successful until French skilled workers were sent to the United States.
2. Positive or negative externalities design the nature of the impact of action of one economic agent on the 'partner' while in the term feedback the term partner loses its involuntary connotation and some sort of reciprocity rule is installed.
3. We should mention here the possibility of a desired technological transfer by the firm that cannot be accomplished. D. Teece (1980) showed that, in some cases, the firm does not necessarily choose to maintain exclusivity over a given technology but the transaction costs of transferring it to its partners may prove to be excessive (precisely because the technology is not an informational commodity).
4. For a comprehensive analysis of the evolution and impact of information technologies see C. Freeman and L. Soete (1987).
5. For a useful comparison of the concepts see Klein (1986).

References

Arrow, K. (1962), 'Economic welfare and the allocation of resources for invention', in Nelson (ed.), *The Rate and Direction of Inventive Activity*, Princeton University Press.

Atkinson, A. and Stiglitz, J. (1969), 'A new view of technological change', *Economic Journal*, 79, pp. 573–8.

Bernstein, J.I. and Nadiri, Ishaq M. (1989), 'Research and development and intra-industry spillovers: an empirical application of dynamic quality', *Review of Economic Studies*, 56, pp. 249–69.

BETA (Bureau d'Economie Théorique et Appliquée) (1988), 'Impact des dépenses de l'Agence Spatiale Européenne', rapport d'étude, Université Louis Pasteur, Strasbourg I.

Callon, M., 'Réseaux technico-économiques et irréversibilité', in Boyer, R. *et al.* (eds) *Figures de L'irréversibilité en Économie*, Editions EHESS, 1991.

Cohendet, P., Ledoux, M.J., Zuscovitch, E. *et al.* (1987), *Les Matériaux Nouveaux*, Economica.

Dasgupta, P. and Stiglitz, J. (1980), 'Industrial structure and the nature of innovative activity', *Economic Journal*, 90.

David, P. (1975), *Technical Choice, Innovation and Economic Growth*, Cambridge University Press.

David, P. (1986), 'New technology diffusion, public policy and international competitiveness', in Landau R. and Rosenberg N. (eds) *The Positive Sum Strategy*, National Academy Press.

Dosi, G. (1988), 'Sources, procedures and microeconomic effects of innovation', *Journal of Economic Literature*, September, pp. 1120–71.

Foray, D. and Mowery, D. (1990), 'L'intégration de la R&D industrielle: nouvelles perspectives d'analyse', *Revue Economique*, no. 3, pp. 501–30.

Foray, D. and Lebas, C. (1986), 'Diffusion de l'innovation dans l'industrie et fonction de recherche technique: dichotomie ou intégration', *Economie Appliquée*, no. 3.

Freeman, C. and Soete, L. (1987), *Technical Change and Full Employment*, Blackwell.

Gaffard, J.L. (1987), 'La création de technologie: stratégies d'entreprises et politique publique', LATAPSES CNRS, Université de Nice.

Griliches, Z. (1979), 'Issues in assessing the contribution of research & development to productivity growth', *Bell Journal of Economics*, no. 10, pp. 92–116.

Griliches, Z. (1991), *The Search for R&D Spillovers*, Harvard University and NBER, September.

Hagedoorn, J. and Schakenraad, J. (1990), 'Strategic partnering and technological cooperation', in Dankbaar, B., Groenwegen J. and Schenk, H. (eds), *Perspectives in Industrial Organization*, KUNER Academic Publishers.

Jacquemin, A. (1987), 'Comportements collusifs et accords en Recherche-Développement', *Revue d'Economie Politique*, I, vol. 97.

Jaffe, A. (1986), 'Technological opportunity and spillovers of R&D', *American Economic Review*, 76, pp. 994–1001.

Klein, B. (1977), *Dynamic Economics*, Cambridge, Harvard University Press.

Levin, R., Klevorick, R., Nelson, R. and Winter, S. (1987), 'Appropriating the returns from individual research and development', Yale Papers on Economic Activity.

Mohnen, P. (1989), 'New technologies and inter-industry spillovers', International Seminar on Science, Technology and Economic Growth, OCDE, DSTI.

Mowery, D. (1983), 'Economic theory and government technology policy', *Policy Studies*, no. 16.

Nelson, R. (1959), 'The simple economics of basic scientific research', *Journal of Political Economy*, 67.

Nelson, R. and Winter, S. (1982), *An Evolutionary Theory of Economic Change*, Cambridge, Harvard University Press.

Pavitt, K. (1984), 'Sectoral patterns of technical change', *Research Policy*, 13(6), pp. 343–73.

Rosenberg, N. (1976), *Perspectives in Technology*, Cambridge University Press.

Rosenberg, N. (1982), *Inside the Black Box*, Cambridge University Press.

Sahal, D. (1981), *Patterns of Technological Innovations*, Addison-Wesley.

Schumpeter, J. (1934), *The Theory of Economic Development*, Oxford University Press.

Shachar, Y. and Zuscovitch, E. (1990), 'Technological learning and efficient organization structure in High Tech. Environment', Communication of the Second International Conference on Management of Technology, Miami, February.

Teece, D. (1980), 'Economies of scope and the scope of the enterprise', *Journal of Economic Behaviour and Organization*, 1.

Zuscovitch, E. (1986), 'The dynamics of technologies development', *Research Policy*, 15, pp. 175–86, December.

4 Technological systems and economic performance: the diffusion of factory automation in Sweden*

Bo Carlsson and Staffan Jacobsson

Introduction

Swedish industry is among the most highly automated in the world. Table 4.1 presents an overview of the usage of flexible automation techniques per employee (referred to as 'density') in various countries in the 1980s. In density of numerically controlled (NC) machine tools, Sweden and Japan were world leaders. There are some problems of standardization of definitions but there is little doubt that the same was true with regard to industrial robots and flexible manufacturing systems (FMS), with Japan clearly ahead in industrial robots and Sweden in FMS. In the area of computer-aided design (CAD), the United States had a significant lead, followed by Sweden, the United Kingdom, France, and West Germany, while Japan and Italy lagged far behind. Thus, Sweden has attained a strong position in the utilization of all of these technologies despite the fact that none or few of the original inventions originated in Sweden. How can we explain the advanced state in the use of factory automation in Sweden, and what lessons can be learned for policy?

The diffusion process is inherently highly dependent on the circumstances in which it takes place. A starting point for us is that the local context refers to much more than 'conventional' aspects, such as relative factor prices, the size distribution of firms, etc. More important, in our judgement, is the institutional set-up and the presence of linkages (market and non-market) between firms and other actors. This is why we have taken a systems view of the determinants of diffusion (see also the previous chapter by Orsenigo).

In an earlier work (Carlsson and Stankiewicz, 1991: 111), we defined *technological systems* as

> *network*[*s*] of agents interacting in a specific economic/industrial area under a particular *institutional infrastructure* or set of infrastructures and involved in the *generation, diffusion, and utilization of technology*. Technological systems are defined in terms of

*The authors would like to thank the participants in the sessions on 'Sweden's Technological System and Future Development Potential' at the 18th annual conference of the European Association for Research in Industrial Economics, Ferrara, Italy, 1–3 September 1991, for valuable comments on an earlier version of this chapter. We would particularly like to acknowledge helpful comments and suggestions by Zoltan Acs, Ellinor Ehrnberg, Lennart Elg, and Franco Malerba. The authors are solely responsible for any remaining errors and omissions.

Table 4.1 Density of flexible automation techniques in various countries (number of units per thousand employees in manufacturing)

Country	NC machine tools* (c.1980)	Industrial robots (1984–5)	FMS** (1988)	CAD (1986)
Belgium		1.13	0.79	
France		0.58	1.52	2.89
W. Germany	11.38	1.16	1.12	2.62
Italy		0.62	0.95	0.31
Japan	22.40	7.53	1.72	0.72
Sweden	22.18	3.57	6.94	3.76
UK	10.51	0.55	1.76	3.17
USA	11.73	1.04	0.72	6.33

* per 1,000 employees in engineering industries
** per 100,000 employees in manufacturing
Sources: Numerically controlled (NC) machine tools: Edquist and Jacobsson, 1988, p. 104
Other equipment: IIASA

knowledge/competence flows rather than flows of ordinary goods and services. They consist of dynamic knowledge and competence networks. In the presence of an *entrepreneur* and sufficient *critical mass*, such networks can be transformed into development blocks, i.e. synergistic *clusters* of firms and technologies within an industry or a group of industries.

A technology system is thus involved with the generation, diffusion and utilization of technology. Much technology policy in the OECD countries would seem to have been concerned mainly with the generation of knowledge (Mowery and Rosenberg, 1990, Jacobsson *et al.*, 1989 and Branscomb, 1991 for the USA). However, the generation of knowledge, in the form of a new invention, and indeed, innovation of a new artefact is the lesser part of the economic picture (Cameron and Metcalfe, 1987). New technologies are significant only insofar as they carry economic weight, i.e. they diffuse extensively and in this process, displace inferior technologies. In another terminology, from the point of view of economic performance of an industry/nation, it is less important to shift the production possibility curve outward (best practice) than to improve the resource allocation of firms and plants within the production frontier (Carlsson and Taymaz, 1991). The main objective of the technological system and of government policy should then be to influence the process of diffusion of new technology so that the entire local industry, including the vast number of small and medium-sized firms, can identify, procure and utilize the new technology earlier than their international competitors.[1]

Of course, the dichotomy between generation and diffusion of technology is too simple-minded. What is diffused is rarely a single innovation but a sequence of innovations, often in the form of a series of generations of the technology (Cameron and Metcalfe, 1987; Ehrnberg and Jacobsson, 1991; Gold, 1981; Metcalfe and Gibbons, 1991; Tryggestad, 1991). As pointed out by Metcalfe and Gibbons (1991), how this sequence develops is closely connected to the diffusion process. Hence, a fast local diffusion may be a function of local innovative activities in developing new generations or varieties of the technology.[2]

By viewing a country's performance in a particular techno-industrial area from a systems point of view, while emphasizing diffusion rather than innovation (keeping in mind their interrelatedness), we are forced to operate in relatively unchartered territory. Few precedents exist – only Freeman (1987) and Mowery and Rosenberg (1990) come to mind. It also means that there is little theory which can provide guidance. Thus, our effort should be viewed primarily as pre-theoretical: the observations and categorizations we make are designed to point out salient features which may constitute building blocks in constructing theory. Only further research, in Sweden and elsewhere, will tell whether the features we identify are indeed the most important – or merely the most prominent to the untrained eye.

The purpose of the present chapter is twofold. The first is to identify the salient features of the Swedish technological system with respect to factory automation which go some way towards explaining the relatively good performance of Swedish industry in this area.[3] Secondly, we are interested in exploring the actual as well as potential role of government policy. Our emphasis is on identifying the key policy issues rather than making policy recommendations.

The chapter is organized as follows. The next section summarizes the results of our study with respect to the composition of the factory automation technology system of Sweden and its historical performance. The third section discusses the policy implications of our analysis.

Salient features of the Swedish technological system for factory automation

In other publications, the characteristics of the Swedish technological system for factory automation have been explored (see Granberg, 1990; Granberg, 1991; Stankiewicz, 1991; Carlsson and Eliasson, 1991; Carlsson and Jacobsson, 1991; Ehrnberg and Jacobsson, 1991; Dahlin, 1991; Tryggestad, 1991). Here we will merely summarize the salient features of that system.

The main components of the technological system are (1) the institutional infrastructure, (2) clustering of resources in the form of networks, and (3) the economic competence of various agents.

Institutional infrastructure

There are four components of the institutional infrastructure: (1) the organization of industrial research and development (R&D); (2) academic infrastructure, particularly the technical universities; (3) other institutions; and (4) government policy. We will limit ourselves here to discussing the academic infrastructure. Government policy and other institutional infrastructures seem to have few unique features in Sweden, if each aspect is treated individually. However, they do seem to be unique if treated as a system. Therefore, these components are dealt with under knowledge/competence networks below, as is the organization of R&D.

Academic infrastructure[4]

The educational system, in particular the Universities of Technology, plays a critical role in the process of generating and diffusing knowledge. As Pavitt has argued in Chapter 1, it is a dual role in that the output of the educational system is both training of people and research results. The latter expand the set of technological opportunities, while training increases the ability to take advantage of the opportunities. Indeed, the quality and quantity of the engineers graduated are among the main determinants of a nation's innovative and absorptive capacity.[5]

Until quite recently, the Swedish educational system in the field of factory automation was largely confined to the training of engineers rather than research. The advanced Swedish research capabilities have rested primarily in firms and in the non-commercial research sector. Before the 1980s, little research was done at the universities, and the curricula lacked a factory automation orientation.[6] It was only during the second half of the 1980s that the universities began to play an important role in research. This is quite remarkable, since the diffusion of microelectronics-based factory automation technology began to be highly visible already in the mid-1970s. Although we do not have data on the adjustment pattern in other OECD countries, it is pertinent to raise the question if a whole decade needs to pass between the point when leading-edge firms[7] identify the new technological opportunities and when the academic system adjusts (or begins to adjust). The persistent shortage of application engineers in the robotics field in the 1980s suggests that a somewhat faster adjustment is warranted.

Moreover, the academic part of the Swedish technology system is still small. The expenditures on factory automation training and research in the Swedish technical universities represent no more than approximately one-half of the R&D budget of ABB Robotics (a subsidiary of Asea Brown Boveri)! Additionally, in spite of the expenditure increase in the 1980s, the academic infrastructure in this area is still underfinanced. The smallness of the system means that it is increasingly difficult for it to cover the whole range of competencies (including both basic technology and applied knowledge) required for the eventual design and implementation of factory automation systems which are becoming ever more complex while, at the same time, not eroding its competence base in other areas. This is, of course, an inherent problem in a small nation, but the underfinanced status of academia makes matters worse.

The academic part of the technology system for factory automation uses its skills primarily to educate engineers, and it also has a strong user orientation in its research. This research is, significantly, not simply oriented to machine design but concerns itself with the broader, practical issues in the implementation of factory automation. This orientation of the research presumably means that the awareness of the newly graduated engineers of these issues is heightened. The applied character of the research strengthens the diffusion orientation of the Swedish technology system for factory automation in spite of the above-mentioned weaknesses of the academic part.[8]

There may be another reason for the relatively applied character of research in factory automation in Sweden. Henderson and Clark (1990) have suggested a distinction between 'component knowledge' and 'architectural knowledge'. In the area of factory automation it may well be that it is primarily the 'component knowledge' (i.e. how a machine or process functions in detail – the realm of the engineer rather than the researcher) which is important. In other fields, such as biotechnology, 'architectural knowledge' may be more important, and hence also the role of academic research.

Knowledge/competence networks (clustering)

There are three aspects of networks which seem noteworthy in regard to Sweden's technological system for factory automation. The first is the existence of a set of 'bridging' institutions. The second is the importance of user–supplier linkages, and the third is the question of critical mass.

'Bridging' institutions

As indicated earlier, the academic sector has played a supportive rather than a leading role. Its emphasis has been on training rather than research, and its research emphasis has been on application/implementation rather than basic research.

By contrast, the business sector has distinguished itself by what seems to be the highest level of R&D expenditure in relation to sales or value added anywhere in the world. According to OECD statistics, Sweden's R&D expenditures have been the highest in the world in recent years. The vast bulk of these expenditures are in the engineering industries. This is clearly important. But it is inherent in such spending that it is highly skewed: a few firms are responsible for most of the R&D. This is probably true everywhere, not just in Sweden. In our view, what makes Sweden unique – and what may partly explain the high degree of automation in Sweden – is the existence of well-functioning networks consisting not only of these few large R&D spenders but also of smaller firms, academic institutions, government agencies, and perhaps even more importantly, institutions which provide bridges among these various types of units.

These 'bridging institutions' are NUTEK and its various sub-units, as well as Mekanförbundet and IVF. NUTEK[9] is the main government agency for implementation of technology policy. Mekanförbundet is the branch organization of the Swedish engineering industry. In addition to the duties normally performed by a branch organization, Mekan plays a significant role in the diffusion of technology as well as in the formulation of technology policy in Sweden. IVF (the Swedish Institute of Production Engineering Research), a private organization, is a collective research institute operated jointly by Mekan and NUTEK (Stankiewicz, 1991.) It conducts technology scanning, monitoring, adaptation, and diffusion, as well as contract research and testing for individual firms.

There is cross-representation of all these organizations on numerous committees and task forces, and small as well as large firms are represented. Academic institutions are represented, primarily in NUTEK and IVF. Thus,

IVF and NUTEK provide links between academia and business, while Mekan bridges the gap between government and industry. Through their organization and interaction they scan the world for new technology, evaluate it and rapidly disseminate information on it, thus fostering a high level of technical awareness at all levels of Swedish industry, an awareness which is central to explaining a rapid diffusion of technology. In addition, the bridging institutions help to accumulate and integrate the results of innovative activities which otherwise tend to be highly firm-specific and make them useful and available to other firms as well, thus internalizing them within the technological system.

The role of these organizations is not only to disseminate know-how but also to provide a compensating mechanism for weaknesses within other parts of the technological system, particularly the academic system. IVF may serve as an example here: it has an important monitoring role; it builds up specialized skills very early in the technology's life; it collaborates with the leading edge firms in the development and application of new technology and it acts strongly as a diffuser of the new technology to small and medium-sized firms. By playing this role, IVF has, for instance, compensated for the lack of a strong Swedish supplier of CAD/CAM (see Stankiewicz 1991).[10]

User–supplier linkages

It is clear that in spite of the large and growing international trade in factory automation products, access to new technology is not the same for all potential customers, nor are all firms aware of the existence of new technologies. Inadequate access, information scarcity, and uncertainty limit their diffusion. Only the most advanced users can expect to attract distant suppliers of complex technology; all others are dependent (and increasingly so with declining technical competence on the part of the users) on domestic suppliers – until the technology is mature and generic (Carlsson and Jacobsson, 1991). A local, and advanced, automation industry therefore has a critical role to play in the diffusion of new technology in the local market.

The strategic role of the automation industry implies that it is important that it be broad in the sense of covering many product areas. At the same time, firms in the industry need to follow strategies which emphasize product development as the prime source of competitive strength, in contrast to emphasizing low costs.

So far, the Swedish factory automation industry has been very successful – it includes globally leading firms (centers of excellence) in robots, automatic guided vehicles, measurement systems, CNC grinding machines (including systems) and systems built around CNC lathes.

In part, its success rests on the interaction (market and non-market mediated) that it has had with a set of very advanced local user firms which possess significant problem-identifying and problem-solving capabilities in the field of new production technology as well as in systems integration. Indeed, as we have shown previously (Carlsson and Jacobsson, 1991), it appears that the existence of technologically advanced domestic users is a prerequisite for an advanced and healthy supplier industry. Jointly, these technologically leading firms and the local factory automation industry form a development block which has the strength to influence positively the diffusion of new production technology to the very large number of small and medium-sized firms (see also the chapter by Lundvall).

Public policy has influenced the workings of this development block in several ways. First, as demonstrated by Stankiewicz (1991), a large number of demonstration plants have been partly financed by NUTEK and SIND. The diffusion of FMS technology to small and medium-sized firms was directly stimulated by this policy, which also had a demonstration effect. Moreover, links between the Swedish factory automation industry and Swedish users were stimulated, thus improving the connectivity of the development block. Finally, the Swedish suppliers had an opportunity to gain valuable experience at a fairly early stage of the product's life cycle.

Second, two of the leading machine tool firms were acquired by the state in periods of financial crisis. However, as indicated by Ehrnberg and Jacobsson (1991), the economic competence (see below) of the state has proven to be quite questionable, and the future is now precarious in some central areas of Swedish machine tool industry (lathes and machining centers as well as systems built around these). Holes in the hitherto well-functioning development block may well materialize in the near future. For these firms to survive[11] and continue to be of strategic importance to the local users, they need to (a) stay at the technological frontier and (b) develop a much larger volume of production of the machine tools which form the core of FMS. In both cases, it is feasible to do so but unlikely that a private owner[12] would take the risks of such a venture. In principle, government could revert to a policy of strong intervention, but recent history suggests that there are such information and competence problems involved that it is highly unlikely that the state would be successful in inducing the firms to pursue the right strategic orientation. In other words, the historical government failure would probably be repeated.

Critical mass

In order for a set of relationships such as a network or a technological system to function as a development block (Dahmén, 1989) capable of generating economic activity, two things are needed: an entrepreneur who can formulate a vision of potential economic activity and who can bring it into being, and a certain density or intensity of the relationships among the various agents. We refer to the latter aspect as critical mass (see further Carlsson and Stankiewicz, 1991). The basic idea is that there is a certain minimum number and intensity of interactions required for sustained economic activity to take place. Thus far, Sweden has been in the fortunate position of having a sufficient number of highly advanced users (e.g. Volvo, ABB (formerly ASEA), SAAB, and Electrolux) to sustain a healthy domestic supplier industry. The close networking among various agents in Sweden has made for high technical awareness and rapid diffusion of new technologies. There are, however, signs that the critical mass problem is increasing in importance as a result of rising complexity of technology (i.e. requiring expertise in more fields than before) and weakening domestic manufacturing capability of advanced users as they become more and more internationally spread out and/or globally oriented. Intriguing questions for future research are what is 'sufficient' and 'close' – in other words, what is required to achieve critical mass? Clearly, international comparisons could shed light on this.

Economic competence

The infrastructural and networking aspects of the technological system are clearly important – but without the proper economic competence (defined as the ability to identify, expand, and exploit the opportunity set) on the part of both users and suppliers, they would come to naught.[13] If the users do not demand new technology, or if for some reason they are incapable of using it – or alternatively, if the suppliers are not able or willing to supply it – it may matter little how the system is organized. As illustrated by Tryggestad (1991) and Carlsson Jacobsson (1991), the more complex and customized the technology, the greater the required interaction between supplier and customer for successful transfer and usage. Customized solutions inherently involve adaptation and/or improvement of an existing product or process and normally require far more competence on the part of the buyer both in procurement and in use.

The high technical competence and strongly international orientation of a number of leading Swedish user firms have already been mentioned. It seems clear that these firms have for some time accumulated a considerable competence to generate and utilize business opportunities drawing on the vast technological opportunities of the basic technologies forming factory automation artifacts. In addition to the high degrees of automation in many of these firms, further corroborating evidence is provided by the fact that many of the major international suppliers of automation technology have established local sales and service organizations in Sweden, in spite of the relatively small market size, and by the opinion of several leading German suppliers that Swedish user firms were the most advanced in the world in the late 1970s. Indeed, it would seem that this competence has been greater among users than among suppliers of factory automation. An exception is ASEA Robotics, which flew in one of the first microprocessors from the USA and by opting for microprocessor-based control technology managed to become a technology leader in robotics in the mid-1970s. An early knowledge of the mere existence of the new technologies and, most importantly, an appreciation of their technological and commercial potential at that time was instrumental in shaping a firm strategy which has been very successful hitherto.

The role of government policy

That markets do not always work is not a new proposition. The literature on market failures goes a long way back, to List, Hamilton and John Stuart Mill (Corden, 1974).[14] A whole set of market failures have been identified; public goods problems; external economies; difference in private and social discount rates and risk aversion; access to information on present and future markets and lack of management or economic competence which limits the perceived opportunity set (Alänge and Jacobsson, 1991; Jacobsson and Alam, 1991).

Two points should be made regarding market failures. First, market failures derive from marginalist economic theory which is concerned with static efficiency problems. As we move away from equilibrium analysis to seeing competition as an open-ended process (Metcalfe and Gibbons, 1991), we need to analyze the consequences of market failures in the context of that process, or in the context of the workings of a technology system.

Second, market failure *per se* is not sufficient to justify intervention. Government failure (Krueger, 1990) may, for a whole set of reasons, be even greater than the market failure and consequently, an intervention may not lead to a reduction or elimination of the market failure (as illustrated, e.g., by the failure of government intervention in the Swedish machine tool industry mentioned in the previous section). Indeed, most governments may lack an ability to intervene 'correctly' (Pack and Westphal, 1986).

This inability to rectify market imperfections is not, however, inherent in government bodies; economic competence is certainly not a monopoly of managers. A Japanese senior manager who begins work with MITI does not, of course, lose all his economic competence in the transfer process. Certainly, the experiences of both Japan (Freeman, 1988; Fransman, 1990) and Korea (Jones and Sakong, 1980; Jacobsson, 1986; Pack and Westphal, 1986; Jacobsson and Alam, 1991) suggest to us that the state can have a very positive role to play in economic development.[15]

In periods of rapid technical change, involving a series of technological transitions based on new generic technologies, the state may have a particularly important role to play in the molding of technology systems. This applies to all the constituent parts of the system; clusters/development blocks, economic competence and institutions. However, as we move away from equilibrium analysis to analyzing dynamic problems in a technology system, it is not self-evident that the market failure approach (with its clear policy conclusions) is the appropriate one. Precisely how policy should influence the system is, in fact, unclear.

First, the critical role of clustering of resources and knowledge/competence networks suggests that a well-functioning technology system rests on strong positive external economies (both pecuniary and non-pecuniary) which tie together users, suppliers and competitors.[16] Indeed, given the importance of networks in technology system, it may not be unreasonable to suggest that technology systems and therefore the process of generating and diffusing new technology rests heavily on market imperfections.

But in a properly functioning technological system it would be a mistake to view external economies as 'failures'. Instead, they form an integral part of an economic organization which tends to correct for these 'failures' more or less automatically – the economic organization is in some instances superior to the market. Many of the external economies tend to become reciprocal in nature. In conventional economic theory, such external economies require some coordinated investment decision, *inter alia* by state intervention. However, in a technological system, the exchange of information may well result in a blending of visions (technological expectations) of the future between different actors (Fransman, 1990). Sharing the same vision may then lead to a quasi-coordination of investments between formally independent actors. For instance, the interaction between the technologically leading firms (in terms of factory automation) and the local supplier industry resulted in a blending of their visions of the future production technology. The whole development block then invested in technological capabilities and design developments which were interchanged within the network (cf. the example provided earlier in note 10).

Indeed, the technological system functions much like a large innovating firm which manages to internalize and integrate a variety of more or less unpredictable changes and thus to survive. As Keith Pavitt has put it,

> [l]arge innovating firms are a major source of the world's technology, and in the 20th century have shown great resilience in absorbing successive waves of radical innovations. The key characteristics of these firms derive from the properties of their innovative activities. First, given the specific, differentiated and cumulative nature of technological development, the range of possible choices about both product and process . . . technologies open to the firm depends on its accumulated competence. Second, given functional and professional specialization, the implementation of technological choices requires organization and orchestration across disciplinary, functional and divisional boundaries. Third, given cumulative development and uncertainty, the improvement of these competences requires continuous and collective learning. Fourth, in the light of these characteristics, systems for allocating resources must take into account the benefits of learning by doing, as well as the benefits of outcomes. As a consequence, the technical function in large firms involves not just the implementation of innovations, but also the definition of appropriate divisional objectives and boundaries, the exploration of radical technologies, and the formation of technological expectations about the future. (Pavitt, 1991: 41.)

How well the technological system works is, in part, dependent on how well its constituent parts are connected. The argument here is that the mechanisms which generate variety and also those which select across variety (both basic features in evolutionary economies) are critically determined by the connectivity of the science and technology system. Metcalfe and Gibbons (1991: 13) put this in the following words:

> the rate of experimentation depends on the connectivity of the science and technology system and . . . policy makers can increase connectivity by, for example, promoting the interaction between the higher education and industrial components. The distribution of research funds between institutions, the existence of bridging mechanisms to connect industry to the public science and engineering base, and the degree of public investment in skill formation all play their role in raising creativity in general.

By enhancing the rate at which information flows through the technology system (increasing the connectivity), the general awareness of the technological opportunity set is raised and the visions are blended. Both innovation and diffusion are affected. Moreover, the connectivity of the system affects how quickly learning (from both failures and successes) takes place at the level of the technological system. Policy, it would appear, should therefore aim at enhancing the connectivity rather than correcting for individual market failures in terms of external economies.

Second, economic competence – the ability to identify, expand, and exploit the opportunity set – is critical to the functioning of the technology system but is unevenly distributed among firms. Firms therefore operate with different knowledge bases and under different assumptions concerning technology, markets, competitors etc. Of course, the whole notion of disparate economic competence indicates the existence of market failure in the neoclassical sense.

Rationality, on these assumptions, could well be argued to be restricted by the experience accumulated by management.[17] A particularly important subset of management's experience and competence refers to the firm's technology base. The technological and commercial opportunity set, as perceived by management, is presumably closely connected to this base.[18] When new generic technologies become available, providing vast technological and commercial opportunities, the bounded vision (Fransman, 1990) of managers may very well limit the perception of these opportunities. To a degree, a technological system may manage to blend the visions of some of its constituent parts in a socially optimal way, but in other cases, intervention is required. Fransman (1990: 286), in fact, argues this was the case in the Japanese collaborative research program:

> a number of very senior company research managers . . . pointed to the importance of the blending of bounded visions brought about by the Japanese System . . . these managers were convinced of the important contribution that has been made to long-term economic performance by MITI officials and ETL researchers as a result of their ability to compensate for the shortsightedness and blind spots of the companies.

The observations made by Freeman (1988: 331, 334) underline this role of MITI in Japan:

> The not-so-invisible guiding hand of MITI shaped the long-term pattern of structural change in the Japanese economy and this was largely exerted on the basis of judgement about the *future direction of technical change* and the relative importance of various technologies . . . The 'Visions' of the future produced by STA, MITI, NIRA and other government and private sources do not pretend to be accurate predictions, nor do they commit companies to inflexible plans. They chart the *broad direction of* advance for the economy and for *technology* and give companies sufficient confidence in this vision to make their long-term investment in research, development, software, equipment and training, . . . Nobody believes that it is possible to eliminate uncertainty, but a thorough discussion serves to mobilize resources, to expose difficulties, and bottlenecks, and above all to energize the participants, secure consensus and heighten *awareness*. (Our emphasis.)

The role of MITI could thus, in part, be seen as helping firms to improve their economic competence and expand their perceived opportunity set, in particular in periods of rapidly increasing technological opportunities. Such a role is, of course, not restricted to MITI.[19] Compensating for inadequate economic competence can constitute a major objective of policy in the process of molding a technology system.

Third, institutions matter. This refers to a whole set of features of a technological system. One of the more important institutions is the educational system, especially the training of engineers. In particular, the system needs to be pro-active and flexible: pro-active in order to be able to supply industry with specialized skills and new knowledge in emerging new technological fields; flexible in order to adjust the orientation of education from old technologies (e.g. shipbuilding technologies) to new (e.g. microelectronics). A second critical institutional aspect is that of 'bridging' institutions which act as information exchanges within the technological system. These institutions may well need initial support to get started – once they function they can be financed by those who discover the usefulness of their 'bridging' work.

A third critical institution is, as always, the functioning of the financial market. In particular, this matters for innovative small firms which will eventually become the new Volvos of the third millenium if the technological system works well.

Technology development strategy of the state

In conclusion, one can identify a set of conditions for reaching the objective of rapid and extensive diffusion:[20]

1. A speedy and wide diffusion of basic engineering competence to all the components of the production system (large and small companies) which sets strict demands on the functioning of the educational system.
2. A mechanism for monitoring the technological development globally where the emerging features of new production systems are identified at a very early stage.
3. A well-functioning communication system in industry (large and small companies) which is tied into strategic R&D planning to build up the necessary competence among users.
4. Centers of excellence for the integration of the various technologies. These can be located in both supplier and user companies or even outside industry, but the know-how produced in these centers must be accessible to the other actors in the industrial network.

Jointly, this would make up a functioning technology system. It is the molding of this system which ought to be the prime objective of policy – not to rectify individual market failures. This implies that all of the constituent parts of the technology system can be subjected to the influence of policy. It also means that intervention is not always necessary just because there is a market failure in the neo-classical sense. The best policy shaping the technological system can therefore not be stated in precise terms. It *might* be necessary to increase the connectivity of some of the constituent parts; the state *might* have to rescue a central component of the system or find compensating mechanisms; it *might* have to influence the visions of management and, most of all, it *might* have to get some institutions and networking started. The precise nature of intervention would be expected to vary from case to case, both between different technological systems in a country and between countries. But it is worth emphasizing that an essentially diffusion-oriented technology development strategy of the state does not preclude but, indeed, may require close attention to the status of the domestic supplier industry. As was argued by Carlsson and Jacobsson (1991), in the earlier phases of the diffusion of a new technology, the market failures can be very significant and can give the local supplier industry a key role to play in the local diffusion process.

While it is difficult to specify the *content* of a sensible public technology policy for the reasons just mentioned, it is clear that taking a systems perspective (i.e. viewing the system as a whole rather than simply as a set of individual activities) means that such a policy should not be limited to R&D support of individual firms or projects. In fact, that may not even be an appropriate part of such a policy. If one adds to the systems perspective an evolutionary instead of a static approach, recognizing the difficulties of identifying future technological opportunities, the emphasis shifts from the pursuit of individual technologies or approaches to either pursuit of alternative strategies simultaneously or emphasis on early identification of important developments and on increasing the absorptive capacity of the economy. While it may seem paradoxical at first glance, such an evolutionary and systems approach may actually impose far less stringent competence requirements on the policy-makers than a more static, less flexible approach which requires in-depth knowledge of how firms operate and compete in individual product areas.

Notes

1. We recognize that under some conditions (e.g. lumpy investments combined with fast product changes) it may be advantageous not to be the first to invest in a new technology.
2. For further discussion of user–producer interaction and the role of geographical and cultural proximity, see Lundvall, 1988 and Tryggestad, 1991.
3. This is not to say that other factors are not important, only that we have elected to focus here on what we deem to be most imortant, namely institutional aspects. More conventional explanatory factors have been explored elsewhere. For example, Carlsson *et al.* 1979, pp. 21–3, suggest that the conditions for factory automation in Sweden may have been particularly favorable in Sweden in the late 1970s because of a combination of factors (including a wage explosion exacerbated by wage subsidies to declining industries) which reinforced the persistent labor shortage (reflecting both demographics and a strong preference among young people for non-industrial jobs) in Sweden in the postwar period. In addition, the relatively high level of education of the labor force and the favorable attitude of labor unions towards automation constituted positive factors. On the other hand, one can show that neither the distribution of value added by industry in manufacturing nor the size distribution of firms can provide much explanatory power. Other possible factors yet to be investigated are the traditional orientation of Swedish engineering industry towards highly specialized producer goods typically manufactured in small batches (as distinct from mass-produced consumer goods) and extremely small wage differentials.
4. Unless otherwise indicated, this section is based on Granberg (1991).
5. For a discussion of the notion of absorptive capacity in firms and the dual role of industrial research and development as representing both innovation and learning, see Cohen and Levinthal, 1989 and 1990.
6. A relevant question, to which we do not yet know the answer, is that if little research was done in this rapidly expanding area, what was the quality of the education provided? How important is first-class education for application (as distinct from generation) of technology?
7. The President of the highly successful Japanese firm Fanuc (global leader in both CNC units and robots) explained once that 'We applied the technical innovations in the semiconductor field to machine tools earlier than the computer industry' (*Metalworking Production*, 1975, cited in Jacobsson, 1986). In particular, Fanuc introduced microprocessor-based CNC units in 1975, a fact of great significance in

explaining Fanuc's global leadership today.

8. One of the reasons for the emphasis on training rather than research may be the decentralization of Swedish universities via growth of new universities which took place during the 1970s and 80s. Was this policy more beneficial to technological diffusion than a more centralized, research-oriented strategy might have been?
9. On July 1, 1991, the Swedish National Board for Technical Development (STU) was merged with the Board for Industrial Development (SIND) and the Energy Agency (Energiverket) to form NUTEK, the National Board for Industrial and Technical Development.
10. The following example (based on *Teknik i tiden*, No. 1, 1991, p. 20) illustrates how the system works. In 1983/4, NUTEK (then STU) started a five-year support program in CAD/CAM designed to stimulate the construction of reference systems by supporting technical development with regard to application within user firms. CAD/CAM was not a new technology when the program started, but it was not widely diffused. During the operation of the program, there was rapid technical development of the equipment, making it difficult for firms interested in the technology to 'link up'. At the same time, there were early users who wanted to proceed from an experimental phase to more thought-out, long-run investments.

    The design of the program reflected the strong need for both increased user competence and technical development of the systems on the supplier side. As a result, the practical application has been speeded up by several years in most participating companies. In many cases, the program was the decisive factor influencing investment decisions even in large firms like SKF. Participating firms report having had a valuable exchange of ideas and experiences helpful in their own development. The participating group of firms initiated technology diffusion to other firms (via 26 conferences on various topics where participating firms presented their experiences), arranged seminars for teachers of technical subjects, and influenced standardization (via IVF) at the international level. Through the participation of academic researchers in company projects, a basis was established for exchange of know-how and experience across the academic research–industry line.

    The program was directed by a person from IVF, enabling IVF to further develop, expand, and disseminate its know-how within the CAD/CAM area.
11. See Ehrnberg and Jacobsson (1991) for a detailed discussion.
12. The two firms were recently sold to private firms.
13. For a discussion of the nature and importance of economic competence, see Carlsson and Eliasson, 1991.
14. These authors put forward the infant industry argument which rests on market failures.
15. Lack of economic competence, corruption, etc., has led to disastrous government failures in many other countries.
16. Porter (1990) makes a similar observation.
17. 'By being specialized the firm develops restricted technological vision. Here lies a major evolutionary paradox. To specialize is to be efficient but it is also necessarily self-limiting in terms of the ability to respond creatively to a change in technological possibility' (Metcalfe and Gibbons, 1991: 20).
18. A technologically specialized firm would then tend to have management with a more bounded rationality than a firm with a broader technology base.
19. In the literature on Korea's phenomenal industrial development, a key role of the state has been to enlarge the perceived opportunity set by 'field manipulation' (Jones and Sakong, 1980).
20. The following four points were taken from an early draft by Stankiewicz (1991).

References

Alänge, S. and Jacobsson, S. (1991), 'Industrial policy for a declining industry – the case of the Swedish textile and clothing industry', mimeo, Department of Industrial Management and Economics, Chalmers University of Technology.

Branscomb, L.M. (1991), 'Toward a U.S. technology policy', *Issues in Science and Technology*, Summer 1991, pp. 50–5.

Cameron, H.M. and Metcalfe, J.S. (1987), 'On the economics of technological substitution', *Technological Forecasting and Social Change*, 32, pp. 147–62.

Carlsson, B. and Eliasson, G. (1991), 'The nature and importance of economic competence', paper presented to the EARIE conference, Ferrara, Italy, 1–3 September 1991.

Carlsson, B. and Jacobsson, S. (1991), 'What makes the automation industry strategic?', *Economics of Innovation and New Technology*, 1(4) (forthcoming).

Carlsson, B. and Stankiewicz, R. (1991), 'On the nature, function, and composition of technological systems', *Journal of Evolutionary Economics*, 1(2), pp. 93–118.

Carlsson, B. and Taymaz, E. (1991), 'The role of technological progress and economic competence in economic growth: a micro-to-macro analysis', paper presented to the EARIE conference, Ferrara, Italy, 1–3 September 1991.

Carlsson, B., Dahmén, E., Grufman, A., Josefsson, M. and Örtengren, J. (1979), *Teknik och industristruktur – 70-talets ekonomiska kris i historisk belysning* (Technology and Industrial Structure – The Economic Crisis of the 70s in Historical Perspective), IUI and IVA, Stockholm.

Cohen, W. and Levinthal, D. (1989), 'Innovation and learning: the two faces of R&D', *The Economic Journal*, 99(397), pp. 569–96.

Cohen, W. and Levinthal, D. (1990), 'Absorptive capacity: a new perspective on learning and innovation', *Administrative Science Quarterly*, 35, pp. 128–152.

Corden, W.H. (1974), *Trade Policy and Economic Welfare*, Clarendon Press, Oxford.

Dahlin, K. (1991), 'Diffusion and industrial dynamics in the robot industry', working paper, Department of Industrial Management and Economics, Chalmers University of Technology, March 1991.

Ehrnberg, E. and Jacobsson, S. (1991), 'Technological discontinuities, industry structure and firm strategy – the cases of machine tools and flexible manufacturing systems', paper presented to the EARIE conference, Ferrara, Italy, 1–3 September 1991.

Fransman, M. (1990), *The Market and Beyond*, Cambridge University Press, Cambridge.

Freeman, C. (1987), *Technology and Economic Performance: Lessons from Japan*, Pinter Publishers, London.

Freeman, C. (1988), 'Japan: a new national system of innovation?', in Dosi *et al.* *Technical Change and Economic Theory*, Pinter Publishers, London.

Gold, B. (1981), 'Technical diffusion in industry: research needs and shortcomings', *Journal of Industrial Economics*, 29(3), pp. 247–69.

Granberg, A. (1991), 'Mapping an evolving technology cluster: the composition and structure of factory automation', working paper, Research Policy Institute, University of Lund, January 1991.

Granberg, A. (1991), 'A study of the academic infrastructure of factory automation', working paper, Research Policy Institute, University of Lund.

Henderson, R.M. and Clark, K.M. (1990), 'Architectural innovation', *Administrative Science Quarterly*, 35, pp. 9–30.

Jacobsson, S. (1986), *Electronics and Industrial Policy. The Case of Computer Controlled Lathes*, Allen and Unwin, London.

Jacobsson, S., Granstrand, O., Sjölander, S. and Alänge, S. (1989), 'Strategic technology issues in U.S. manufacturing industry: a critical analysis', CIM-Working Paper no. 1989: 3, Department of Industrial Management and Economics, Chalmers University of Technology.

Jacobsson, S., Oskarsson, C. and Bångens, L. (1989), 'Industriell elektronik i Sverige – Aktörsstruktur och tillväxtmönster', Report submitted to the Swedish National Board for Technical Development.

Jones, L. and Sakong, I. (1980), *Government, Business, and Entrepreneurship in Economic Development: The Korean Case*, Harvard University Press, Cambridge, Mass.

Krueger, A. (1990), 'Economists' changing perceptions of government', *Weltwirtschaftliches Archiv*, 126(3), pp. 417–31.

Metcalfe, S.J. and Gibbons, M. (1991), 'Technology policy in an evolutionary world', working paper, University of Manchester, May 1991.

Mowery, D. and Rosenberg, N. (1990), 'The U.S. national innovation system', working paper no. 90–3, Consortium on Competitiveness and Cooperation, University of California, Berkeley, September 1990.

Pack, H. and Westphal, L. (1986), 'Industrial strategy and technological change. Theory versus reality', *Journal of Development Economics*, pp. 87–128.

Pavitt, K. (1991), 'Key characteristics of the large innovating firm', *British Journal of Management*, 2, pp. 41–50.

Porter, M.E. (1990), *The Competitive Advantage of Nations*, The Free Press, New York.

Stankiewicz, R. (1991), 'The role of the R&D infrastructure', working paper, Research Policy Institute, University of Lund, January 1991.

Tryggestad, K. (1991), 'Customers as designers of advanced manufacturing technology: the case of the Swedish located engineering industry', paper presented to the EARIE conference, Ferrara, Italy, 1–3 September 1991.

Part II
The models revolution: cumulative learning, irreversibility and diversity of trajectories

Introduction to Part II

Robert Boyer

The paradoxes of the 1980s: conventional economic theories at bay

In recent years, research on technical change has been stimulated by the puzzling evolutions observed in advanced industrialized countries. How to interpret the slowing down of productivity after 1967 in the United States and in other OECD countries after the two oil shocks? Is there any explanation to the Solow paradox: everybody observes a speeding up of innovations, whereas productivity growth is still sluggish in most sectors and economies? Why have technological and organizational innovations become key components in the structural competitiveness of firms and countries?

The initial strategy has been to use the conventional models inherited from the golden age. For example, growth accounting has desperately looked for an explanation of the residual term of total factor productivity . . . without really succeeding. Similarly, the neo-classical growth theory built upon the seminal Robert Solow paper (1956) implied a progressive catching up of the leading country by the followers: this phenomenon has taken place but does not explain American productivity slow-down, nor the absence of development in very poor countries. Of course, research about the determinants and impacts of R&D expenditures (Griliches, 1979) has put the emphasis upon the advances of knowledge, as a complement to physical capital accumulation. Nevertheless, sophisticated econometric studies on panel data now show that R&D has a significant but limited impact upon productivity and growth (Mairesse and Sassenou, 1991).

Clearly, a renewal of micro and macro researches upon the interaction between innovation, competition, productivity and growth was called for by the puzzles of the 1980s. Part II of this book provides a significant sample of the alternative strategies which have been followed by a new vintage of economists, whatever their background, let it be neo-classical, Keynesian, Schumpeterian or evolutionary. All the chapters included do share some common features, but of course exhibit rather conflicting concepts or methodologies.

Once upon a time, micro and macro theories used to be totally separated, with almost no bridge between sophisticated, but not very relevant, general equilibrium models and very crude but enlightening Keynesian macro models. On the other side, growth theory and analyses about technical change were rather disconnected, with few interactions between theoretical investigations and more empirically oriented researches. This period is now over, since after the semi-seminal paper by Paul Romer and the dynamism of new industrial economics, this has become a very active field for research. Most researches, including the chapters of this book, exhibit much more realistic hypotheses, which partially fill the gap between microeconomists, experts in management and growth theorists.

Now economists try to formalize really existing economies . . .

Basically, theoreticians now try to capture the fundamental stylized facts about the role of innovation upon *the wealth of firms and nations*. At odds with most of previous researches, at least ten genuine features are to be found in this new literature.

1. First of all, innovation is at the core of *the strategy of the firms* in order to get and preserve their competitive edge. There is no more any helicopter dropping innovations stochastically from the sky . . . without any deliberate effort from individuals to find out new products or processes. This is a far more realistic hypothesis, which gives a definite role to managers, contrary to the old neo-classical theory of the firm.
2. Symmetrically, consumers are better off when they face more *diversified goods*, since they ask for differentiation in the context of the renewal of durable consumption goods. Consequently, price, quality and diversity are jointly to be considered to explain the market share of any economy. Basically, innovation is a method for continuously creating oligopolistic or monopolistic rents.
3. Therefore, the smooth and anonymous process of competitive equilibrium is replaced by the conflict between *strategic behaviours* of firms themselves, or between firms and consumers. Again, this is evidence for much more realism than in past researches. The authors consequently develop new analytical tools, whether game theory, micro dynamic simulations or evolutionary models.
4. In neo-classical growth theory, technological change used to shift the whole production function or globally affect the bias of innovation between capital and labour. In contemporary researches, quite on the contrary, *all innovations are initially local* and only modify the productivity and capital per capita at the margin of the operating productive system. Here comes the very important idea that only specific paths along a given technological paradigm are explored, in such a way that no general property can be derived from the success or the failure of a given local innovation.
5. Nevertheless, the diffusion of some generic technologies delivers spillover from one innovation to another, from one sector producing the technology to others using it. In this respect, some key innovations are finally interdependent since they suppose strong complementarities. *Network externalities* give one polar example of the *dynamic increasing returns to scale*, which are associated with such a clustering of innovations, investments, information and knowledge.
6. *The variety of innovation* mechanisms and components is a salient feature from any review of these new models: R&D spillovers in new neo-classical theories (Romer, 1986), role of education and/or on-the-job training as public goods for other models (Lucas, 1988), impact of transport and telecommunication infrastructures, or learning by doing, interacting or using in the tradition of Kenneth Arrow (1962) are alternatively selected by the various authors. Nevertheless, whatever these variants, similar mechanisms and externalities seem to prevail.
7. Symmetrically the shift from micro to macro modelling implies a greater focus upon the precise characteristics of the goods, their objective or perceived qualities, the number of varieties or models among the same product. *Imperfect competition* precisely takes place due to this continuous change in the upgrading and diver-

sification of goods. If extrapolated at the economy wide level, this framework explains better some key factors in national competitiveness in the contemporary world. The realism of the hypotheses and the processes formalized is bringing a common language to both managers and economists, theoreticians and applied researchers. This is good news for the relevance and applicability of these emerging theories.

8. As a consequence, *innovation becomes endogenous* within each country and all the firms are competing to get the same domestic and foreign markets. Consequently, the relative growth of *open economies* is closely related to the ability to innovate successfully and gain market shares among identical consumers. On the contrary, most neo-classical models used to consider only closed economies with mostly exogenous technical change.
9. Along the related steady growth path, the economy was quasi-stationary when all the variables were deflated by total factor productivity. To the contrary, the new models are mainly *dynamic and sequential*, since the economy is no more at its full long-run Marshallian equilibrium, but evolves from one temporary equilibrium to another, due to the very renewal of innovations as a result of the competitive process itself.
10. Last but not least, *no one unique equilibrium growth path* usually exists: small differences in the initial endowments or in the parameters representing individual behaviour might trigger diverging development patterns. In some extreme cases, the rapid growth of one country can be obtained at the cost of a relative decline of another, at odds with conventional international trade theory. The potential consequences about the role of markets and public interventions is important indeed.

Evolutionary models versus a sophistication of neo-classical tools

Even if most of contemporary researches share a large fraction of these features, economists have explored alternative strategies and methodologies.

1. Limited versus full rationality

On the one hand, the inherent complexity of technological innovations is assumed to *exceed the informational and computational abilities* of economic agents, especially when strong sectoral or intertemporal externality prevails. This was precisely the starting point for Nelson and Winter (1982) and their alternative to Solow's growth theory. This research agenda has been renewed in order to cope with the specificities of the socio-technical system which is currently emerging. In the present Part Francesca Chiaromonte and Giovanni Dosi, and J.L. Gaffard extend and refine the distinction between imitation and innovation, propose to model various price, wage and innovation regimes and assess their impact upon the dynamics of aggregate variables. Consequently, *evolutionary models* are conceived as a complete alternative to neo-classical theory.

On the other hand, the inaccurate prediction of this theory is essentially related to unsatisfactory hypotheses about the innovation process. Therefore, all the tools about *economic rationality, intertemporal optimization, and general equilibrium* can be extended to a configuration with imperfect competition, strategic behaviour and significant externalities associated with innovation, infrastructure and education. Paradoxically, rather Schumpeterian hints can be encapsulated into genuine micro models, the properties of which are far richer than those associated with a steady growth path, defined by totally exogenous total factor productivity trends. The chapter by Philippe Aghion and Peter Howitt clearly belongs to this category and has to be compared with recent advances by Helpman and Grossman (1991).

2. Plug increasing returns into post-Keynesian and neo-Schumpeterian models

But typical evolutionary models and extended neo-classical new growth models are not the only paths explored by contemporary researches. At least two other traditions do provide worthwhile insight into the links between innovation and growth: the post-Keynesian tradition has investigated the issue of dynamic increasing returns to scale for at least two decades, whereas the neo-Schumpeterian modelling of the diffusion process can be extended to dynamic externalities.

During the 1970s Nicholas Kaldor (1981) and Anthony Thirlwall (1979) investigated the consequences of the trade balance constraint upon sustainable growth: basically each country growth is limited by its specialization as captured by import penetration and export share in world trade. In his chapter Bruno Amable is extending this framework in two directions. On the one hand, *increasing returns to scale* are formalized within a Kaldor–Verdoorn relation between productivity and growth. On the other hand, exports are related to *quality*, which is obtained by the accumulated experience represented by aggregate production.

According to the Schumpeterian analysis, the impact of any innovation has to be related to the characteristics of its diffusion process, which depends upon the relative profitability of this innovation with respect to the existing routines. In a later chapter, C. Antonelli proposes an extension of this framework to a more complex world in which product, process and user innovations are closely interrelated, as they are in the telecommunication industry. Consequently, *network externalities* appear and deliver a genuine dynamics of the joint system of telecommunication, equipments and related services.

Each approach has its pro and con

The four approaches share the same concern for dynamic increasing returns to scale, but assume different mechanisms and generally provide quite varied methodologies in dealing with the existence of an equilibrium path or a dynamic evolution of the economic system. Each approach has its own merits and limits.

1. Microfoundations or/and empirical relevance?

The new endogenous growth theory in the tradition of neo-classical methodology exhibits clear *microeconomic foundations* for a dynamic growth model and contributes to close the gap between general equilibrium theory and macro modelling. Still more, the chapter by Aghion and Howitt gives a *rigorous formalization* of the creative destruction process, a striking novelty which combines some elements of Schumpeterian dynamics with the full rationality of behaviours and expectations. Nevertheless, this assumes *a large dose of rationality and computational ability*, far in excess of what is observed in really existing economies: can firms actually have rational expectations about innovations, which by definition cannot be dealt with in the conventional probability approach?

The evolutionary approach has symmetrical advantages and drawbacks. Gaffard as well as Chiaromonte and Dosi adopt *a limited or situated rationality principle*, according to which the firms elaborate routines in order to deal with the radical uncertainty inherent to the innovation process. Both insist upon the idea of regime, defined as the conjunction of routines about imitation, price and wage formation, access to the financial system and so on. A learning process built upon past experience is replacing the forward optimization which is adopted by the new endogeneous growth theorists. But this realism has a cost: given the dynamic and sequential features of these models, it is quite hard to derive *any clear analytical and general result* which would be equivalent to the long-run equilibrium in the neo-classical approaches.

2. *Cumulative causation and/or innovation push growth*

The post-Keynesian modelling can claim some anteriority in the representation of *growth as a cumulative process*. V. Norman, commenting on a 1989 paper by R. Baldwin, has pointed out how close the first Paul Romer model was to Nicholas Kaldor's pet ideas about the existence of dynamic increasing returns to scale. Bruno Amable belongs to this tradition in macroeconomic modelling and points out the interaction between structural competitiveness captured by quality and the ability to sell on high growth markets and finally the long-run growth of a national economy. Potentially, this can answer the old question by Denison and Poullier (1967): 'Why do growth rates differ?' with better arguments than the new endogenous growth theory. But some will object that the macro modelling lacks any firm micro foundation and that some key hypotheses are *ad hoc*. Clearly, contemporary researches exhibit a tension between the need for micro behaviour full formalizations and their relevance with respect to major stylized facts. After all, is it not the job of economists to build theories and models delivering useful insight about a given problem for a given period, i.e. to be *cleverly ad hoc*?

The neo-Schumpeterian dynamics benefits from a long legacy in the explanation of capitalist growth and cycles (Schumpeter, 1911): in the absence of sufficiently powerful and numerous innovations providing large but transitory monopoly rents, the accumulation process would stop. Therefore, the process of innovation and its interaction with competition is *the real engine of capitalist growth* (Rosenberg and Birdzell, Jr, 1986). The major contribution of Antonelli's chapter in Part III is to extend this approach to the case of a cluster of innovations which are providing increasing returns to scale due to network externalities. Consequently, some cumulativeness is brought into the conventional diffusion model. Given the very meso approach adopted, two symmetrical criticisms might be addressed. On the one hand, the microeconomists will complain about the lack of any explicit formalization of the *individual behaviour* of firms. On the other, the macroeconomist will call for a complete model analyzing the links between the externalities in the telecommunication industry and the growth at the *economy wide* level.

Some unsettled methodological issues

A general view of the four chapters in this Part makes clear some difficult and not already solved questions, even if they might yet converge upon some common conclusions.

1. Equilibrium or dynamics?

The first issue concerns the choice between *equilibrium, disequilibrium or dynamics* concepts. Clearly, evolutionary approaches and equilibrium intertemporal models exhibit opposite advantages and drawbacks: realism but lack of any general and strong results versus nice theorems obtained via questionable hypotheses. But the relative *indeterminacy of the conclusions* is not specific to evolutionary models: J.L. Gaffard points out that the best experts in the new field of industrial organization confess that the same feature is taking place: 'the principal result of theory is to show nearly anything can happen' (Fisher, 1991).

2. The representative agent against the inherent heterogeneity of production techniques

Most economic theories are built upon the simple but questionable assumption that a *representative agent* is sufficient to derive general results for the whole system. This hypothesis is specially inadequate for models dealing with innovations which by definition continuously create heterogeneity among firms, workers, goods . . . (Dosi, Orsenigo, 1988). In this respect, Philippe Aghion and Peter Howitt capture only a part of the Schumpeterian legacy: even if firms are heterogeneous through time, there exists only one firm incorporating a given product and technology. On the contrary, Francesca Chiaromonte and Giovanni Dosi build the macroeconomic evolution out of the aggregation of heterogeneous agents. They conclude that: '*the form* of the distribution of micro characteristics (technological competences and parameters of decision rules) of the agents affects macro dynamics, even for *unchanged mean values*'. This is one of the more challenging issues in contemporary researches about possible bridges between micro and macro modelling. For two decades, mathematical economists have, rather unsuccessfully, tried to use the first order conditions in order to derive an aggregation method from the individual optimization; now, Grandmont (1991) convincingly argues that on the contrary the existence of any aggregate function relies upon a *sufficient diversity* among the characteristics of the individual economic agents. Paradoxically enough,

Chiaromonte and Dosi get the same conclusion for sequential analysis: 'diversity among agents is a crucial ingredient of system dynamics. (. . .) Unlike common theoretical wisdom, more micro diversity tends to yield simple macrodynamics.'

3. *R&D versus learning by doing and using*

The third issue relates to the inner characteristics of innovation. Is it the result of a rational *ex ante* assessment or the unintended and joint result of economic activity itself? What is the influence of the private appropriability of innovation upon the competition and growth patterns? In the neo-classical tradition, following Griliches (1979), even if R&D expenditures deliver uncertain outcomes, a probability distribution is supposed to be known by the firms. Aghion and Howitt provide a very clever formalization combining both a stochastic part and a positive influence of R&D expenditure upon the probability of success. Nevertheless, they suppose *rational expectations and a full computation* of the long-run consequences of any innovation. Basically, the same framework could be applied to any investment decision. On the contrary, for Bruno Amable, productivity and quality improvements come from *learning by doing*, and this feature is present as well in Gaffard's model and that of Chiaromonte and Dosi. Following the tradition of Arrow (1962), technical progress is the joint outcome of production itself: it has not any specific and separable cost. No doubt it is crucial to assess which of the two assumptions are the best suited in the emerging socio-technical system. It would be nice if model builders could refer more frequently to Pavitt's (1984) taxonomy, according to which technical change can occur via the advances of basic science, the supply of sophisticated intermediate or equipment goods or by continuous learning by doing or learning by using processes. The emerging endogenous growth theory would gain a lot from such a comparison with the stylized facts worked out by the experts in science and technology (Romer, 1990).

4. *Knowledge is more than information*

The final issue emphasizes the differences between information and knowledge: even if the relevant information about an innovation is freely available, a firm might be unable to use it productively for its own profit, given the lack of competence and specific skills. Chiaromonte and Dosi insist upon the partly tacit knowledge, specially in problem-solving, which is necessary for an innovation born elsewhere to be successfully implemented in the given firm. On the contrary, the strategy of the new industrial organization theory is to imagine sophisticated contracts or incentive mechanisms in order to solve the dilemma inherent to any socio-technical system (Nelson, 1988): on one side, to make the innovation possible, i.e. giving sufficient return to the innovator; on the other, to foster its subsequent diffusion . . . without destroying the incentive to launch new research and development projects. The evolutionary and neo-Schumpeterian models suggest that a part of innovation is easy to diffuse, whereas another is idiosyncratic to a given process, product and firm (see also Chapters 2 and 3 in Part I). This would explain the coexistence of different productive organizations and the persistence of heterogeneity. This explanation exhibits a clear superiority with respect to old fashioned neo-classical models. Nowadays, the theory of innovation and endogenous growth is in a state of flux and calls for further empirical and methodological researches. Nevertheless, the rapidly spreading literature has already provided a series of new and stimulating results.

Cumulativeness, diversity and the productivity slow-down

The chapters under review provide fresh answers to the questions which used to be paradoxical within conventional growth theory. Six major stylized facts can now be distinguished.

1. The innovation triggered by competition may feed long-run endogeneous growth

In Solow's model, long-term growth was set according to two exogeneous trends, associated respectively with working population and technical change. In the new endogenous growth theory, even a constant population can propel a *cumulative growth*. At the micro level, innovation is the consequence of firms' strategies in order to build some market power, and consequently at the macro level, spillover effects from one firm to another, if they are sufficiently strong, will induce a continuous growth process. This is a common result for new neo-classical models (Aghion and Howitt), as well as for post-Keynesian formalizations (Amable). Still more, evolutionary models provide the same result, while preserving a large diversity in the production technologies used by firms: Chiaromonte and Dosi push a step forward the path-breaking contribution by Nelson and Winter (1982).

Finally, modern theoreticians are delivering rigorous models which encapsulate one of the major stylized facts about capitalist growth: in the long run development is the outcome, not only of capital accumulation, but of a continuous pressure toward innovation (Rosenberg and Birdzell Jr, 1986; Maddison, 1991). Clearly now Schumpeter has a wide following . . . even in the neo-classical and post-Keynesian school!

2. There exists an optimum degree of competition for innovation and growth

The adhesion to the concept of *creative destruction* is not without consequence. In the previous analyses, the knowledge of an economy was always augmenting with the size of R&D expenditure. In the new models, if competition is very sharp, then each firm may overinvest in R&D, just to cope with the pace set by its competitors. At the macro level, a possible and surprising outcome is that competition may *destroy* much more knowledge than it creates. Consequently, growth and employment can be adversely affected by an excess of innovation, at odds with the standard conception according to which innovation is always good for the firms and the economy as a whole.

This is the strong and rather unconventional and new conclusion obtained by Aghion and Howitt. But by quite different concepts and methods, Chiaromonte and Dosi get an equivalent result: 'Variations in appropriability above a certain minimum threshold have ambiguous effects on growth for individual innovators, but may also slow-down diffusion, and through that aggregate growth in income and productivity.' Both chapters give new insights about the possible mechanisms delivering creative destruction and technological unemployment. Again, economic theory becomes more relevant and can help in interpreting the interwar as well as the present periods.

3. *Free markets do not guarantee an optimum investment in technical change*

The core argument of new industrial organization economics and endogenous growth theories has definite consequences concerning the ability of markets to monitor the complex process of innovation. Basically, the social returns of most large innovations usually exceed their private returns; consequently, again and again, contemporary researches confirm the seminal conclusion of Arrow (1962): under pure and perfect competition, firms will underinvest in innovation and by extension in skill formation, as far as they cannot capture totally the corresponding benefits.

From an intellectual standpoint, Aghion and Howitt deliver quite unconventional results out of mainstream tools. Whereas most governments *subsidize R&D expenditures* in order to enhance national competitiveness and growth, there might exist configurations in which the competition is so intense that it would be socially desirable *to tax the innovators*. Two other authors give a similar appraisal of the relative inefficiency of pure market mechanisms. For Gaffard 'the market-oriented system makes a process of change viable if the human resource constraint is never the dominant one (. . .) that supposes the need for new skills can be satisfied on the external market'. Consequently, market-oriented economies might suffer from an underinvestment in skills and even in R&D and non-physical capital. Amable shows that a country which would select its specialization according to purely static comparative advantage might be stuck into a low growth trajectory.

4. *Conflicts between static and dynamic efficiency are possible*

In most standard neo-classical models, the steady path was supposed to be unique for a given country. If new advances in basic and applied sciences were common knowledge, then the long-run trend for the income per capita would be the same in any country. Quite on the contrary, new endogenous growth models with increasing returns to scale might exhibit *multiple equilibria*. Consequently, some strategic choice during an early phase of the innovation process can determine the long-run trajectory.

Amable gives a good example of such a configuration. A country might stick to a static analysis of its competitive advantage and therefore get initially higher income results, but over a longer period, a specialization in markets with low growth might induce a lower growth rate. On the contrary, the acceptance of lower performance in earlier periods but the search for high growth markets might deliver a faster long-term growth. This is a typical example of the new results obtained by modern international theories (Krugman, 1990).

5. *The growth process can exhibit path-dependency: history matters*

The importance of increasing returns to scale delivers the possibility of multiple equilibria. Consequently, according to the *initial conditions* and the precise characteristics of a given economy, the growth pattern may take different configurations: fast and steady growth for some countries, instability and/or stagnation for others. This has long been recognized as a major stylized fact concerning capitalist development (Maddison, 1991). The models by Aghion and Howitt as well as Amable shed some light upon such phenomena: the growth might be replaced by cycles for the first authors, stability by instability for the second.

The corresponding *path-dependency* is still more striking within evolutionary models, submitted to random shocks. Chiaromonte and Dosi's various simulations show that closely similar economies might evolve quite differently, if for example the distribution of firms and their heterogeneity are different. In some respects a larger diversity provides a better ability to cope with an unexpected disturbance and therefore deliver a steadier if not faster growth (Eliasson, 1987). The same philosophy is encapsulated into the model elaborated by Gaffard. All these analyses provide quite interesting macro models, with similar results to those obtained by simpler and unidimensional stochastic models of path-dependency (Arthur, 1988).

Therefore, growth theory finally captures some key stylized facts put forward by economic historians. Still more it opens formalization to the very characteristics of historical processes (Romer, 1990; David, 1988): economic history can teach a lot to growth theoreticians.

6. *The Solow paradox reconsidered: the learning process associated with a new socio-technical system takes time*

The basic neo-classical model according to which productivity increases are related to capital labour substitution and the advance of knowledge, captured by the equivalent of the capital stock for physical equipment, has turned out to be quite unable to explain the productivity slow-down: since the end of the 1970s R&D expenditures per capita have boomed . . . but total factor productivity has not recovered the trends of the 1960s, especially in the United States.

The new approaches allow a stimulating even if quite provisional interpretation: 'the dynamics of labour productivity illustrate (. . .) the possibility that the evolutionary dynamics may induce absolute decreases in, average productivity even if best practice techniques may only improve.' This conclusion drawn from the chapter by Chiaromonte and Dosi confirms a previous result derived from a rather different model (Silverberg, Dosi, Orsenigo, 1988). When a paradigm in production and technology is emerging, one observes an obsolescence of previous explicit and tacit knowledge. Symmetrically, the new technologies and organizations only become efficient after a rather long learning period. During all the intermediate phase, the dynamism of invention and innovation can therefore coexist with a sluggish productivity growth, since new equipments are not fully nor efficiently used and all the required skills are not available. It may be that contemporary economies will experience with the computer the equivalent of the diffusion of the dynamo at the beginning of this century (David, 1991).

An agenda for further researches

Very likely the 1990s will experience a quasi-*paradigmatic change* in the theory of innovation and growth. The theoreticians have taken into account most of the specificities associated with R&D expenditures spillovers, learning by doing or by using, and the collective good aspects associated with an efficient and diversified educational system. Consequently, the gap between management practices and economic theory has been reduced and the modelling of organizational and technological change exhibits a new relevance.

Similarly, the *globalization of competition* drastically alters the sources of *national or regional competitiveness*. New conceptions about economic policy, as well as technological and industrial strategies will probably emerge during the next few years. It would be a challenging task for the economists, the historians of technology and firm organization and of course for the experts in science and technology to work out new concepts, methods and models in order to interpret such a complex transition from the Fordist golden age to a renewal of mass production of differentiated and quality goods.

Understanding the origins of the wealth of nations will long be a key issue in the agenda of economists and policy-makers. Nowadays, almost everybody agrees that this complex process derives from *the interaction of technological and organizational innovation*. Quite modestly, all the chapters of this book tentatively provide tools and hints in this direction.

References

Amable, B. (1991), 'National effects of learning, international specialization and growth paths', this volume.

Antonelli, C. (1991), 'The dynamics of technological interrelatedness', this volume.

Aghion, P. and Howitt, P. (1991), 'A model of growth through creative destruction', this volume.

Arrow, K.J. (1962), 'The economic implications of learning by doing', *Review of Economic Studies*, vol. XXIX, no. 2, pp. 155–73.

Arthur, B. (1988), 'Competing technologies: an overview', in Dosi G. *et al.* (eds) *Technical Change and Economic Theory*, Pinter Publishers, London.

Baldwin, R. (1989), 'The growth effects of 1992', *Economic Policy*, no. 9, October, pp. 248–81.

Chiaromonte, F. and Dosi, G. (1991), 'The microfoundation of competitiveness and their macroeconomic implications', this volume.

Denison, E.F. and Poullier, J.P. (1987), *Why Growth Rates Differ?*, Brookings Institution, Washington DC.

David, P. (1988) (1991), 'Computer and dynamo. The modern productivity paradox in a not-too-distant mirror', mimeograph prepared for the International OECD Seminar on Science, Technology and Economic Growth, Paris, 6–8 June.

Dosi, G. and Orsenigo, L. (1988), 'Coordination and transformation: an overview of structures, behaviours and change in evolutionary environments', in Dosi G. *et al.* (eds) *Technical Change and Economic Theory*, Pinter, London.

Dosi, G., Freeman, C., Nelson, R., Silverberg, G. and Soete, L. (1988), *Technical Change and Economic Theory*, Pinter, London.

Eliasson, G. (1987), 'The knowledge base of an industrial economy', *Research Report* no. 33, Industrial Institute for Economic and Social Research, Stockholm.

Fisher, F. (1991), 'Organizing industrial organization. Reflections on the Handbook of Industrial Organization', *Brookings Papers on Economic Activity*, pp. 201–25.

Gaffard, J.L. (1991), 'Towards a theory of the creation of technology as an out-of-equilibrium process', this volume.

Grandmont, J.M. (1991), 'Transformations of commodity space, behavorial heterogeneity and the aggregation problem', WP Cepremap, Paris.

Griliches, Z. (1979), 'Issues in assessing the contribution of R&D to productivity', *The Bell Journal of Economics*, vol. 10, pp. 92–116.

Helpman, E. and Grosman, I. (1991), *Endogeneous Growth Theory*, MIT Press, Cambridge, Mass.

Kaldor, N. (1981), 'The role of increasing returns, technical progress and cumulative causation in the theory of international trade and economic growth', *Economie Appliquée*, vol. 34, no. 4.

Krugman, P. (1990), *Rethinking International Trade*, MIT Press, Cambridge, Mass.

Lucas, R. (1988), 'On the mechanics of economic development', *Journal of Monetary Economics*, 22, pp. 3–42.

Maddison, A. (1991), *Dynamic Forces in Capitalist Development. A Long-Run Comparative View*, Oxford University Press, Oxford.

Mairesse, J. and Sassenou, M. (1991), 'Recherche-développement et productivité: un panorama des études économétriques sur données d'entreprise', *Science/Technologie/Industrie Revue*, no. 8, April, pp. 9–45.

Nelson, R.R. (1988), 'Institutions supporting technical change', Chapter 15 in Dosi G. *et al.* (eds) *Technical Change and Economic Theory*, Pinter, London.

Nelson, R.R., Winter, S.G. (1982), *An Evolutionary Theory of Economic Change*, Harvard University Press, Cambridge, Mass.

Pavitt, K. (1984), 'Sectoral patterns of technical change: towards a taxonomy and a theory', *Research Policy*, vol. 13, No. 6, pp. 343–75.

Romer, P. (1986), 'Increasing returns and long run growth', *Journal of Political Economy*, vol. 94, no. 5, pp. 1002–37.

Romer, P. (1990), 'Future researches on economic growth', CIAR WP.

Rosenberg, N. and Birdzell, L.E. Jr. (1986), *How the West Grew Rich. The Economic Transformation of the Industrial World*, Basic Books, New York.

Schumpeter, J. (1911), *Theorie de l'évolution économique*, first edition 1911, French Edition, Dalloz 1935 and 1983, Paris.

Solow, R.M. (1956), 'A contribution to the theory of economic growth', *The Quarterly Journal of Economics*, vol. 70, no. 1.

Thirlwall, A. (1979), 'The balance of payments constraint as an explanation of international growth rate differences', *Banca Nazionale del Lavoro Quarterly Review*, pp. 45–53.

5 The micro foundations of competitiveness and their macroeconomic implications

Francesca Chiaromonte and Giovanni Dosi

1. Introduction

This chapter discusses the relationship between the modes of microeconomic innovation, competition, and the dynamics of aggregate variables such as income and average productivity. The impressive transformation undergone by modern industrial economies in the last century or two – with their secular growth in income and inputs efficiencies, their changes in organization of production and patterns of consumption, and so forth – is metaphorically attributed by economic historians to the powers of the 'Unbound Prometheus', rooted in capitalism-based competition and innovation.[1]

In turn, the intuitive notion of 'competition', held by several historians and businessmen alike, is of an uncertainty-ridden process which mobilizes individual creativity and is checked by some interaction that establishes rewards and penalties to individual behaviours. This idea of competition as a process full of discoveries, mistakes and differential rewards to the various agents closely relates to some intuitive notion of 'competitiveness' ('. . . firm A is more competitive than firm B'; '. . . What can we do in order to improve our competitiveness?'; . . . etc.). In this common-sense meaning of competitiveness, a crucial role is played by technological and organizational innovation.

In the economic discipline, that innovation is a fundamental factor in long-term growth emerges also as indirect evidence from the application of equilibrium growth theory to empirical data (or, more correctly, from the failure in the exercise of 'explaining growth' simply in terms of accumulation, population growth and interfactoral substitution). However, in the standard (Solow-type) growth theory, any process of discovery and competition is strikingly absent: the implicit micro foundation is a General Equilibrium with an exogenous drift in technical possibilities.

The exogeneity restriction has been considerably relaxed in recent growth models with endogenous R&D, and/or increasing returns and externalities in the generation or use of technical knowledge (see Romer, 1986 and 1990; Aghion and Howitt, 1989 and Chapter 7 of this book; Grossman and Helpman, 1991a and 1991b). For a survey see Verspagen (1990). In those models, growth theory incorporates also parts of the Schumpeterian insights on the 'creative destruction' associated with endogenous innovation (Aghion and Howitt, 1989). Moreover, along with Schumpeter (and, earlier, Marx), a necessary condition for innovation by profit-motivated agents is the private appropriability of some quasi-rents stemming from their successful discoveries.

Still, the theoretical representation remains relatively far from the messy but powerful and disequilibrating force of the Capitalism Prometheus – with its story of trials and errors, long-lasting successes and miserable failures. Of course, it may well be that the essential features of contemporary growth processes can be theoretically reconstructed simply by means of (pre-analytical) endowments, some auxiliary assumptions on the nature of non-convexities in the technology possibility set, jointly with quite strong assumptions on the 'rationality' and forecasting abilities of individual actors. Naturally, if all this holds, the *process of competition* and the seemingly common-sense notion of *competitiveness* recalled earlier, must not carry any theoretical implication. They could simply be epiphenomena: transient appearances of an underlying process of adjustment to a more deeply rooted steady-state.

Here, however, we want to explore a different analytical approach. We start by assuming a wide set of notional opportunities of innovation, which is permanently expanding via an endogenous process. Further, we model individual agents that are motivated in their innovative search by economic incentives, but also constrained and 'locked' by their competences. Finally, we represent an 'out of equilibrium' environment wherein: (i) Individual agents, characterized by various and permanent forms of diversity, compete on the grounds of their specific technological achievements and behavioural rules (in each 'industry'), and (ii) Market interactions determine also intersectoral adjustments in demand, prices and, ultimately, in the levels and changes of macroeconomic variables. Such a micro foundation of the process of coordination and change – broadly in an 'evolutionary' perspective – as pioneered by Nelson and Winter (1982) – is plausibly nearer sheer intuition on the nature of innovation and firms' competitiveness in decentralized economies. This notwithstanding, does the different approach make an interpretative difference? Can one show that 'disequilibrium behaviours' – such as technological 'mistakes' and what *ex post* turn out to be 'irrational' decisions – permanently affect system-level changes? We shall argue that it generally does. The distributions and changes in micro level 'competitiveness' bear macro-level effects. Developing and modelling some of the ideas discussed at greater length in Coricelli, Dosi and Orsenigo (1991), we shall show that the possible regularities in the dynamics of aggregate variables (such as income and productivity growth) are *emergent properties* of a system that self-organizes far from equilibrium. The model presented here concerns a closed economy case: however, one can easily conceive extensions to interdependent economies which may indeed reinforce the conclusions achieved here.

In section 2 we outline the basic hypotheses and the structure of the model. Section 3 presents its formal structure. Its properties are explored via simulation, and results are discussed in section 4.

2. Innovation, competition and growth[2]

Ideally, a satisfactory model of growth should account for, or be consistent with, a series of 'stylized facts' on macro dynamics and on the microeconomics of innovation and competition. Several of these empirical regularities are discussed in Kaldor (1974), Simon (1984), Solow (1984), Maddison (1982), Greenwald and Stiglitz (1988), Coricelli, Dosi and Orsenigo (1991). Let us briefly recall the stylized facts that are more directly relevant to the model which follows.

At the macro level:

(i) Output per unit of employment grows at relatively stable rates over rather long periods, and so do per capita incomes.
(ii) There are persistent economy-wide fluctuations (i.e. so-called 'business cycles').
(iii) One can identify 'phases' of development intertwined by 'crisis' and discontinuities in the average rates of growth of income and labour productivity.
(iv) The levels and rates of growth of income and labour productivity significantly differ across countries.
(v) There is no systematic trend in capital/output ratios, nor are there systematic correlations across countries between such ratios and the level of development.
(vi) Significant levels of unemployment may persist for quite a long time.

Moreover, at the micro level:

(vii) Innovation in products and processes is to a good extent endogenous to the activity of the business sector, via firms' R&D activities and also via more informal mechanisms of learning by doing and by using.
(viii) Firms (and countries) systematically differ in their commitment to innovation and in their ability to innovate.
(ix) Forms of market interactions different from pure competition are prevalent. Market structures are endogenous to the process of innovation and competition, with a lot of variability in market shares, relative costs and relative profitabilities of individual firms.

The theoretical task is to model a process of innovation, competition and growth whereby these 'stylized facts' can in principle hold together. The

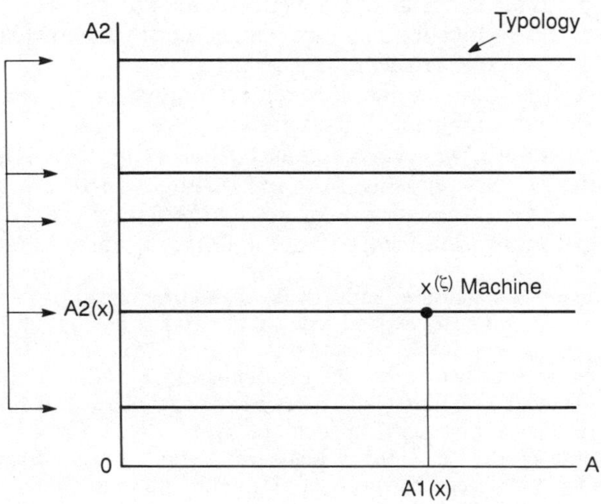

Figure 5.1

model presented here consists of two vertically connected sectors. Sector 1 produces heterogeneous (durable) production inputs ('machines') with labour alone. Sector 2 produces an homogeneous consumption good with labour and 'machines'. The bi-sectoral structure is clearly more apt than one-sector models to capture the transmission of demand and technological impulses between agents that might not have any competitive interaction with each other (i.e. it may be taken as a metaphor for the input–output structure of the economy). Moreover, it makes easier the representation of both processes of endogenous innovation and diffusion. Labour is an homogeneous input. Each sector is composed of a finite number of firms. Their behaviour is represented as routinized, and they are characterized by a persistent variety in behavioural rules and technological capabilities. Firms' market shares endogenously change as a function of their relative 'competitiveness' (defined below), that, in turn, is connected with behavioural rules and technological capabilities. Firms of sector 1 innovate by developing new ('more efficient') types of machines and by finding more efficient techniques to produce them. Firms of sector 2 innovate by adopting and learning how to use the new types of machines. The collective outcomes of individual disequilibrium decisions determine also aggregate variables such as income growth, employment, wages, etc.

Innovative opportunities

The very possibility of innovation must rest on the existence of some unexploited technological opportunities, or the continuous emergence of new ones. In line with a growing body of evidence from the microeconomics of innovation (Freeman, 1982; Rosenberg, 1976 and 1982; Dosi, 1988), we assume that unexploited opportunities permanently exist and that what is actually explored is much smaller than the set of notional possibilities.

Furthermore, we draw a fundamental distinction between *knowledge* and *information*. Innovation, we suggest, does not imply only some processes of information acquisition about new products or techniques, but is grounded in pre-existing, partly tacit, forms of problem-solving knowledge embodied in the agents themselves. Innovation, in turn, augments this knowledge for individual agents or collections of them.

All this implies that, unlike the representation of technical change in conventional growth or business cycle models, agents are not constrained by 'nature' in their innovative achievements, but rather by their own capabilities. More emphatically, paraphrasing Milton Friedman, there are always (semi-) 'free lunches', whose exploitation is primarily limited by agents abilities.[3]

Our model of innovation incorporate the following major characteristics:

1. Innovation involves changes in both products and production processes.
2. Opportunities to innovate expand over time as a function of income growth (plausibly carrying with it more basic research, more scientific advancements, etc.).
3. Technical change is of two types, 'incremental' and 'radical'.
4. Abilities to innovate and imitate are firm-specific and depend on their past innovative record (i.e. innovative learning is cumulative).
5. Although some of the economic benefits from innovation and adoption of new products and processes can be appropriated by the innovators themselves, there are learning externalities: the easiness of imitation augments with the number of

incumbents already producing a certain commodity, and the skills in using a particular type of equipment partly 'leaks' from individual adopters to the whole industry.

In the model, innovative opportunities in sector 1 correspond to 'machines' (x), each of which in turn corresponds to a technique of production for the final good defined by the labour productivity coefficient $A_1(x)$, and to a technique of production for the machine itself, defined by its own labour productivity coefficient $A_2(x)$ ('machines' are measured in terms of productive capacity, so that the machine input coefficient per unit of output is equal to one for each technique in sector 2).

In analogy with several empirical studies on innovation pointing out the difference between 'incremental' change – on the grounds of an unchanged knowledge base (a given 'paradigm') – and a 'radical' change (a new technological 'paradigm') (Dosi, 1984), we assume that the former is represented by increments in A_1, given A_2, while changes in both A_1 and A_2 correspond to 'qualitative' changes in technological competences and production organization (radical innovations). Hence, although empirically implausible, we assume for modelling simplicity that, within a paradigm, incremental innovation is only a product-innovation for sector 1 agents: i.e. it is the discovery of new machines embodying more efficient techniques for the final good production process – higher values of $A_1(x)$ – but produced with the same technique – constant value of $A_2(x)$. An illustration is in Figure 5.1.

The 'notional opportunities' T(.) are defined by the set of achievable typologies (set of values for A_2) at each time (.). Over time, this set (containing old and new *potential* technological paradigms) grows via a stochastic process dependent on the level of economic (and scientific) development. Note that the opportunity set always grows, or at least remains the same. However, as we shall see, the dynamics of opportunities do not directly influence the actual processes of innovation undertaken by individual agents. Agents' effective explorations generally remain well below the maximum notional opportunities.

Each agent i (i = 1, 2 . . . n) of sector 1 actually 'explores' the set of notional achievable typologies T(.), via an expensive process based on R&D investments, assumed to be measured by the number of R&D workers that it employs. Such a search is represented as a stochastic process: when it turns out to be successful, the newly discovered machine will be a realization ($x\zeta_i(t)$) from a probability distribution dependent on the production competences of the agents (as represented by the machine that is currently manufactured, $x_i(t)$), and on the opportunity set T(t). Incremental progress within a typology of machines (elsewhere, we defined it as the progress within a 'technological paradigm', or along a 'technological trajectory'; Dosi, 1984) is always possible, but subject to a progressive exhaustion of opportunities. This is like saying that any trajectory is characterized, in stochastic terms, by dynamic increasing returns, but at decreasing rates.

Conversely, the more a firm proceeds along a 'trajectory', the more its probability increases to jump on another one characterized by a higher value of A_2 (i.e. to 'discover' a radical innovation). In turn, the discovery of a new 'paradigm' reopens the opportunities for further incremental improvements.

Underlying this formalization, there is the idea that the progressive exhaustion of a trajectory performs as a 'focus device' (Rosenberg, 1982) which triggers the search for new sources of innovative opportunities, while past experience on a trajectory cumulatively builds technological competence useful also for the discovery of new ones.

The exploration process as a whole maintains its increasing returns and cumulative nature with periods of incremental technical progress intertwined with more discrete jumps in both product and process technologies. Moreover, technological knowledge, in most circumstances, is neither a purely public good nor an entirely appropriable blueprint or patent (such as in the formalization of 'patent races') but, rather, can be reproduced and imitated in an expensive way (Mansfield *et al.*, 1981; Pavitt, 1984; Dosi, 1988). Hence, in the model we introduce the possibility for each agent of sector 1 of imitating machines that are known (although not necessarily produced) by other agents.

We formalize imitation as a stochastic process whose success depends on the amount of imitative R&D investment undertaken by each firm. When such a search is successful, we further assume that, first, the probability of imitating any one type of machine is inversely related to the technological distance from the current capabilities of the imitating agent: given the partly tacit nature of technological knowledge, it is easier to imitate machines which are relatively similar to those that the agent already produces or knows. Second, the more widespread the knowledge about a particular machine, the easier might be innovation: thus, we assume that the probability of imitating any one type of machine is directly proportional to the absolute frequency of agents who know it.

Through the search activity of sector 1 agents, the system continuously 'discovers' (and produces, under microeconomic decision rules specified further on) new capital-embodied opportunities for sector 2.

Innovation, for each agent j ($j = 1, 2 \ldots m$) of sector 2 is equivalent to the purchase of new types of machines, via expansion and scrapping decisions (whose rules, again, shall be specified below). However, producers of the (homogeneous) final good must learn how to use each machine type (A_2) up to the optimal use defined by the value of A_1 of each machine belonging to that type (i.e. belonging to any one 'technological paradigm'). Similar to Silverberg, Dosi and Orsenigo (1988), we assume that each firm j embodies a *skill* $s_j(A_2, t)$ (a coefficient between 0 and 1) in the use of machines belonging to the typology A_2, so that the actual labour productivity it can achieve on them at time t might well be less than the optimal value. These skills improve via two learning processes. The first is a learning-by-using activity (as in Rosenberg, 1982), whereby the capabilities of using each typology of machines grow with the actual cumulated use of them. We assume that *private* firm-specific and technology-specific learning follow a logistic path.

However, we assume also a second process of learning, through which part of the private skills 'leak out' to other potential adopters, via information diffusion, personnel mobility, consultancy activities, etc. (Silverberg, Dosi and Orsenigo, 1988); this becomes a *public* skill shared by every firm. Therefore, late-adopters of a particular typology A_2 will start by using only public skills, which in turn depend on (a) the previous number of adoptions; (b) the opportunities of private learning-to-use; and, (c) the appropriability of private capabilities (inversely related to the rate at which private skills become public).

To summarize: in the model, the economic system continuously explores new opportunities of capital-embodied technical progress, via R&D based processes of generation and imitation of 'machines', undertaken by agents in sector 1. Moreover, the flow of new machines represents also a continuous opportunity of innovation for the agents in sector 2, via their adoption, the subsequent process of learning how to use them, and the learning externality that industry-wide diffusion of information and skills provides for late-adopters. Finally, the aggregate income dynamics, specified below, induces an endogenous increment of new potential avenues of innovation. All this defines an expanding universe of *notional* and (partly) *exploited* opportunities of technical progress.

However, in a micro founded model, activities of search, innovation, imitation, adoption by profit-motivated agents must be grounded in some economic incentive to do so, and, more generally, in microeconomic decisions. This is what we shall now discuss.

Microeconomic decision rules

It is a standard practice in the economic profession to assume perfectly optimizing agents (and growth theory is no exception). It is impossible to discuss here the analytical status of that assumption. Suffice to say that such a representation of the decision processes, irrespectively of its empirical plausibility under stationary conditions (i.e. given technologies and preferences) becomes particularly demanding on the computational and forecasting capabilities of the agents when innovations may occur endogenously. As Herbert Simon put it,

> [i]n a formal way, it is perfectly feasible to produce a theory of technical innovation based on the postulate of rationality. Since innovation is presumably produced by investment of human and capital resources, we introduce a new production function for innovation, and equate the value of marginal product of innovation with its cost. From a formal standpoint, we have simply replaced the task of estimating the parameters of a function, the production function for innovation. The only obvious gain from the replacement is that we can now rest comfortable in the knowledge that everything is proceeding rationally. Human rationality is now only bounded by the characteristics of the external environment: the quality of the ore that is mined by the innovation production process. Of course, if we examine the metaphor too closely, we see that 'quality' of the 'ore' is a euphemism for the 'effectiveness of the thought processes of human beings who are doing the innovating'. (Simon, 1984, pp. 41–42)

Indeed, if one sticks to a 'rational' micro foundation of economic dynamics *cum* endogenous innovation, one must subscribe to either (or both) of the following hypotheses, namely, (a) the actual 'thought processes' rather closely resemble optimizing procedures (and generally involve also 'rational' technological expectations, etc.), and, (b) if and when hypothesis (a) does not hold, the competitive process is such that the relative frequency of behaviours that *ex post* turn out to be 'optimizing' rapidly tends to one.

In fact, good empirical and theoretical reasons – related to the nature of the innovation process, the associated uncertainty, the characteristics of innovation-based competition, etc. – strongly suggest that hypothesis (a) does not generally hold (much more on that in Simon, 1984; Nelson and Winter, 1982; Winter, 1987; Dosi and Egidi, 1991; Dosi and Orsenigo, 1988). And, regarding hypothesis (b), its proponents have so far failed to deliver any proof of the uniqueness and stability of rationally microfounded equilibrium dynamics whenever perturbed by 'disequilibrium' behaviours.

Here, we want to explore an opposite behavioural foundation. We assume that agents are characterized by *fixed decision rules* – generally different in their parameters among agents – concerning investments, research, pricing, etc. In principle, these 'rules' must have developed via some *learning processes* over the history of individual firms and of the system as a whole. Here, however, we do not model any such process. We set ourselves the task of showing that some orderly aggregate properties may emerge even in those extreme conditions when *no* individual behaviour can be presumed to be an 'equilibrium' (optimizing) one and, also, when no form of learning is going on. After all, the common modelling assumption of perfect rationality derives a good deal of the purported system properties from the forecasting and computing abilities of 'representative' agent(s). Moreover, whenever one adopts such a compact microfoundation (a single agent representing the whole of them), one also assumes by hypothesis perfect microeconomic coordination and precludes the possibility of analyzing the long-term effects of those non-average behaviours that are typically associated with innovative enterpreneurship. In this respect, our agents are much less able to forecast and compute equilibrium behaviours than commonly assumed, but they are continuously prepared to (imperfectly) adjust to the unexpected, and able to generate and imitate novelty. They are *different* in their decision parameters, although not in the rules themselves. Finally, they differ in their technological abilities and degrees of success in innovation and imitation, in line with the general finding from the innovation literature that technology presents highly firm-specific, tacit and idiosyncratic features.[4] In that, we entirely share and push further Greenwald and Stiglitz's suggestion that

> [i]ndividuals do not have perfect foresight or rational expectations concerning the future. The events which they confront often appear to be unique, and there is no way that they can form a statistical model predicting the probability distributions of the outcomes. There is little evidence that they even attempt to do so. At the same time, individuals are not myopic. They do not simply assume that the future is like the present. (Greenwald and Stiglitz, 1987: 131)

Our decision rules implicitly embody also firm-specific 'theories' on how to cope with a changing world. This implies also that there is no common knowledge of either an 'innovation production function' or even a unique 'production function': both differ across firms and change over time according to agent-specific patterns.

Decision rules: Sector 1

Machine producers do know that notional opportunities for better machines always exist, and they also know the general procedure of discovering them, i.e. by investing in R&D. Finally, they expect to derive some economic benefit from successful innovation or imitation: new machines are partly appropriated, due to various forms of knowledge specificity (with or without a patent system), and indeed it turns out that some innovators increase their market shares and their profits. Similarly, successful imitation yields a sort of 'business stealing'.[5] The fact that, *for some*, innovation and imitation result in differential profits and increasing market shares, implies what we call *weak incentive compatibility* of search activities. However, agents do not know the actual probabilities of innovation or imitation, or the precise technical features of what they will discover. Relatedly, market shares and profits of each firm depend on the performances and prices of every machine produced by all firms, which in turn are a function of the success in innovation and imitation by each of them. In such circumstances we make the extreme assumption that agents determine the level of their decision variables via fixed and firm-specific rules.

In sector 1, each firm is characterized by the following decision rules:

(i) *Desired* allocation of resources to R&D, via a fixed percentage of previous-period turnover (note that here and throughout, individual agents may be unable to achieve their 'desired' rates of investment and production, given their maximum available liquidity determined by cash flow plus bank credit, the latter being a fixed proportion of the former).
(ii) A parameter distributing R&D between innovative and imitative search.
(iii) Desired prices on each new machine that is or could be manufactured, via a mark-up on direct costs of production.
(iv) An evaluation function comparing prices and performances of currently produced machines with those stemming from innovation and imitation, if any, in order to decide which one to manufacture (we assume single-product agents).
(v) Desired level of production for whatever machine is chosen.

Sticking to our general endeavour of exploring the possible orderly characteristics of system coordination and change in the presence of highly routinized microbehaviours, we assume a very simple adaptive expectation which extrapolates from the demand of firm i over the previous two periods. Given these five decision rules also the desired demand for labour can be automatically derived.

Decision rules: Sector 2

The industry producing the final good is somewhat similar to what Pavitt (1984) calls a 'supply-dominated' sector: innovations are generated elsewhere in the system and introduced via decisions of scrapping and/or expansion. The levels of production and scrapping, jointly with the choice of the type of machine to purchase, influence production efficiency and, thus, unit costs of production.

Firms are characterized by the following decision rules:

(i) Desired level of production. As in sector 1, it is derived from adaptive expectations on demand, based on the own-demand of the previous two periods. (Given the existing capital stock, the rule allows also the automatic determination of desired net investment.)
(ii) Desired price, again, via a mark-up over direct unit costs.

(iii) A choice function which compares prices and performances of the new machines to be purchased, if any.
(iv) A scrapping rule, based on a pay-back period criterion.

As in sector 1, these rules are sufficient to determine also the desired demand for labour.

Competition, coordination and aggregate dynamics

Firm-specific decision rules, together with the history of innovation, imitation and learning by individual agents, account for a permanent diversity among them. Firms differ in (i) the quality of products they offer (sector 1); (ii) the prices they charge (both sectors); (iii) the parameters of the decision rules concerning their control variables (again in both sectors). All this, jointly with the fact that no agent is capable of correctly computing *ex ante* equilibria where individual actions are reciprocally consistent, implies that competition is a *permanent disequilibrium process*. Its outcomes are also variations in the market shares and profits of individual firms. Hence markets perform as *selection mechanisms*, acting upon the relative competitiveness of individual actors.

In the model, as in common language, *competitiveness* is a *relative* concept: it implies a comparison among different agents. In sector 1, the competitiveness of each firm i is determined by (i) the production efficiency for the users of the machine offered on the market (as expressed by the value of $A_1(x)$); (ii) its price (connected to the value of $A_2(x)$); and (iii) the degree to which the firm might 'ration' the market since it might not be able to fulfil all demand, given forecasting errors and/or liquidity constraints. Similarly, in sector 2, the competitiveness of each firm j is determined by (i) the price of the output (note that, despite the homogeneous good, we allow for prices that are different among firms); and (ii) the possible degrees of 'rationing' exercised by the firm – as described above for sector 1.

Firms with above-average competitiveness expand their market shares, while firms below average shrink or even die. However, the selection criteria themselves are endogenous: they are the collective outcome of the dynamics of the competitiveness of each firm. Consider the selection process in sector 1. It is endogenous in two ways. First, it is determined by the individual dynamics of product-innovation (i.e. the changes in A_1 – and in the typologies A_2 – achieved by each i), in prices and in production capacity. Second, it is also determined by the evolution of the users sector – via learning by using and expectations of machine-users. Similarly, selection in sector 2 is shaped by the individual dynamics of capital-embodied process innovation, by learning by using and by the externality that all this entails.

In a biological analogy, the selection processes of the model represent the equivalent of a 'fitness function'. However, the 'landscape' over which 'fitness' is defined *co-evolves* with the behaviours and technological characteristics of individual agents in both sectors.

In such an environment, individual abilities of forecasting the technology and the behaviours of the others are highly imperfect. Each agent can be expected to face unfulfilled expectations and lack of correspondence between *ex ante* plans and *ex post* realizations. Therefore, the agents must possess some adjustment rule in their control variables. In particular, with regard to prices, we model an (imperfect) adjustment rule whereby each firm changes *actual* prices trying to strike some balance between desired (mark-up) prices and its relative competitiveness on the market (as revealed by the dynamics of market shares).

The system coordinates and evolves via competition among heterogeneous agents which continuously introduce, adopt, imitate technological innovations. In turn, the outcome of the competitive process feeds back on individual behaviours, directly, via price adjustments, and indirectly, via the effects that relative competitiveness exerts on firm size and cash flow (and thus, also, on investments in R&D and machines).

Aggregate ('macro') variables are the endogenous outcomes of such processes of innovation and competition. Aggregate gross and net investments in machines sum, *ex post*, over the investments of individual firms. So does employment.

In line with the unorthodox spirit of this work, we further add the following assumptions:

(i) Labour supply is unlimited at any positive wage, and wages change, in the most general formulation, as a function of (a) previous-period variations in labour demand; (b) changes in average (system-level) labour productivity; (c) changes in final good prices.

Indeed, we represent different *labour market regimes* by allowing varying combinations of 'institutional' mechanisms of indexation of wages on variations of both cost-of-living and productivity of labour, and a 'classical' adjustment on net variations in labour demand.

(ii) In an extreme and out-of-fashion 'Keynesian' simplification,[6] all wages are consumed (in the final good), profits are either invested (in R&D, labour, or machines) or saved in the form of interest-yielding deposits. The interest rate is exogenously fixed, and does not have any effect on investment decisions. There is no credit ceiling at a system level.[7]

Indeed, these extreme assumptions, jointly with the earlier ones on the nature of technological change and on rule-driven behaviours, define a system that is highly *unconstrained*. It is not bounded by 'nature' in its dynamics: unlike standard models, one cannot ultimately find a principle of order either in the scarcities of primary inputs or in some objective technological constraint. Moreover, highly institutional (rule-driven) behaviours cannot provide by themselves those ordering virtues which a perfectly rational 'representative' agent with foresight is assumed to possess.

These features of our model are instrumental in investigating the conditions under which a system that continuously explores a wide and expanding set of possible states can self-organize despite, *or because of*, disequilibrium micro behaviours. Ideally, the non-linear dynamics of the system should be able to generate the 'stylized facts' recalled at the beginning of this section: here, we shall simply present some highly preliminary results. However, before doing that, we shall present more formally the structure of the model.

3. A self-organization model

Innovation, imitation and production: Sector 1

Machines are identified by couples of integers (A_1, A_2), corresponding to points in a non negative subset of a bidimensional space ($\mathbb{R}^2_+$).[8]

We define each 'typology' of machines belonging to the same paradigm as a set of points $x = (A_1, A_2)$ in $\mathbb{R}^2_+$, given by

$$(x \in \mathbb{R}^2_+ : A_1 \in N, \text{ and } A_2 = a_2),\ a_2 \in N$$

in correspondence with various values of a_2 (see Figure 5.1).

The set of achievable typologies T(.), augments via a two-stage stochastic process, the first stage determining the access to the second (cf. Nelson and Winter, 1982). In the first one, the stochastic event 'access to the creation of a new paradigm' is determined as a draw from a Bernoulli distribution with probability

$$\Pr(AT(t) = 1) = 1 - \exp(-r \cdot Y_m(t)) \qquad (1)$$

where $Y_m(t)$ represents a moving average of aggregate income up to time t.

If access occurs, the system adds a new typology to the set, T, of the achievable ones, via an equiprobability draw in the set of the integers within the closed interval

$$[A_2^*(t-1); (1+h) \cdot A_2^*(t-1)]$$

where $A_2^*(t-1)$ is the maximum value in the set $T(t-1)$. r in eq. (1) and h are parameters that implicitly capture the general state of 'scientific' opportunities and the effectiveness of their exploitation, yielding new *potential* paradigms. In order to *actually* access new 'machines', each agent i in sector 1 undertakes a two-step stochastic procedure, similar to the above one, through which it may generate a new machine, in t, with probability

$$\Pr(AG_i(t) = 1) = 1 - \exp(-r_g \cdot I_i^g(t)) \qquad (2)$$

where I_i^g represents firm i's investment in innovative search, in terms of number of R&D workers, and r_g is a system parameter. If the random access occurs, the new machine is drawn from a uniform probability distribution whose support is defined by the set of points with second coordinate (A_2)

$$A_{2i}^g(t) = \max(T(t) \cap [A_2(x_i(t)), A_2(x_i(t)) + \lambda \cdot \theta_i(t)]) \qquad (3)$$

and first coordinate given by the integers in

$$[\max(0; A_1(x_i(t)) - (1/\lambda) \cdot (A_{2i}^g(t)/A_1(x_i(t))); \\ A_1(x_i(t)) + (1/\lambda) \cdot (A_{2i}^g(t)/A_1(x_i(t)))] \qquad (4)$$

where λ is a system parameter and

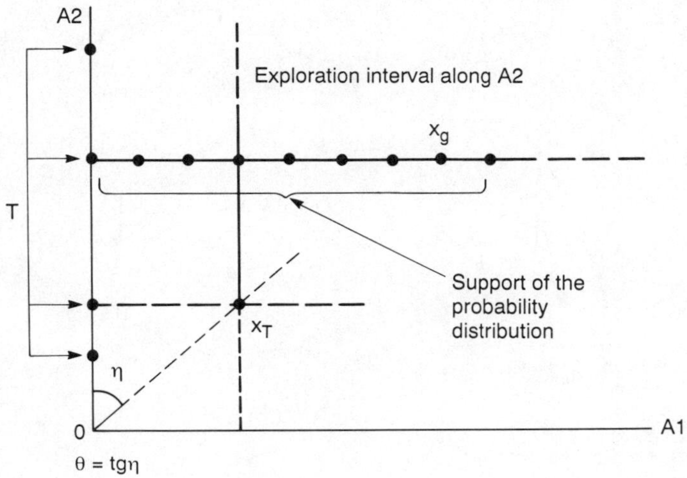

Figure 5.2

$$\theta_i(t) = A_1(x_i(t))/A_2(x_i(t)).$$

This formulation implies that exploration along a trajectory (holding A_2 constant) induces a relative shrink of further opportunities of incremental innovation, but also expands the set of attainable machines on other trajectories. An illustration is in Figure 5.2.

Agents can also imitate each others machines, again, via a two-stage stochastic process.

The probability of access to imitation is

$$Pr(AM_i(t) = 1) = 1 - \exp(-r_m \cdot I_i^m(t)) \tag{5}$$

$I_i^m(t)$ is the investment in imitative search of agent i, and r_m is a system parameter which is lower the higher is the appropriability of innovations.

If access to imitation occurs, the imitated point (machine) will be a realization ($x_i^m(t)$) from a probability distribution defined over the set

$$Z_i(t) = X(t) \setminus X_i(t)$$

where X(t) is the set of the points known by the whole population of agents of sector 1, and $X_i(t)$ is the set of the points known by i (one rules out the possibility of imitating oneself!). The probability, for firm i, of imitating any one type of machine in Z_i is inversely related to the distance, expressed in an orthogonal metric, between each point $z \in Z_i(t)$ and the set $X_i(t)$.

$$D_{MET}(z, X_i(t)) = \min_{x \in X_i(t)} [(z - x)' \cdot MET \cdot (z - x)]^{1/2}$$

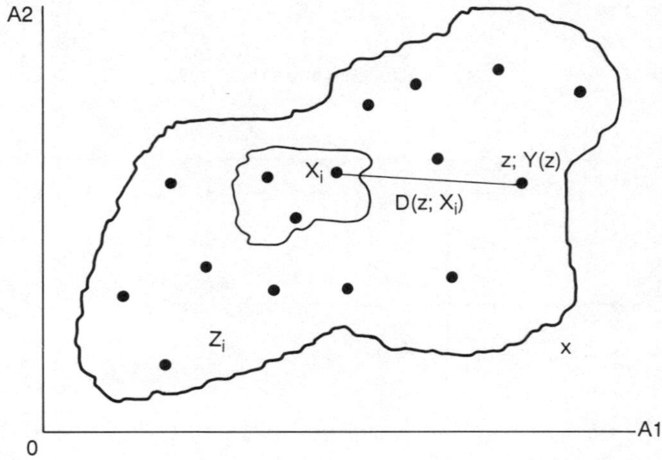

Figure 5.3

where MET is the matrix 2 × 2

$$\text{MET} = \begin{pmatrix} \text{met1} & 0 \\ 0 & \text{met2} \end{pmatrix}$$

The probability is also directly related to the frequency of agents knowing any one kind of machine in Z_i, $\gamma(z, t)$. Thus, the probability of imitating a machine $z \in Z_i(t)$, normalizing, is given by

$$\Pr(x_i^m(t) = z) = \frac{\gamma(z, t) \cdot [D_{MET}(z; X_i(t))]^{-1}}{\sum_{x \in Z_i(t)} \gamma(x, t) \cdot [D_{MET}(x; X_i(t)]^{-1}} \tag{6}$$

An illustration is presented in Figure 5.3. Each firm i determines its total desired allocation in R&D according to the rule

$$I_i^d(t) = \mu_i \cdot S_i(t - 1)/w(t) \tag{7}$$

where $S_i(t - 1)$ is the previous period turnover, μ_i is a firm-specific parameter and $w(t)$ is the current wage (hence, R&D investments are defined in terms of number of research workers). Actual investments, $I_i(t)$, as well as the actual values of the other decision variables of the firm, correspond to the desired ones up to the liquidity constraint.

A firm-specific parameter ζ_i distributes research between innovative and imitative activities, so that

$$I_i^g(t) = \zeta_i \cdot I_i(t) \quad \text{and} \quad I_i^m(t) = (1 - \zeta_i) \cdot I_i(t).$$

At each time, each agent has to choose which machine to manufacture among the variety which it already produces, x_i; the machine discovered via innovative R&D, if any, x_i^g; and that imitated, if any, x_i^m. That choice involves an evaluation of the 'quality' of each machine for the users – approximated by the technically minimum cost of production on the machines, given current wages – and also of the supply price of the machines themselves (p_i):

$$V_i(z, t) = \phi_1 \cdot \ln(w(t)/A_1(z)) + \phi_2 \cdot \ln(p_i(z, t)) \tag{8}$$

V_i is the notation for the evaluation functional, and it has to be applied to $z = x_i; x_i^g; x_i^m$.

For 'new machines' (candidates for production coming from innovation and/or imitation, and never manufactured before by the agent), prices are *desired* prices, calculated via a mark-up rule

$$p_i^d(z, t) = m_i \cdot (w(t)/A_2(z)) \tag{9}$$

(m_i being the firm-specific mark-up parameter) while for the machine already in production, the price is the *actual* price applied by the agent in the previous period (which, as we shall see, may well be different from the desired one). If all three candidates to production belong to the same typology of machines (that is $A_2(x_i) = A_2(x_i^g) = A_2(x_i^m)$) the choice is straightforward, and will satisfy a simple minimization of $V_i(.)$. However, if one or both the new candidates belongs to typologies that differ from that previously manufactured ($A_2(x_i)$), then their evaluation v_i is multiplied by a 'prudential' firm-specific coefficient $\chi_i/w(t)$. The prudential correction requires that, in order to switch production to a new paradigm, with the consequent market uncertainty and learning costs for the users, the expected superiority of the new machine *vis-à-vis* the previously manufactured one must exceed a certain threshold, influenced by the 'animal spirits' of individual entrepreneurs.

Having selected which machine to produce (the new x_i for the current period), each agent defines its desired level of production in relation to its demand expectations and its carried-over inventories. Expected demand is extrapolated as

$$D_i^e(t) = \delta_i(D_i(t-1), D_i(t-2)) \tag{10}$$

where δ_i is a firm-specific function expressing demand expectations derived from the level of actual demand in the previous periods.[9]

Thus, the level of desired production is

$$Q_i^d(t) = D_i^e(t) - O_i(t-1)$$

$O_i(t-1)$ being the inventories from the previous periods of production. As for research investment, actual production level ($Q_i(t)$) is equal to the desired one, unless adjusted via the liquidity constraint. Production occurs with labour alone, under constant returns to scale, at unit costs $w(t)/A_2(x)$.

Diffusion, learning and production: Sector 2[10]

The machines are purchased and used by sector 2 producers of the final good. Each agent j (j = 1, 2 . . . m) utilizes each machine with a skill level $s_j(A_2, t)$ (bounded between 0 and 1) specific to the agent and to the machine type (A_2). Thus, actual labour productivity on each machine x whereby $A_2(x) = \bar{A}_2$, is

$$s_j(A_2, t) \cdot A_1(x) \leqslant A_1(x) \tag{11}$$

Let us define the set of typologies actually produced at each time (.) by sector 1 as $T^s(t)$. On the new typologies offered on the market for the first time in t, the initial using skills are assumed to be identical for all potential adopters of sector 2, and equal to the system parameter σ. An agent that adopts any one type (A_2) of machines will grow in its using skill according to

$$\begin{aligned} s_j(A_2, t) - s_j(A_2, t-1) = \alpha_1 \cdot [Q_j(A_2, t)/(CQ_j(A_2, t) + \alpha_2)] \\ \cdot [s_j(A_2, t-1) \cdot (1 - s_j(A_2, t-1))] \end{aligned} \tag{12}$$

where Q stands for the production undertaken by j in t with machines of typology A_2; CQ is the correspondent cumulated production,

$$CQ_j(A_2, t) = \sum_{t=0..t} Q_j(A_2, t)$$

and α_1 is a system parameter capturing learning-by-using opportunities.

Private skills s_j tend to partly diffuse to the whole industry. The *public* machine-using skill for each typology $A_2(s_p(A_2, t))$, which starts at level σ when a typology is first introduced, grows over time according to

$$s_p(A_2, t) - s_p(A_2, t-1) = \beta_1 \cdot (s_m(A_2, t-1) - s_p(A_2, t-1)) \tag{13}$$

where s_m is the average of private skills, weighted with market shares, and β_1 is a parameter which is higher the lower is the private appropriability of machine-using skills. Any new late-adopter of machines of a certain type (A_2) will start with public skill levels $s_p(.)$.

Each firm determines its levels of production based on a rough expectation rule on its own demand

$$D_j^e(t) = \delta_j(D_j(t-1), D_j(t-2)) \tag{14}$$

(with a meaning analogous to that of equation (10) for sector 1), yielding desired production

$$Q_j^d(t) = D_j^e(t) - O_j(t)$$

O being, again, the inventories. The level of production allowed by the current capital stock is identical to the number of (heterogeneous) machines retained by the agent. Calling $\Gamma_j(t)$ the set of them (and recall that they are measured in terms of units of production capacity), we have

$$K_j(t) = \sum_{x \in \Gamma_j(t)} g_j(x, t)$$

where $g_j(x, t)$ are the absolute frequencies of machines. The desired net investment (expansion) is

$$IE_j^d(t) = \max(0; \hat{K}_j^d(t)),$$

$\hat{K}_j^d$ being the desired variation in capital stock, calculated as

$$\hat{K}_j^d(t) = v_j \cdot [Q_j^d(t) - u_j \cdot K_j(t)] \cdot (1/u_j) \tag{15}$$

where u_j is the firm-specific level of desired capacity utilization ($0 \leqslant u_j \leqslant 1$), and v_j is a firm-specific parameter that can be taken to capture expectations about future levels of demand as compared to the past (note also that throughout the model own-demand expectations are based on both industry-wide levels and one's own market share).

With a non-binding liquidity constraint both levels of actual production and actual expansion investment will equal the desired ones (otherwise, they are proportionally corrected). Each j must choose within the set of machines currently offered by sector-1 producers ($X^s(t)$). It follows from the foregoing representation of technological learning that each agent j, whenever choosing among different typologies, must account also for its specific skills in using them, $s_j(A_2, t)$. These skills imply a form of dynamic appropriability of learning by using. However, precisely because of such an (partly) appropriable learning opportunity, expectations on potential skill improvements enter the choice rule (like Silverberg, Dosi and Orsenigo, 1988). Call a_j the '*expectation bonus*'. The adjusted skills used for machine selection are

$$s_j^e(A_2, t) = \begin{cases} s_j(A_2 t) + a_j & \text{if } s_j(A_2, t) < 1 - a_j \\ 1 & \text{otherwise} \end{cases} \tag{16}$$

Moreover, as we shall see below, nothing guarantees that markets for each type of machine clear: the degrees of 'rationing' by machine suppliers, $l_i(x, t)$, negatively influence preferences.

The choice of the machine depends on its price, the labour productivity that it entails and its possible levels of rationing. The chosen machine will be the one that maximizes the evaluation functional

$$M_j(x, t) = \phi_3 \cdot \ln(s_j^e(A_2(x), t) \cdot A_1(x)) - \phi_4 \cdot \ln(p_i(x, t)) - \phi_5 \cdot \ln(l_i(x, t)) \tag{17}$$

Call $x_j^*(t)$ the chosen machine.

Note that more than one producer may manufacture the same machine, but possibly offer it at a different price. The choice of sector 2 agents is therefore among 'couples' machines-producers which identify both the machine characteristics (typology A_2 and 'quality' A_1), on the one hand, and the supply price and 'rationing' conditions (p_i and l_i, respectively), on the other. Since machine-rationing can occur, each j may be forced to 'second best' options.

As regards scrapping, we assume that it follows a payback-period rule. The current capital stock $K_j(t)$ is made of different kinds of machines x associated with *actual* labour productivities $\pi_j(x, t) = s_j(A_2(x), t) \cdot A_1(x)$, and *actual* unit labour costs $c_j(x, t) = w(t)/\pi_j(x, t)$. ('Optimal' unit labour costs, given the machine stock, are simply $w(t)/A_1(x)$.)

Correspondingly, call $c_j^*(t)$ the actual unit labour cost associated with $x_j^*(t)$.[11] The subset of the capital stock that j desires to scrap will be

$$OS_j(t) = (x \in \Gamma_j(t)\colon [p(x_j^*(t))/(w(t)/A_1(x)) - c_j^*(t)] \leq b_j) \tag{18}$$

where b_j is a firm-specific payback period parameter. Desired substitution investment will thus be

$$IS_j^d(t) = \sum_{x \in OS_j(t)} g_j(x, t)$$

Like producers of sector 1, also each j may be unable to meet its desired plans due to a liquidity constraint, in which case, in sector 2 we assume that firms will satisfy, in order, current production, expansion investments and replacement.

Production is characterized by constant returns to scale and the notional average productivity depends on the frequencies of different types of equipment

$$\pi_{mj}^n(t) = \sum_{x \in \Gamma_j(t)} (1/A_1(x)) \cdot (g_j(x, t)/K_j(t)) \tag{19}$$

However, actual average productivities, and average unit labour costs (c_m) depend on such frequencies, but also on typology-specific and firm-specific skills[12]

$$c_{mj}(t) = \sum_{x \in \Gamma_j(t)} (w(t)/\pi_j(x, t)) \cdot (g_j(x, t)/K_j(t)) \tag{20}$$

Finally, desired prices are determined via a mark-up over average labour unit costs, $p_j^d(t) = m_j \cdot c_{mj}(t)$.

Market interactions and system dynamics

Each sector 1 producer competes in the machine market for the demand stemming from expansion and replacement decisions of sector 2 manufacturers. Similarily, the latter compete on the final good market whose size is determined by the real wages of all employed workers.

In sector 1 the competitiveness of agent i, $(E_i(t))$, manufacturing x, is defined as

$$E_i(t) = -\omega_1(t) \cdot \ln(w(t)/A_1(x)) - \omega_2(t) \cdot \ln(p_i(x, t)) \\ -\omega_3(t) \cdot \ln(l_i(x, t-1)) \tag{21}$$

where $A_1(x)$ is the labour productivity of machine x for its users, $p_i(x, t)$ is the price charged for it by manufacturer i, and $l_i(x, t - 1)$ is the unfulfilled demand in the previous period.[13] As in Silverberg (1987), the selective benchmark is the industry average competitiveness:

$$E^1_m(t) = \sum_i E_i(t) \cdot f_i(t)$$

that is, the average of individual levels of competitiveness weighted by market shares (f_i).

In sector 2, individual levels of competitiveness are

$$E_j(t) = -\omega_4 \cdot \ln(p_j(t)) - \omega_5 \cdot \ln(l_j(t - 1)) \tag{22}$$

and $E^2_m(t)$ is the analogous average competitiveness in the final good industry.[14]

Moreover, notice – not surprisingly – the similarity between eq. 21 (the competitiveness of machine producing firms) and eq. 17 earlier, expressing the evaluation functionals of different machines by would-be customers. In fact, the arguments of both are 'objective' supply-side parameters (labour productivity allowed by the machine, prices and degrees of 'rationing') which differentiate producers. However, the actual competitiveness of each producer is also the outcome of varying distributions of *heterogeneous users' skills and users' technological expectations*. These distributions, weighted by the demand shares of each customer, determine the $\omega(t)$'s in eq. 21.[15] Each market share f_i changes according to a non-linear process

$$f_i(t) - f_i(t - 1) = F^1(E_i(t - 1) - E^1_m(t - 1)) \tag{23}$$

Indeed, in the structure of the simulation exercises presented here the F function itself is constructed to be only the *ex post* aggregate outcome of (mistake-ridden) evaluation and choice rules by individual buyers.

Similarly, market shares for the final good change as a function of the difference between individual and average competitiveness

$$f_j(t) - f_j(t - 1) = F^2(E_j(t - 1) - E^2_m(t - 1)) \tag{24}$$

In analogy with Silverberg *et al.* (1988) we assumed the following adjustment dynamics of individual prices. In sector 1

$$\begin{aligned} \ln(p_i(x, t)) - \ln(p_i(x, t - 1)) = {} & \psi_1 \cdot (\ln(p^d_i(t)) - \ln(p_i(x, t - 1)) \\ & + \psi_2 \cdot (E_i(t - 1) - E^1_m(t - 1)) \end{aligned} \tag{25}$$

and similarly in sector 2

$$\begin{aligned} \ln(p_j(t)) - \ln(p_j(t - 1)) = {} & \psi_3 \cdot (\ln(p^d_j(t)) - \ln(p_j(t - 1)) \\ & + \psi_4 \cdot (E_j(t) - E^2_m(t)) \end{aligned} \tag{26}$$

That is, changes in actual prices – expressed in logs – (different among firms, *also in the homogeneous good industry*) are influenced by desired (mark-up) prices but they also track the relative competitiveness of each firm in the market. The ψ's are system parameters whose relative values capture the nature of the *product-market regime*: for example, a highly competitive regime is defined by relatively high ψ_2 and ψ_4 and near zero ψ_1 and ψ_3. The opposite applies to quasi-monopolistic markets.

Aggregate demand variables are emergent properties resulting from the microeconomic decision rules and interactions described so far. However, *ex post*, they are simply the sum of the corresponding realized decisions of individual units. Aggregate gross and net investments in machines (respectively $IE^T(t) + IS^T(t)$ and $IE^T(t)$) are the sum in j of individual investments. The same applies to employment. With fixed coefficient technologies and constant returns to scale, employment in the final good sector ($L^2(t)$) sums over the actual outputs compounded by labour productivities of each agent. In sector 1, employment is given by the sum of production employment and research employment (recalling that $I_i(t)$'s – investments in research – are measured in labour) ($L^2(t)$). Of course, total employment is simply the sum over the two sectors $L^T(t) = L^1(t) + L^2(t)$.

Aggregate ('national') income $Y^T(t)$ follows the standard accounting identities (it equals total production which equals total expenditure plus 'involuntary' changes in inventories). Finally, on the labour market we assume that wages (expressed in logs) change according to

$$\Delta\ln(w(t)) = \xi_1 \cdot \Delta\ln(p^2_m(t-1)) + \xi_2 \cdot \Delta\ln(\Pi^T(t-1)) + \xi_3 \cdot \Delta\ln(L^T(t-1)) \tag{27}$$

where p^2_m is a price index for the final good and Π^T is the average labour productivity in the economy.

4. The properties of the model: some preliminary results

Clearly, the structure of the model prevents analytical solutions and inferences on its properties can only be drawn from simulation exercises. In the particular simulation that we selected to present here as an illustration, we assume ten agents in each sector that start with identical technological competences but different parameters in their search rules (research investments, and propensity to innovate and imitate in sector 1) and in their technological expectations ('confidence' in new products-machines for the agents of both sectors, and learning expectations in sector 2). (See Appendix 1 for a description of some parameters of the simulation.) In the competitive process, market interactions determine variation in the relative size and profitability of each firm (they may also 'die' and, in sector 1 only, 're-enter' with some probability, which is actually equivalent to the birth of new firms).

Figure 5.4 presents the dynamics of notional opportunities (those generated at aggregate level by income changes that yield new discoveries by 'basic-science') and that of opportunities that actually become technological innovations. As the simulation shows, the innovative exploration of individual firms 'floats' in an expanding set of potential new technological paradigms. However, their path-dependent processes of technological learning make them

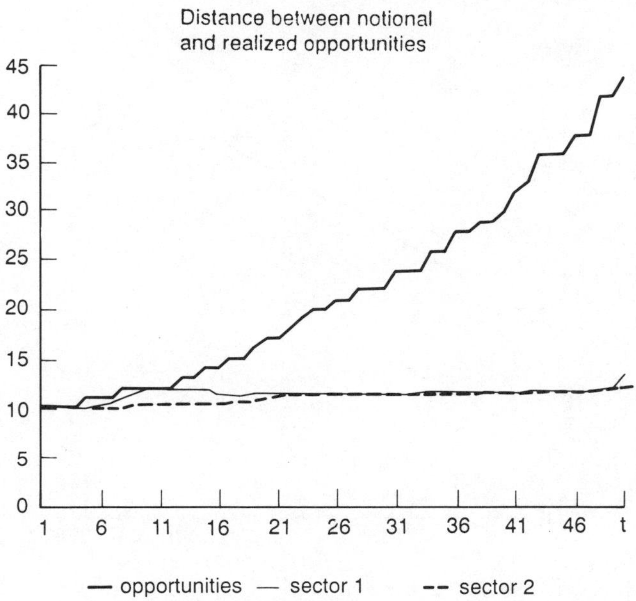

Figure 5.4 Distance between notional and realized opportunities

discover and develop only three of them (corresponding to the vertical steps along the otherwise horizontal line). All this illustrates also our earlier proposition that agents are not limited in their innovativeness by some exogenously given 'natural' constraint, but rather by the history of their competences.

At this stage, one can only say that the generated series of income and average productivity seem 'plausible' (Figures 5.5 and 5.6): we conjecture that the aggregate dynamics might show econometric properties similar to the empirical ones.

Note also that, as in 'Real Business Cycle' models, one cannot distinguish between transitory (cyclical) and permanent (trend) components in the generated time series. However, unlike the former models, innovations do not take the form of exogenous stochastic shocks but, rather, are generated endogenously by agents themselves.

Moreover, the transmission mechanism from innovation to aggregate dynamics is neither the intertemporal adjustment of consumption plans by a rational agent (cf. Long and Plosser, 1982), nor the optimal intertemporal distribution of investment plans (Schleifer, 1986). Indeed, in our model, innovations influence aggregate dynamics via two interrelated processes, namely: *first*, via time-consuming diffusion among producers and users, and *second*, via the demand impulses that innovation and diffusion generate. The latter, 'Keynesian' feature of the model implies that, although (*endogenous*) 'shocks' may have a technological nature, the actual aggregate impact is determined by related *demand* shocks (via investment decisions, and levels of consumption).

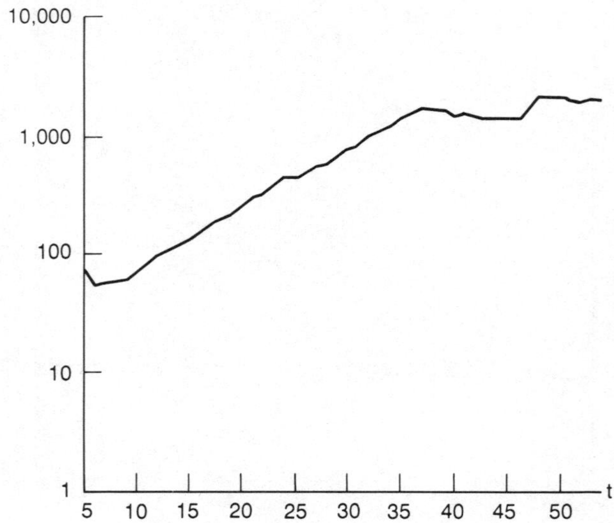

Figure 5.5 Income time series (logscale)

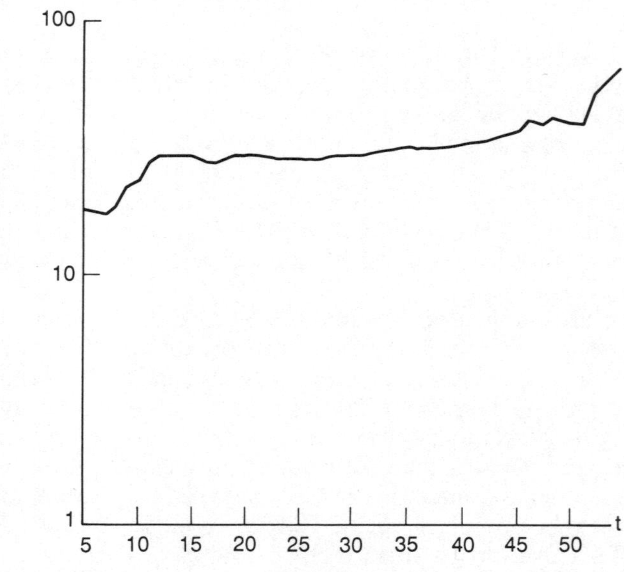

Figure 5.6 Labour productivity time series (logscale)

Figure 5.6 shows the dynamics of labour productivity and illustrates also the possibility that the evolutionary dynamics may induce absolute decreases in average productivity even if best practice techniques may only improve (or remain the same): this is due to the learning lags of users and the possible fall in the market shares of early innovators. As with the results obtained by Silverberg, Dosi and Orsenigo (1988), innovators may sometimes be 'lambs' whose 'sacrifice' produces a learning externality for the whole system.

Corroborating the Schumpeterian intuition, our simulations show that appropriability (as captured by the imitation costs in sector 1 and by the 'private learning' in sector 2) is a necessary condition for innovation. However, *variations* in appropriability, above a certain minimum threshold, have ambiguous effects on growth: higher appropriability is beneficial for individual innovators, but may also slow down diffusion, and through that aggregate growth in income and productivity.

The *form* of the distribution of micro characteristics (technological competences and parameters of decision rules) of the agents affect macrodynamics, *even for unchanged mean values*. For example, repetitions of the simulations shown in the above figures assuming identical agents (with parameters equal to the mean values of the distributions as from the previous case), lead to very little technical progress and long-term fall in aggregate income (see Figures 5.7 and 5.8).

Other simulation experiments, not shown here, appear to confirm the

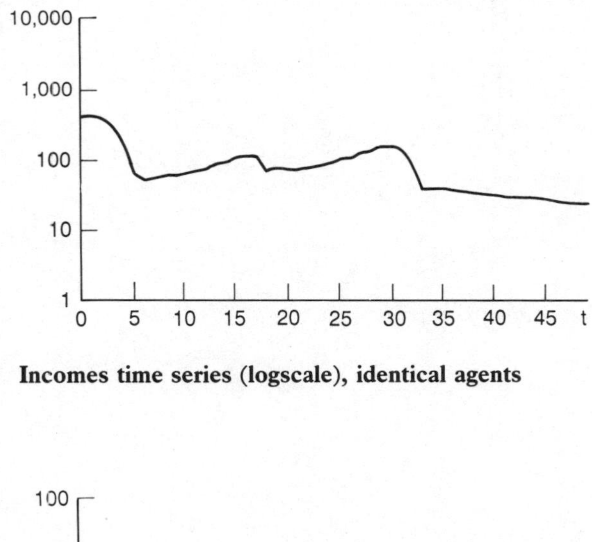

Figure 5.7 Incomes time series (logscale), identical agents

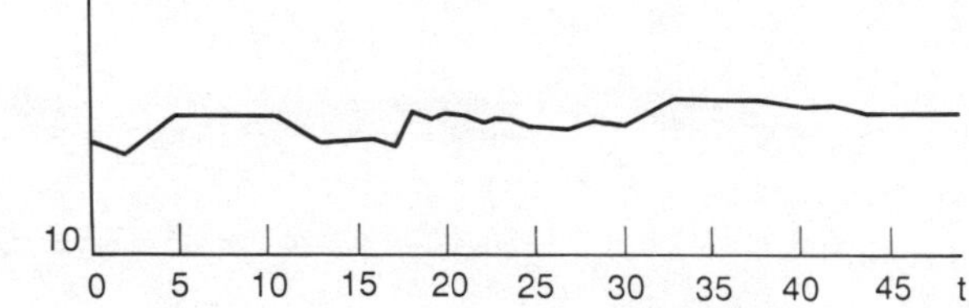

Figure 5.8 Labour productivity time series (logscale), identical agents

importance for aggregate dynamics of the *variance* and *the form of the distributions* of micro parameters. If these results turn out to withstand a more sophisticated investigation, this will also confirm the misleading implications of modelling exercises microfounded in 'representative agents'. Moreover, in our view, our results seem to highlight the importance of behavioural variety and of non-average behaviours in determining technical change and growth.

Of course, even accepting our results, one could interpret them as the outcome of an equilibrium game in mixed strategies. However, this would imply also some unspecified coordination process based on incredible demands on forecasting and computing abilities of individual actors. On the contrary, we suggest that behavioural variety is permanently associated with non-equilibrium behaviours, 'irrational' beliefs, unforecastable entrepreneurial innovations, etc. In fact, the microstructure underlying the growth process does not resemble any equilibrium dynamics.[16] Rather the relative com-

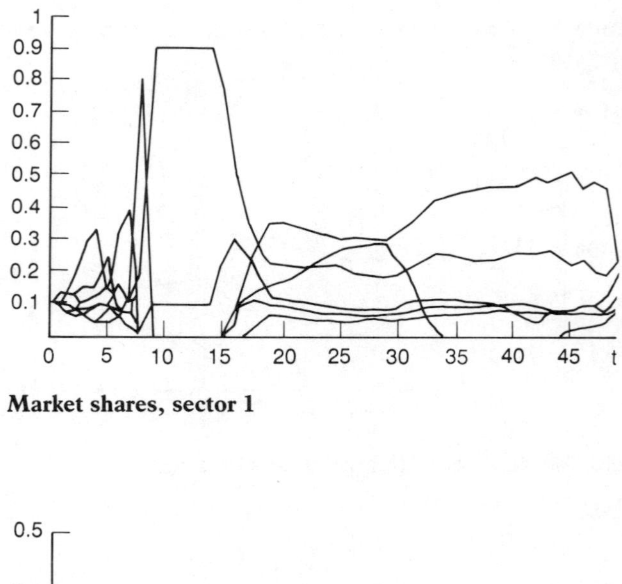

Figure 5.9 Market shares, sector 1

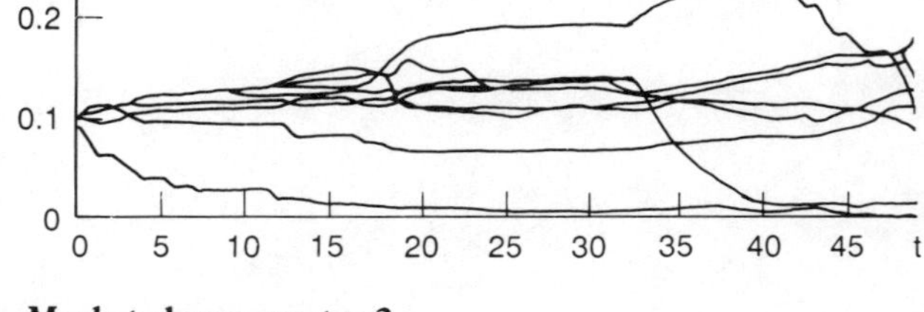

Figure 5.10 Market shares, sector 2

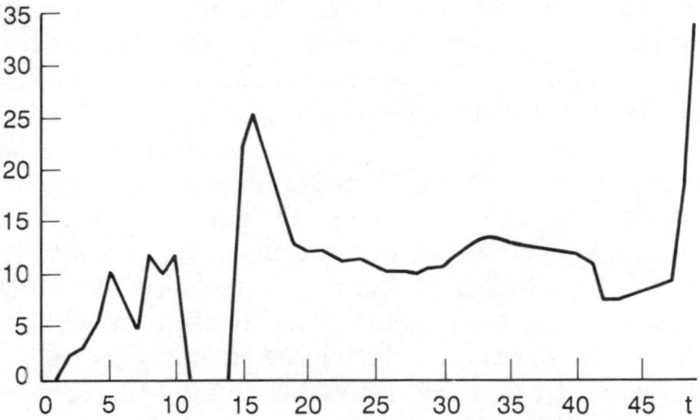

Figure 5.11 Technological asymmetry, sector 1

petitiveness and the market shares of the individual firms continuously change (see Figures 5.9 and 5.10).

Market structures are endogenous and bear a bi-univocal link with innovation. In one direction, their dynamics depend on the different degrees of innovative success of individual agents. In the other direction, they affect the access of individual producers to innovative opportunities. The link between endogenous generation of innovation and market structure appears particularly evident in sector 1, and contributes to explain the difference in market-shares dynamics between sector 1 (in which agents innovate in both production processes and products) and sector 2 (in which innovation relates only to production techniques).[17]

Technological asymmetries among the agents are also endogenous, and appear to be a permanent phenomenon (see Figure 5.11 for an illustration from sector 1: asymmetries are measured as standard deviations in performances of the machines that are sold).

5. Conclusions

The model presented here bases the microfoundation of productivity and income growth in an explicit disequilibrium process of innovation, imitation and competition. In that, we make a direct theoretical use, and show the crucial importance, of those characteristics which account for the differential competitiveness of individual agents. Some highly preliminary simulation results are encouraging for our hypothesis that the basic features of empirical aggregate time-series can indeed be 'theoretically reconstructed' and interpreted as the outcome of a self-organization process driven by *endogenous learning* and *market selection*. Coordination and change cannot be meaningfully distinguished: the process of change in the 'fundamentals' (in our case, technology and endowments) is the same one that produces an (imperfect) reciprocal adjustment of individual behaviours.

One cannot generally define any steady state for the system. Rather, the model yields a highly path-dependent dynamics both at system level and at the level of individual agents, since their rates and directions of learning depend on their history of competence accumulation and relative market success.

Moreover, behavioural and technological *diversity* among agents is a crucial ingredient of system dynamics. In fact, our conjecture, not falsified by the circumstantial evidence stemming from simulations, is that, unlike common theoretical wisdom, *more* micro diversity tends to yield *simpler* macro dynamics. This might well be so because in an opportunity-ridden system, variety assures that with higher probability some 'superior' innovation will be discovered and some (time-contingent) 'fitter' behaviours will be present.

As mentioned at the beginning, this closed economy model can in principle be promisingly extended to open systems trading with each other. Indeed, we speculate, irreversible processes of international differentiation may be triggered by those patterns of variation in micro 'competitiveness', which, it has been shown, produce system-level effects also in the closed economy case.

Appendix 1

Values attributed to the parameters defining the innovative micro behaviours in the simulation presented in section 4:

Parameters	Means	Distributed between
sector 1		
χ	0.15	(0; 0.3)
μ	0.1	(0; 0.2)
ζ	0.5	(0.2; 0.8)
sector 2		
a	0.2	(0; 0.4)
b	5	(3; 7)

The random distributions are truncated normal around the above means.

Recall the meaning of the parameters:

for sector 1 agents

χ is a firm-specific 'prudential' coefficient in the decision rule about the choice to manufacture radical innovations in manufactured machines

μ is the propensity to invest in technological search

ζ is the parameter that distributes research investments between innovation and imitation

for sector 2 agents

a is the firm-specific learning expectation

b is the payback period used in the scrapping rule

Notes

1. The classic reference is Landes (1986). See also Mokyr (1990).
2. In this and the following sections we shall present only some basic features of the model: more details are in Chiaromonte (1990).
3. Compare this view with that of an exogenous random arrival of 'technological shocks' as in Long and Plosser (1982), or the steady flow of innovations in Schleifer (1986). Nearer to our approach, *prima facie*, are Romer (1990) and Aghion and Howitt (1989), where the rates of innovation depend on the 'human capital' invested in exploration. However, in our model, the *abilities* in exploring are themselves an endogenous variable, and are, to some extent, firm-specific.
4. Cf. Freeman (1982), Winter (1987), Pavitt (1987), Dosi (1988).
5. Cf. Aghion and Howitt (1989) who, however, use the expression to refer only to the substitution among different generations of innovations.
6. Of course, the 'Keynesianism' of Cambridge, England!
7. Although there is for individual borrowers, as a function of their cash flows.
8. The restriction to the integer values of A_1 and A_2, while not influencing any of the conclusions, allows a positive probability that two or more different agents may produce/use the same machine (see below).
9. In the simulations of section (4) we have distinguished demand expectations on new typologies of machines, correcting them with a firm-specific parameter of 'confidence' in new technologies.
10. The following part of the model draws significantly from the earlier model in Silverberg, Dosi and Orsenigo (1988).
11. In its calculation, as in Silverberg *et al.* (1988), one uses the actual skill levels: it would be too adventurous to throw away good equipments simply on learning expectations. Moreover, if the new machine belongs to a typology not yet adopted by j, the initial skill level is the public one $s_p(x, t)$.
12. In the model presented here there are no variable inputs other than labour and no overhead costs.
13. Supposed equal to 0 if the currently manufactured machine is a new one.
14. Note however that, while the coefficients of the 'competitiveness' function of sector 1 agents are themselves functions of time, for sector 2 agents they are parameters (see below).
15. One cannot presume each producer to know the $\omega(t)$'s. Rather, they result by aggregation over a heterogeneous population of customers.
16. Unlike 'new growth theory', where one often assumes monopolistic competition equilibria (cf. Romer (1990), Aghion and Howitt (1989), Grossman and Helpman (1991a,b).
17. Innovation *cum* product differentiation in sector 1 tends to yield higher levels of market concentration than in sector 2.

References

Aghion, P. and Howitt, P. (1989), *A Model of Growth with Creative Destruction*, Cambridge, Mass., MIT, Dept. of Economics, Discussion Paper.

Chiaromonte, F. (1990), *Processi di Innovazione Microeconomici e Dinamiche Macroeconomiche: un Modello Evolutivo Bisettoriale*, Rome, Department of Economics.

Coricelli, F., Dosi, G. and Orsenigo, L. (1989), 'Microeconomic dynamics and macro-regularities: an evolutionary approach to technological and institutional change', *Technology and Productivity*, Paris, OECD.

Day, R. and Eliasson, G. (eds) (1986), *Market Dynamics*, Amsterdam, North-Holland.

Dosi, G. (1984), *Technological Change and Industrial Transformation*, London, Macmillan.

Dosi, G. (1988), 'Sources, procedures and microeconomic effects of innovation', *Journal of Economic Literature*.

Dosi, G. and Egidi, M. (1990), 'Substantive and procedural uncertainty. An exploration of economic behaviours in changing environments', *Journal of Evolutionary Economics*.

Dosi, G. and Orsenigo, L. (1988), 'Coordination and transformation: an overview of structures, behaviours and change in evolutionary environments', in Dosi *et al.* (1988).

Dosi, G., Freeman, C., Nelson, R., Silverberg, G. and Soete, L. (eds) (1988), *Technical Change and Economic Theory*, London, Frances Pinter and New York, Columbia University Press.

Freeman, C. (1982), *The Economics of Industrial Innovation*, London, Frances Pinter, 2nd edn.

Greenwald, B. and Stiglitz, J. (1987), 'Keynesian, New-Keynesian and new-classical economics', *Oxford Economic Papers*.

Greenwald, B. and Stiglitz, J. (1988), 'Examining alternative macroeconomic theories', *Brooking Papers on Economic Activity*.

Grossman, G. and Helpman, E. (1991a), 'Quality ladders and product cycles', *The Quarterly Journal of Economics*.

Grossman, G. and Helpman, E. (1991b), *Innovation and Growth*, Cambridge, Mass., MIT Press.

Kaldor, N. (1974), *Essay on Stability and Growth*, London, Duckworth.

Landes, D. (1969), *The Unbound Prometheus*, Cambridge, Cambridge University Press.

Long, J.B. and Plosser, C.I. (1982), 'Real business cycle', *Journal of Political Economy*.

Lucas, R. (1988), 'On the mechanics of economic development', *Journal of Monetary Economics*.

Maddison, A. (1982), '*Phases of Capitalist Development*', Oxford, Basil Blackwell.

Mansfield, E. (1961), 'Technical change and the rate of imitation', *Econometrica*.

Mokyr, J. (1990), *The Lever of Riches*, Oxford, Oxford University Press.

Nelson, R. and Winter, S. (1982), *An Evolutionary Theory of Economic Change*, Cambridge, Mass., The Belknap Press of Harvard University Press.

Nelson, C.H. and Plosser, C.I. (1982): 'Trends and random walks in macroeconomic time series', *Journal of Monetary Economics*.

Pavitt, K. (1984), 'Patterns of technical change: towards a taxonomy and a theory', *Research and Policy*.

Romer, P. (1986), 'Increasing returns and long-run growth', *Journal of Political Economy*.

Romer, P. (1990), 'Endogenous technical change', *Journal of Political Economy*.

Rosenberg, N. (1976), *Perspective on Technology*, Cambridge, Cambridge University Press.

Rosenberg, N. (1982), *Inside the Black Box*, Cambridge, Cambridge University Press.

Schleifer, A. (1986), 'Implementation cycles', *Journal of Political Economy*.

Silverberg, G., Dosi, G. and Orsenigo, L. (1988), 'Innovation, diversity and diffusion: a self-organization model', *Economic Journal*.

Simon, H. (1984), 'On the behavioural foundation of economic dynamics', *Journal of Economic Behaviour and Organization*.

Solow, R. (1984), *Growth Theory*, Oxford, Oxford University Press.

Verspagen, B. (1990), *'New' Neo-Classical Growth Model and Their Relations to Evolutionary Theory of Economic Growth: an Interpretative Survey of some Recent Literature*, Maastricht, University of Limburg, MERIT, Research Memorandum.

Winter, S. (1987), 'Natural selection and evolution', *The New Palgrave: a Dictionary of Economics*, London, Macmillian.

6 Towards a theory of the creation of technology as an out-of-equilibrium process*

J.L. Gaffard

What we might consider by now as the traditional interpretation of the process of innovation stipulates that every technological advance is fully defined and developed at its first appearance and generally consists in new productive processes embodied in durable capital equipment. The role of the firm is then reduced to decide what amount of resources is to be devoted to R&D expenditures and thereby determining the degree of technological advance. Thus the only problems faced from an economic point of view are that of allocation of generic resources and of the social or industry efficiency. They are resolved by market forces or, in case of failure of the latter, by organizational coordination.

Nevertheless, what we can name 'stylized facts', point to the essentially economic nature of innovation which is to be regarded as a pervasive process through which new forms of production and consumption are continuously brought into being. In this perspective, among the economic problems faced the main one is that of new specific resources. This is essentially a dynamic problem as what finally matters is how to determine a learning process can be viable. Viability depends on the way markets and organizations are structured together.

This paper is dedicated to examining this problem. It is organized as follows. Section 1 briefly stresses research of allocative efficiency as the main aspect of the theory of innovation, in Arrow's formulation as well as in the modern refinements brought about by the analysis of game theoretic interactions among rival firms. Section 2 defines innovation as a process of creation of technology and introduces some reflections about creative efficiency in relation to the appearance of sunk costs. Section 3 refers to industrial organization, defined as a set of markets and organizations getting structured together, and focuses on its role in the processes of creation of technology. Finally, section 4 proposes some conjectures on particular forms of industrial organization that make a process of innovation viable, with reference to an original modelling based on a temporal representation of production.

*This paper owes much to my collaboration with M. Amendola in recent years. However he is not responsible for the way in which I have taken advantage of it.

1. Research of allocative efficiency

The modern revisitation of the economics of technological change begins with the seminal paper by Kenneth Arrow (1962). In this paper Arrow interprets invention as the production of knowledge. Technology is no more exogenous and becomes a produced resource. The field of economic activities is therefore extended to include research and development (or invention) activity. Correlatively, the list of goods is enlarged so as to make possible the reference to a market. The new activity is completely disconnected from the others and, in particular, from the production activity *stricto sensu*. It can be described by means of a production function, in which some types of capital goods and labour skills are listed as inputs and the output is a codified knowledge (i.e. information). The new output is a commodity which can be traded on a market. In this way the innovation process is seen as a linear process.

The main problem lies not in the characterization of technology-creating production processes but in the realization of optimal resource allocation for invention which depends on the properties of technology classed as information. In fact, in the case of invention all the reasons for the failure of the market to achieve optimality in resource allocation exist. It is uncertainty that makes technology such a specified commodity. Besides this commodity is neither divisible nor thoroughly appropriable. Given that there is no incentive to allocate some resources to R&D activity, the real value of its results (information) cannot be fixed by the market forces.

The conclusion of this analysis is twofold. On the one hand it appears that a free market economy underinvests in R&D. On the other hand, it is admitted that an incentive to invent can exist which results from profits earned on product markets by firms using invention, when, for example, the latter is production cost reducing. In other words organizational coordination is necessary to make up for market failures, but it implies a distortion in the allocation of resources from a social point of view.

In recent modelling developed from the same line of thought, various aspects of R&D activity and their relationships to market structure have been analyzed. In the Dasgupta–Stiglitz model (1980), for example, the aim is to compare different market positions in relation with various benchmarks. The typical outcome of these comparisons is that 'the market economy may be characterized both by excessive expenditures on R&D and too low a rate of technical progress as compared to the socially managed one' (*ibid.*: 279). In other models the issue has been the dependence of R&D investment incentives on the strategic consideration of preemptive innovation. The essence of these models is that a firm always has the opportunity to make commitments which alter its environment, i.e. the conditions in which an oligopoly game takes place. These commitments can be R&D expenditures that are decided in the first step of a multistage game, where a strategy for a firm is a commitment, or a bid (spending on R&D) and a Nash equilibrium is a vector of commitments (or bids). As F. Fisher (1991) has pointed out, such models have produced a sort of taxonomy of particular cases, but have failed to provide a robust analysis of real situations: 'the principal result of theory is to show that nearly anything can happen' (Fisher, 1991: 207).

At this stage of our argument it is important to emphasize what sort of assumption is made in all these models. In fact the key assumption consists in admitting no real resources are expended until the market structure is determined, i.e. until the number of firms with non-negative profits is determined. The firms bid what they would spend, but only the firms which finally participate in the oligopoly game actually develop the invention, i.e. spend resources on R&D (see Reinganum, 1988). In other words, the models, that have been worked out to analyze R&D activity, do not remove a fundamental axiom of the Arrow–Debreu model, that is the axiom of firm productive efficiency. The strategies these models describe do not involve the appearance of sunk costs, as these costs are defined as resulting from a dissociation in time between costs and proceeds, causing the financial position of the firm to acquire an important role. Each firm is efficient, not at a particular point of time, but along a sequence of successive periods considered as a whole. Thus time becomes actually irrelevant.

Now, the innovation process, as Schumpeter (1934) has emphasized, is a creative destruction process whose main aspect is made up of repeated changes in industrial organization which inevitably result in the appearance of sunk costs. That sets a specific problem to be solved from the efficiency viewpoint. In fact, when markets and organizations are not given, the efficiency analysis, as it is understood in the standard theory, loses its meaning, as comparison and choice analysis is not the kind of analysis to be carried out.

A comparison analysis concludes, as we have seen, that a free market economy does not allocate given resources for invention efficiently. Nevertheless observation of stylized facts suggest that a market system, as it is organized in developed countries, is the best system for making innovation processes viable. This calls for a reconsideration of the analysis of innovation processes in order to identify what is the real problem to be solved by the real firms which are involved in such processes and are operating in real time. As we shall see, this problem concerns the productive efficiency of the firm rather than the allocative efficiency at the social or industry level.

2. A definition of innovation

The innovation processes which are observed in modern industrial systems are sequential learning processes. These processes may appear to be little more than examples of learning by doing, aimed at making minor improvements in equipment and manufacturing processes and at reducing production costs. They have gradually incorporated learning by interaction with customers (users) and by interaction between research and development, manufacturing and marketing of product activities. They characterize an empirical (or concrete) model of innovation that S. Kline and N. Rosenberg (1986) have named the chain-linked model. In this context, the distinction between major (or radical) and minor innovations no longer has any meaning, whereas it is now essential for the analysis to focus on the relationships between the various actors involved in a process in which all component phases are interwoven. In other words if we are to understand the phenomenon of innovation, stress is to be laid on the composition of the network and the way it works.

However, it would be improper only to consider factual features and not to perceive that the understanding of this new empirical model of innovation makes it necessary to focus on some general analytical points. First, in this model, markets and organizations are not distinct institutions that would alternately perform the same allocative task depending on the nature of commodities that are traded or the kind of information that each economic agent withholds. On the contrary, they are considered together as complementary elements of a unique process. Indeed this aspect is an analytical feature (and not only an empirical one) of a dynamic model of creation of resources – be it a chain-linked model, a linear model, or whatever. The second point that the reference to the new empirical model of innovation helps to clear up, concerns the real dimension of the production process. Focusing on vertical relationships between successive phases, such as R&D, manufacturing and marketing, implies the taking into account of a construction phase distinct from an utilization phase. In this way it appears, on the one hand, that production takes time, and, on the other hand, that markets don't exist, at each period of time, for most intermediate products which characterize the various stages and which are intertemporal complements to one another. As a consequence, costs and proceeds cannot be contemporaneous, except in the case of a complete synchronization of production; that is, in a steady state, when elementary production processes are arranged in line in the factory system and enough of them are carried out so that all stages take place at the same time, and that the vertical ordering of the successive phases of the process itself fades away (Georgescu-Roegen, 1965). This appearance of sunk costs (in a proper sense) characterizes systematically an economy out of equilibrium whether it is organized, once again according to the chain linked, the linear, or whatever model.

Summing up, what is at the heart of the analysis is innovation as a process in time and not R&D as a product. And coordination and organization of innovation are different from coordination and organization of R&D.

3. Industrial organization

The successful implementation of innovation processes requires that firms solve the problem posed by the learning processes and the existence of sunk costs which are always associated with them, that is by the conflict between short- and long-term requirements with regard to profitability. How to deal with the time profile of costs and proceeds is what matters for the innovation processes to be viable.

The task of the firm is still then that of coordination, no longer though in the sense of allocating given resources but rather of organizing the process of change so as to make it viable. This is a specific task for which only the firm (or a group of firms) as an organization is suited. The reason is that dealing with the sunk costs implied by a change interpreted as a learning process requires establishing a wide range of cooperative relationships over time: not only internally, but also between different firms, between firms and banks, and so on.

In this perspective, the environment is no longer an exogenous constraint. The firm itself must organize the process of change; that is, it must develop innovation loci and procedures. The environment becomes a resource or an option, in fact a strategic variable to be moulded according to and in function of the strategy pursued (Amendola and Bruno, 1990).

The theory of the firm becomes that of a complex system of intra- and inter-firms' relationships. Accordingly, a firm is characterized by coherent organizational design, i.e. by some combination between a mode of coordination (horizontal or hierarchical) and a scheme of incentives (market-oriented contracts or rank hierarchies) (Aoki, 1990). It is obvious that a coherent organizational design tends to favour innovative processes, i.e. to generate learning processes and specific quasi-rents that are associated with them. These quasi-rents are, on the one hand, organizational quasi-rents that result from the mutual commitments of employers and employees, and on the other hand, relational quasi-rents that are group-specific economic returns attributable to the cooperation between firms (Aoki, 1988). From this point of view an organization that would be intrinsically superior to any other (in terms of whatever general criterion) doesn't really exist. In fact, different combinations between modes of coordination and schemes of incentives can be carried out which correspond to several efficient systems.

How to make an innovation system (i.e. a firm, a group of firms, or a national economy) efficient (or viable) is a problem where solution encompasses the building up of a satisfactory set of complementarities (via coordination and incentives modes) among internal as well as external factors which become specific resources of the transformation process (Amendola and Bruno, 1990). However, this solution is never unique. Small changes in these factors involve a different articulation over time between the decisions and the constraints and then determine a new profile of the transformation process which can always be viable.

M. Aoki is the author who, having most advanced in this direction, proposes a comparative analysis which brings forward the hypothesis that the relative efficiency of the two mentioned systems depends on values taken by different environmental parameters. These parameters concern the characteristics of market demand, the mechanism of price (and wage) determination, the nature of technology involved in the production processes, and institutional (or historical) traits. In doing so, Aoki establishes a link between organizations and markets. He underlines that what is important is not to determine where the frontiers between firms and markets are, but to understand the reasons why costs (as it is the efficiency criterion which is adopted) differ among systems, each of them being characterized by some organizational design coupled with a particular market working. Markets and organizations are no longer substitutes that perform allocative tasks. They become complements in a process of creation of resources. However Aoki's analysis remains a comparative static one that only permits the setting of an 'exemplifying' (an expression proposed by F. Fisher, *ibid.*) theory of the Japanese firm in contrast with the agency theory reputed to represent the American firm.

4. A viable process of innovation

The ingredients of Aoki's analysis can be turned to treat processes of creation of resources. What we need for this to be done, however, is to shift the focus of the analysis. This implies, in the first place, interpreting sunk costs as the outcome of a true sequence of a production process conceived as a scheme for transforming 'in time' a flow of labour inputs into a flow of final output. In fact this requires explicit consideration of a phase of construction and (following it) a phase of utilization of fixed capital goods. The neo-Austrian model (Hicks, 1973) with its full vertical integration gives a satisfactory answer to the above requirement. Being completed by introducing specific human resources as inputs in such a way that learning can be contemplated (Amendola and Gaffard, 1988), this type of model permits bringing to light the consequences of the breaking of an equilibrium configuration.

Let us start by defining an elementary process of production showing the above mentioned features as follows:

$$A = [A^c, A^u];\ b$$

That is, by means of an input matrice A partitioned into the submatrices $A^C \equiv [a^c_{hk}]$ and $A^u \equiv [a^u_{hk}]$ whose elements represent the quantities of the different types of labour (h = 1, 2, . . . j) required by the elementary process in the different periods of the phase of construction (k = 1, 2, . . . , n) and the phase of utilization (k = n + 1, . . . , n + N) respectively, and a vector of final output $b \equiv [b_k]$ whose elements refer to the amounts of final output obtained in different periods of the phase of utilization.

Production capacity at time t is given by the number of processes x of type j and age k both in phase of construction (x^c) and in phase of utilization (x^u) actually carried on. This number is determined as the minimum between the number of processes inherited from the past and what is required for investment and for current production, respectively, that is:

$$x^c_{jk}(t) = \min[x_{jk-1}(t-1), \bar{x}^c_{jk}(t)]$$

$$x^u_{jk}(t) = \min[x_{jk-1}(t-1), \bar{x}^u_{jk}(t)]$$

where $\bar{x}^c_{jk}(t)$ and $\bar{x}^u_{jk}(t)$ are determined by the model whose framework is hereafter exposed.

Suppose that there are two kinds of processes (j = 1, 2) in the economy-labelled routine (j = 1) and innovative (j = 2) processes. Firms receive at date t the proceeds from the production of the previous period. The level of their activity depends on these proceeds or, alternately, on labour supply inherited from the past. Let m(t − 1) be the mentioned proceeds, h(t − 1) be the unspent money carried over from previous periods, f(t) be the external finance, w(t) be the wage fund, c(t) be the intermediary consumption (i.e. the 'take out'), h(t) be the idle balances (desired or not). Then the firm finance constraint is given by:

$$m(t-1) + h(t-1) + f(t) = w(t) + c(t) + h(t)$$

$$w(t) \leqslant \bar{w}(t)$$

where $\bar{w}(t)$ is the maximum wage fund as it is determined by the human resource constraint in the current period.

Suppose now that no price change occurs within each period of the sequence. Changes can only take place at the junction from one period to the next. That implies that the firms' money proceeds are given by:

$$m(t-1) = \Sigma \min[p_j(t-1) \cdot s_j(t-1), p_j(t-1) \cdot d_j(t-1)]$$

where s_j and d_j are the current supply and demand of commodity j, respectively, and p_j its price.

In this way, stock changes substitute for price changes at least inside the elementary period of time. An excess of supply, e.g. is added to stocks.

$$o_j(t) = s_j(t) - d_j(t) > 0$$

where o_j is the accumulated stock of commodity j.

Let the operator g be defined by $gz(t) = [z(t) - z(t-1)]/z(t-1)$ for any variable z(t). Then prices and wages, that can only change at the junction of successive periods, are given by:

$$gp_j(t) = k_j\ \xi_j(t-1)$$

where $\xi_j(t-1) = [d_j(t-1) - s_j(t-1)]/s_j(t-1)$ stands for the rate of excess demand (supply) of commodity j, and,

$$gw_h(t) = k_h\ \xi_h(t-1)$$

where $\xi_h(t-1) = [l_h^d(t-1) - l_h^s(t-1)]/l_h^s(t-1)$ stands for the rate of excess demand (supply) of labour skill h. Let us write the demand for the firms as:

$$\log d_j(t) = \epsilon_{jj}\log p_j(t) + \epsilon_{ij}\log p_i(t) + \log g_j[s_j(t-1)]$$

where the last term indicates a learning in consumption effect.

The supply confronted with this demand is given by:

$$s_j(t) = [\delta_j(t)m^*(t)]/p_j(t)$$

where $m^*(t)$ is the expected money value of final demand for the firms and $\delta_j(t)$ is the proportion of existing productive capacity devoted to produce the commodity j.

Firms may either run out of stocks or be accumulating them. Then the production $q_j(t)$ of the commodity j is given by:

$$q_j(t) = s_j(t) - \eta o_j(t)$$

where η is a parameter which determines the proportion of stocks to be eliminated.

Firms' investment, i(t), is constrained by the wage fund once finance of current production has been taken care of; so it is given by:

$$i(t) = w(t) - w_u(t)$$

where $w_u(t)$ is the wage fund which has been spent to carry out productive processes in their utilization phase.

Given the range of the processes of production that the skills of the existing human resource make it possible to devise, the labour supply vector sets a constraint both on the type and the number of the processes that can actually be carried out in each given period. However, this constraint is not strictly rigid. Learning affects the structure of existing skills rendering it step by step more consistent with the actual requirements. The new labour supply vector to be confronted with the labour demand at each period, is given by:

$$\log l_h^s = \epsilon_{hh} \log w_h(t) + \log(l^{se} + l^{si}(t))$$

where l^{se} stands for the exogenously determined supply of labour and l^{si} for the supply of labour which is determined as an effect of learning by employees on the innovative process.

Suppose we start in a situation where there are only routine processes with $p_1 = 1$ and $s_j(t) = d_j(t) < t$ (i.e. in a steady state). At date $t = 1$ the demand for commodity 1 is less than it would be in a steady state as a consequence of a determinist shock. Stocks of this commodity are involuntarily accumulated ($o_1 > 0$). This event gives a signal to carry out innovative processes at date 2. At this date proceeds of firms have decreased. This is assumed to decrease their investment expenditures. From this moment onwards what happens depends on the way the finance constraint interacts with the human resource constraint. It is this interaction which determines a more or less stringent distortion in the age structure of productive capacity, i.e. the viability of the process of change.

It would not be easy (if it were possible) to obtain an analytical solution even when we had given all the functions a suitable parametric form. We shall not try to do this. But it may be feasible to analyze some out of equilibrium paths corresponding to different values of parameters or to different parametric forms of the functions, by the means of a step-by-step numerical calculation.

The main relevant functions concern external finance, intermediary consumption (that includes purchases from subcontractors), prices and wages determination, learning in consumption and learning in production. Different parametric forms of these functions and different values of parameters refer to different modes of articulation between markets and organizations, i.e. to what we could name different systems of innovation.

We can make some conjectures about the way in which different systems affect the interaction between the financial and the human resource constraint over the process of change.

Two polar cases will be singled out. One polar case is characterized by a structureless labour market, a market-oriented financial system, and non-cooperative relationships between firms and their subcontractors. The other polar case on the contrary, is characterized by a structured labour market, a bank-oriented financial system, and cooperative relationships between firms and their subcontractors.

With a market-oriented system one can assume that a finance constraint generally prevails as a consequence of the breaking of an equilibrium configuration. The main reason lies in the behaviour of those (shareholders, bondholders, banks and other financial intermediaries) whose function is to provide external finance. In fact, this behaviour implies a redistribution of financial assets towards the firm when things go well, and away from it when the internal source of finance shrinks. This results in a strengthening of the existing finance constraint. If, for a short period of time, the human resource constraint should have to be the dominant one, then wages go up, being determined through spot market contracting, and hence that would result in a stronger finance constraint. Moreover, non-cooperative relationships between firms and their subcontractors prevent the sunk costs being divided. However, the finance constraint could be made smaller in such a way that the process of change is left viable. On the one hand, as a consequence of unemployment, wages would decrease which would permit the maintenance, at a certain level, of the intensity of the activity of construction of new productive processes, and hence limit the distortion in the age structure of the productive capacity. On the other hand, the same effect could be obtained when the externalization by firms of some activities is chosen with the aim of reducing labour costs.

The market-oriented system makes a process of change viable if the human resource constraint is never the dominant one for a long period of time. This supposes that the needs for new skills can be satisfied on the external market.

With organized markets for labour, finance and subcontracting a process of innovation can also be viable. In this different context, the human resource constraint prevails in most cases. In fact, the contracyclic intervention of a main bank contributes to release the financial constraint in such a way that it is less strong than the human resource constraint. Wages do not fluctuate in response to excess demand or supply of labour but are linked to the growth of the proceeds of the firm. Hence the wage bill goes up or down with the internal financial resource. That prevents the fluctuations of the resources available for new investments from being too strong. Moreover, subcontracting appears to be a valid option when shifting the available human resources to other tasks is relatively more productive than keeping them in productions that can be decentralized to other firms (subcontractors). That makes it possible to intensify the learning process associated with the innovative project being developed (Amendola and Bruno, 1990).

We have gone through this essentially methodological exploration in order to focus on a general feature of innovation dynamics which is a sequential dynamics by nature. Thus the real analytical problem is not to endogenize rate of growth specifying individual behaviours, particularly R&D behaviours. It is to determine how markets and organizations have to interact along the way to make a process of change viable. Such an analysis results in a sort of taxonomy of innovative systems which deserves a more detailed exploration than it is possible to do in this chapter.

References

Amendola, M. and Bruno, S. (1990), 'The behaviour of the innovative firm: relations to the environment', *Research Policy*, 19(5), pp. 419–33.

Amendola, M. and Gaffard, J.L. (1988), *The Innovative Choice*, Oxford, Basil Blackwell.

Aoki, M. (1988), *Information, Incentives and Bargaining in the Japanese Economy*,

Cambridge, Cambridge University Press.

Aoki, M. (1990), 'Toward an economic model of the Japanese firm', *Journal of Economic Literature*, XXVIII, pp. 1–27.

Arrow, K. (1962), 'Economic welfare and the allocation of resources for invention' in NBER conference no. 13, *The Rate and Direction of Inventive Activity. Economic and Social Factors*, Princeton, Princeton University Press.

Dasgupta, P. and Stiglitz, J. (1980), 'Industrial structure and the nature of innovative activity', *Economic Journal*, 90, pp. 266–93.

Fisher, F. (1991), 'Organizing industrial organization. Reflections on the Handbook of Industrial Organization', *Brookings Papers on Economic Activity*, pp. 201–25.

Georgescu-Roegen, N. (1965), 'Process in farming versus process in manufacturing. A problem of balanced development', repr. in *Energy and Economic Myths* (1976), Pergamon Press.

Hicks, J.R. (1973), *Capital and Time*, Oxford, Clarendon Press.

Reinganum, J. (1988), 'The timing of innovation: research, development, and diffusion', in R. Schmalensee and R. Willig (eds), *Handbook of Industrial Organization*, Amsterdam, North-Holland.

Schumpeter, J.A. (1934), *The Theory of Economic Development*, Cambridge, Mass., Harvard University Press.

7 A model of growth through creative destruction*

Philippe Aghion and Peter Howitt

1. Introduction

This paper presents a model of endogenous stochastic growth, based on Schumpeter's idea of creative destruction. It starts with the same premises as other endogenous growth models, namely that growth in the long run is attributable primarily to the accumulation of knowledge rather than factors of production, that this accumulation is endogenous to the economic system, and that the process underlying it involves significant external economies. The particular process it focuses on is research aimed at producing innovations in the form of better products.

This focus highlights three factors that have been relatively neglected in recent literature. The first is imperfect competition. Monopoly rents are needed to induce research. Perfectly competitive models[1] can generate technological progress only through the incidental spillover effects of learning by doing; purposive research has no reward.

The second factor is obsolescence. Technological progress creates losses as well as gains, by rendering obsolete old skills, goods, markets, and manufacturing processes. This has both positive and normative implications for growth. In positive terms, the prospect of a high level of research in the future can deter research today by threatening the fruits of that research with rapid obsolescence. In normative terms, obsolescence creates a negative externality from innovations, and hence a tendency for *laissez-faire* to generate too much growth.

Obsolescence does not fit well into existing models of endogenous growth, which emphasize positive externalities in the form of technology spillovers.[2] The closest in spirit to the present model is that of Romer (1990), in which innovations produce new intermediate goods, neither better nor worse than existing ones. Once invented a good remains in production forever. This model does a good job of capturing the division-of-labor aspect of growth. But adding obsolescence, by allowing old goods to be displaced by the introduction of new goods, may eliminate growth (see Deneckere and Judd, 1986).

* The authors wish to acknowledge the helpful comments and criticisms of Roland Bénabou, Olivier Blanchard, Patrick Bolton, Mathias Dewatripont, Dick Eckaus, Zvi Griliches, Elhanan Helpman, Rebecca Henderson, Louis Phaneuf, Nancy Stockey, and Patrick Rey. The second author also wishes to acknowledge helpful conversations with Louis Corriveau.

The third factor is discontinuity. The chapter adopts the view taken by many economic historians that individual technological breakthroughs have aggregate effects, that the uncertainty of innovations does not average out across industries. This view is supported by empirical evidence of authors such as Nelson and Plosser (1982) and Campbell and Mankiw (1987) that the trend component of an economy's GNP includes a substantial random element. Recent authors like Cochrane (1988) have challenged these results. But whatever one's views on unit roots there is considerable evidence that aggregate productivity is random (see, for example, Prescott, 1986).

Existing models of endogenous growth do not produce a random trend, unless exogenous technology shocks are added. But technology shocks should not be regarded as exogenous in an analysis that seeks to explain the economic decisions underlying the accumulation of knowledge. In this chapter they are entirely the result of such decisions. Statistical innovations in GNP are produced by economic innovations, the distribution of which is determined by the equilibrium amount of research.[3]

In highlighting these factors the paper follows Schumpeter (1942: 83, his emphasis):

> The fundamental impulse that sets and keeps the capitalist engine in motion comes from the new consumers' goods, the new methods of production or transportation, the new markets, . . . [This process] incessantly revolutionizes the economic structure *from within*, incessantly destroying the old one, incessantly creating a new one. This process of Creative Destruction is the essential fact about capitalism.

In Schumpeter's view, capitalist growth is inherently uncertain. Fundamental breakthroughs are the essence of the process, and they affect the entire economy. However, the uncertainty is endogenous to the system, because the probability of a breakthrough depends upon the level of research, which in turn depends upon the monopoly rents that constitute 'the prizes offered by capitalist society to the successful innovator' (1942: 102). Thus, 'economic life itself changes its own data, by fits and starts' (1934: 62).

The present chapter develops a simple model that articulates these basic Schumpeterian elements. Although it is quite special in some dimensions we believe it is also simple and flexible enough to serve as a prototype model of growth through creative destruction.[4]

2. The basic model

a. *Assumptions*

There are three classes of tradable objects: labor, a consumption good, and an intermediate good. There is a continuum of infinitely-lived individuals, with identical intertemporally additive preferences defined over lifetime consumption, and the constant rate of time preference $r > 0$. The marginal utility of consumption is assumed constant; thus r is also the rate of interest.

There is no disutility from supplying labor. There are three categories of labor: unskilled labor, which can only be used in producing the consumption good; skilled labor, which can be used either in research or in the intermediate sector; and specialized labor, which can be used only in research. Each individual is endowed with a one-unit flow of labor. Let M, N and R denote respectively the mass of unskilled, skilled and specialized individuals.

The consumption good is produced using the fixed quantity M of unskilled labor, and the intermediate good, subject to constant returns. Since M is fixed the production function can be written as:

$$y = AF(x) \tag{2.1}$$

where $F' > 0$, $F'' < 0$, y is the flow output of consumption good, x the flow of intermediate input, and A a parameter indicating the productivity of the intermediate input.

The intermediate good is produced using skilled labor alone, according to the linear technology:

$$x = L \tag{2.2}$$

where L is the flow of skilled labor used in the intermediate sector.

Skilled labor used in research produces a random sequence of innovations. The Poisson arrival rate of innovations in the economy at any instant is $\lambda\Phi(n, R)$, where n is the flow of skilled labor used in research, λ a constant parameter, and Φ a constant-returns, concave production with positive marginal products. Both λ and Φ are given by the technology of research. There is no memory in this technology, since the arrival rate depends only upon the current flow of input to research, not upon past research. (The 'linear' case where $\Phi(n, R) = n$; that is, where $R = 0$, will be used frequently.)

Time is continuous, and indexed by $\tau \geqslant 0$. The symbol $t = 0, 1 \ldots$ denotes the interval starting with the tth innovation (or with $\tau = 0$ in the case of $t = 0$) and ending just before the $t + 1^{st}$. The length of each interval is random. All prices and quantities are assumed to remain constant within each interval. If n_t is applied to research in interval t, the length of the interval will be exponentially distributed with parameter $\lambda\Phi(n_t, R)$.

Each innovation consists of the invention of a new intermediate good, whose use as input allows more efficient methods to be used in producing the consumption good. Real-world examples include such 'input' innovations[5] as the steam engine, the airplane, and the computer, whose use made possible new methods of production in mining, transportation, and banking. An innovation need not, however, be as revolutionary as these examples, but might consist instead of a new generation of intermediate good, similar to the old one.

Specifically, use of the new intermediate good increases the productivity parameter A in (2.1) by the factor $\gamma > 1$. There are no lags in the diffusion of technology.[6] The most modern intermediate good is always produced, so that:

$$A_t = A_0\gamma^t; \quad t = 0, 1, \ldots \tag{2.3}$$

where A_0 is the initial value given by history. (Of course, it is always possible to produce the consumption good using an old technology, with a correspondingly old intermediate good.)

A successful innovator obtains a patent which it sells to an intermediate firm that becomes a monopolist until the next innovation occurs. (Assume perfect competition in all markets other than those for the intermediate good and for the patents.) The innovator offers its patent for sale at a price equal to the expected present value of the monopoly rents accruing to the patent. The buyer pays that price as a lump sum in exchange for exclusive use of the patent.

b. Research

There are no contemporaneous spillovers in research; that is, a research firm employing the amounts z, s of the two factors will experience innovations with a Poisson arrival rate of $\lambda\Phi(z, s)$, independently of the inputs of other research firms. Thus the firm's flow of expected profits will be:

$$\lambda\Phi(z, s)V_{t+1} - w_t z - w_t^s s \tag{2.4}$$

where V_{t+1} is the value of the $t + 1^{st}$ innovation, and w_t and w_t^s are the respective wage rates of skilled and specialized labor. All values and prices are expressed taking the consumption good as numéraire.

Because Φ has constant returns, and because the total flow of specialized labor must equal R in equilibrium, it follows from the Kuhn-Tucker conditions of maximizing (2.4) that:

$$w_t \geqslant \varphi'(n_t)\lambda V_{t+1}, \quad n_t \geqslant 0, \text{ with at least one equality} \tag{2.5}$$

where $\varphi(n_t) \equiv \Phi(n_t, R)$, and n_t is the economy-wide flow of skilled labor used in research during interval t.

The value V_{t+1} is the expected present value of the flow of monopoly profits (π_{t+1}) generated by the $t + 1^{st}$ innovation over an interval whose length is exponentially distributed with parameter $\lambda\varphi(n_{t+1})$:

$$V_{t+1} = \frac{\pi_{t+1}}{r + \lambda\varphi(n_{t+1})}. \tag{2.6}$$

There is an important intertemporal spillover in this model. An innovation raises productivity forever. It allows each subsequent innovation to raise A_t by the same multiple γ, and with the same probability $\lambda\varphi(n_t)$, but from a starting value A_{t-1} that will be higher by the multiple γ than it would otherwise have been. The producer of an innovation will capture (some of) the rents from that productivity gain, but only during the next interval. After that the rents will be captured by other innovators, building upon the basis of the present innovation, but without compensating the present innovator.[7] This intertemporal spillover will play a role in the welfare analysis of section 4 below.

The model thus embodies Schumpeter's idea of 'creative destruction'. Each innovation is an act of creation aimed at capturing monopoly rents. But it also destroys the monopoly rents that motivated the previous creation. Creative destruction accounts for the term $\lambda\phi(n_{t+1})$ in the denominator of V_{t+1} in (2.6). More research will reduce the expected tenure of the current monopolists, and hence reduce the expected present value of its flow of rents.[8]

c. *The intermediate monopolist's decision problem*

The intermediate monopolist's objective is to maximize the expected present value of profits over the current interval. When the interval ends so do the rents. The only uncertainty concerns the length of the interval. Except in section 6 below, the monopolist is assumed to take as given the amount of research at each time, and hence also takes as given the length of the interval.[9]

Let x_t be the quantity of the intermediate good produced by the monopolist in interval t. By (2.2), x_t also equals employment of skilled labor in manufacturing. The inverse demand curve facing a monopolist charging the price p_t is the marginal product

$$p_t = A_t F'(x_t) \tag{2.7}$$

Thus the monopolist chooses x_t to maximize $[A_t F'(x_t) - w_t]x_t$, taking A_t and w_t as given. The necessary first-order condition is

$$\omega_t = F'(x_t) + x_t F''(x_t), \tag{2.8}$$

where $\omega_t = w_t/A_t$ is the productivity-adjusted wage rate.

Assume that the monopolist's marginal-revenue schedule is downward sloping:

Assumption A.1: $2F''(x) + F'''(x) < 0 \quad \forall x > 0$.

This condition holds automatically when $F''' < 0$; we show in Appendix 1 that it also holds when F comes from a CES production function. It follows from (2.8) and A.1 that there is a unique solution to the decision problem:

$$x_t = \tilde{x}(\omega_t) \tag{2.9}$$

where $\tilde{x}$ is strictly decreasing. The monopolist's flow of profits can be expressed as:

$$\pi_t = A_t \tilde{\pi}(\omega_t)$$

where $\tilde{\pi}(\omega) \equiv [F'(\tilde{x}(\omega)) - \omega]\tilde{x}(\omega)$ is strictly decreasing in ω.

The monopolist's decision problem has not been constrained by potential competition from the holder of the previous patent. There are two ways to rationalize this assumption without assuming that innovations are always drastic in the usual case (see, for example, Tirole, 1988: ch. 10). The first is to note that if the innovations were nondrastic then each patent would be of greatest value to holders of the previous patent, who would face no such competition; thus non-drastic innovations would be implemented by incumbent producers, in accordance with the evidence of Scherer, 1980.[10] The second is to note that if there were even small fixed costs of entry that had to be repaid by a previous incumbent to reenter, the credible threat of limit pricing would be enough to deter reentry; the incumbent could thus ignore the competitive fringe.

d. The capital market

Access to a capital market is obviously needed by research firms that must pay their workers before (possibly) becoming successful innovators, and by intermediate firms that pay a lump-sum fee to acquire a patent without knowing for how long they will profitably exploit that patent.

Introducing a capital market that guarantees consistency between all flows of revenue at any instant τ is straightforward, given risk-neutrality. In particular, the model can be reinterpreted assuming a central bank owned by the individuals. During interval t, in the period between τ and $\tau + d\tau$, the research sector borrows $(w_t^s R + w_t n_t)d\tau$ from the bank to pay its workers. The bank, owned by risk-neutral individuals, is willing to lend this amount in exchange for a claim to the t + 1st patent if it is discovered during the period. The expected value of the claim equals $\lambda\varphi(n_t)V_{t+1}d\tau$ which, by profit-maximization (and Euler's theorem) just equals the amount lent. Thus the bank breaks even in its dealings with research firms.

The intermediate firm that acquires the tth patent will actually purchase it from the bank that owns the property rights for that patent. Again, the risk-neutral bank will be indifferent between receiving the lump-sum payment V_t or the flow of monopoly profits π_t during the tth interval. The flow of gross revenues realized per unit of time by the intermediate incumbent producer is $p_t x_t$. This flow is decomposed into wage payments $w_t x_t$, and profits π_t which accrue to the bank as the price of renting the patent. Therefore, the bank also breaks even in its financial relationship with the intermediate sector. The consumption good sector does not need to deal with the bank: its output is deterministic and immediately renumerates the two factors used (intermediate good and unskilled labor).

A final remark concerns the rate of return (the equilibrium dividend to price ratio) paid by the bank to its depositors. The amount of deposits D_t held at the bank during interval t must equal the value V_t of its assets (balance-sheet equilibrium). On the other hand, the financial flows to and from the bank must also permanently be equal

$$R_t D_t + w_t^s R + w_t n_t = \pi_t, \tag{2.10}$$

where R_t is the dividend rate on current deposits. (Depositors also received a capital gain of $(\gamma - 1)D_t$ with each innovation.)

Consider now a stationary path where $\omega_t = \hat{\omega}$ and $n_t = \hat{n}$ for all t.[11] Along such a path the relevant variables of the economy, namely wages, profits and final output, will jump homothetically by the multiple γ each time a new innovation occurs. In particular, $\pi_{t+1} = \gamma\pi_t$. From equations (2.4), (2.6) and (2.10), and the condition that $V_t = D_t$[12]:

$$R_t = r - \lambda\varphi(\hat{n})(\gamma - 1), \tag{2.11}$$

which is strictly less than the rate $(r + \lambda\varphi(\hat{n}))$ at which private firms discount current profits. This discrepancy will be reinterpreted in the following section on welfare analysis, as the formal basis for the intertemporal spillover effect mentioned in (b) above.

e. A reinterpretation of the model

The model could be reinterpreted as one where innovations increase the quality of consumption goods, thereby generating growth in terms of instantaneous utility. More precisely, assume that intertemporal preferences are given by:

$$\int_0^{\infty} u_\tau e^{-r\tau} d\tau,$$

where: $u_\tau = A_t F(x_\tau)$. Let the instantaneous utility u_τ be the numéraire. Then $p_t = A_t F'(x_t)$ is the price of the consumption good, whose output x_t is governed by the production function (2.2). The analysis goes through unaffected by this reinterpretation. (The reinterpretation also facilitates a multisectoral analysis, worked out in Appendix 2, in which an aggregate demand spillover leads to an enhanced level of research.)

3. Perfect foresight dynamics and balanced growth

At any point in time there is only one decision for society to make; namely, how to allocate the fixed flow N of skilled labor between manufacturing (x) and research $(n = N - x)$. This decision is determined by the equilibrium value of ω the productivity-adjusted wage rate, since $x = \tilde{x}(\omega)$. This section shows how ω is determined, and consequently how the entire economy evolves, in a balanced growth equilibrium.

Equation (2.5) can be reexpressed as a first-order difference equation in ω_t. From (2.5) and (2.6):

$$\omega_t A_t / \varphi'(N - \tilde{x}(\omega_t)) \geqslant \frac{\lambda A_{t+1} \tilde{\pi}(\omega_{t+1})}{r + \lambda\varphi(N - \tilde{x}(\omega_{t+1}))}; \ \tilde{x}(\omega_t) \leqslant N; \text{ with at least one equality.} \tag{3.1}$$

Dividing both sides by A_t and using the fact that $A_{t+1} = \gamma A_t$ produces

$$\omega_t = \text{Max}\ \{\Psi(\omega_{t+1}), \underline{\omega}\} \equiv \Psi(\omega_{t+1}) \tag{3.2}$$

where $\underline{\omega}$ is the wage rate at which all labor is employed in manufacturing (i.e. $\tilde{x}(\underline{\omega}) = N$), and Ψ is defined by:

$$\omega_t = \Psi(\omega_{t+1}) \Leftrightarrow \frac{\omega_t}{\varphi'(N - \tilde{x}(\omega_t))} = \frac{\lambda\gamma\tilde{\pi}(\omega_{t+1})}{r + \lambda\varphi(N - \tilde{x}(\omega_{t+1}))}$$

Equation (3.2) states that the expected future wage ω_{t+1} determines the current wage ω_t, where the expected wage ω_{t+1} is used (correctly) to calculate the expected value of the $(t + 1)^{th}$ innovation. Define a perfect foresight equilibrium (PFE) as a sequence $\{\omega_t\}_0^\infty$ satisfying (3.2) for all $t \geqslant 0$.

Characterization of PFE is simplified by:

Lemma 1: The mapping: $\omega \to \Psi(\omega)$ is decreasing in ω.

Lemma 1 follows from the fact that $\tilde{x}$ and $\tilde{\pi}$ are decreasing functions, φ is increasing, and φ' is non-increasing. The economic interpretation is as follows: A foreseen increase in ω_{t+1} will reduce the reward V_{t+1} to the next innovator: on the one hand it raises the flow of monopoly rents $\tilde{\pi}(\omega_{t+1})$ and on the other hand it increases the amount of research and hence the amount of creative destruction next interval (the effect in the denominator of (2.6)). The decrease in V_{t+1} will reduce the value of labor in research, which decreases the equilibrium wage ω_t this period (if $x_t < N$).

The dynamic equation (3.2) is pictured by Figure 7.1.

There exists a unique stationary equilibrium[13] (SE) defined by: $\hat{\omega} = \Psi(\hat{\omega})$. In this steady state the economy experiences balanced growth in the sense that the allocation of employment between manufacturing and research remains constant over time. Growth is positive if $\tilde{x}(\hat{\omega}) < N$; i.e. if $\hat{\omega} > \underline{\omega}$, and it is zero if $\hat{\omega} = \underline{\omega}$, because there will be innovations if and only if a positive amount of labor is allocated to research. Except in section 8 the rest of the paper focuses on positive balanced growth equilibria, defined by the equation:

$$\hat{\omega} = \frac{\lambda\gamma\tilde{\pi}(\hat{\omega})}{r + \lambda\varphi(N - \tilde{x}(\hat{\omega}))}\varphi'(N - \tilde{x}(\hat{\omega})). \tag{3.3}$$

This equation defines $\hat{\omega}$ as a function of the parameters r, λ, γ, N. The equilibrium level of research is $\hat{n} = N - \tilde{x}(\hat{\omega})$, which depends positively on $\hat{\omega}$.

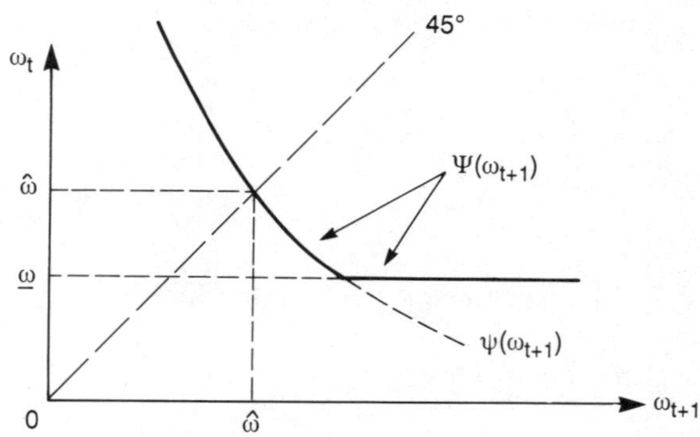

Figure 7.1

A particularly simple case is the linear, Cobb-Douglas case: $\varphi(n) = n$ and $F(x) = x^{\alpha}$, $0 < \alpha < 1$. In that case $\hat{n}$ is defined by the equation:

$$\lambda\gamma\left[\frac{1-\alpha}{\alpha}\right](N - \hat{n}) = r + \lambda\hat{n} \tag{3.4}$$

For growth to be positive, it is necessary and sufficient to have:

$$\lambda\gamma\left[\frac{1-\alpha}{\alpha}\right]N > r. \tag{3.5}$$

In particular, for a fixed value of γ, λ, r, N, positive balanced growth is possible if and only if the parameter α is sufficiently small: $\alpha < \alpha^* \equiv \lambda\gamma N/(\lambda\gamma N + r) < 1$.

To interpret this result note that α is an inverse measure of monopoly power in the intermediate sector. Specifically, $1 - \alpha$ is the Lerner (1934) measure of monopoly power (price minus marginal cost divided by price), $(1 - \alpha)^{-1}$ is the elasticity of demand faced by an intermediate monopolist, and $1 - \alpha$ is the fraction of equilibrium revenue in the intermediate sector accruing to the monopolist,

$$\frac{\pi_t}{\pi_t + w_t x_t}.$$

Thus, if monopoly power is too weak ($\alpha \geqslant \alpha^*$) then the flow of monopoly rents from the next innovation will not be enough to induce positive research aimed at capturing those rents even if they can be retained forever, with no creative destruction in the next interval.

Equation (3.4) shows how a change in the parameters r, λ, γ, N, α affects the SE level of research $\hat{n}$. In particular, an increase in monopoly power (reduction in α) increases $\hat{n}$. The effects of a change in the remaining parameters r, λ, γ, N turn out to be quite general, as shown by the following Proposition:

Proposition 1: The amount of research in a positive balanced growth equilibrium, defined by equation (3.3), will:

(a) decrease with the rate of interest r,
(b) increase with the arrival parameter λ,
(c) increase with the size γ of each innovation, and
(d) increase with the total endowment N of skilled labor.

Proof: Proposition 1 follows immediately from the same argument as Lemma 1.

This proposition is intuitive:

(a) an increase in the interest rate r means an increase in the required rate of return on research, whose effect will be to reduce investment in R&D;

(b) an increase in the arrival rate λ will have both the positive effect of raising the speed at which research pays (the effect on the numerator of 3.3), and the negative effect of increasing creative destruction (effect on the denominator). The first effect dominates;[14]
(c) an increase in the size γ of each innovation also increases research by increasing the size of the next interval's monopoly rents relative to today's research costs;
(d) an increase in skilled labor endowment N increases research by increasing the supply of one factor used in research.

One of the benefits of endogenizing technical change is to endogenize the average rate of growth (AGR) of the economy. Rather than AGR depending upon exogenous population growth and/or exogenous technical progress as in the neo-classical growth model, it depends upon the various factors that affect the incentive to do research and the fruitfulness of research, as in other endogenous growth models.

The volume of real output (i.e. the flow of consumption goods) during interval t is:

$$y_t = A_t F(\tilde{x}(\hat{\omega})) = A_t F(\hat{x}), \tag{3.6}$$

which implies:

$$y_{t+1} = \gamma y_t. \tag{3.7}$$

Thus the time path of the log of real output $ln\ y(\tau)$ will be a random step-function starting at $ln\ y_0 = ln\ F(\hat{x}) + ln\ A_0$, with the size of each step equal to the constant $ln\ \gamma > 0$, and with the time between each step $\{\Delta_1, \Delta_2, \ldots\}$ a sequence of iid variables exponentially distributed with parameter $\lambda\varphi(\hat{n})$. This statement, together with (3.3) and $\hat{n} = N - \tilde{x}(\hat{\omega})$, fully specifies the stochastic process driving output, as a function of the parameters of the model.

Not surprisingly, this stochastic process is non-stationary. Suppose observations were made at discrete points in time 1 unit apart. Then from (3.7):

$$ln\ y(\tau + 1) = ln\ y(\tau) + \varepsilon(\tau); \quad \tau = 0, 1, \ldots \tag{3.8}$$

where $\varepsilon(\tau)$ is $ln\ \gamma$ times the number of innovations between τ and $\tau + 1$. It follows from the above discussion that

$$\left\{\frac{\varepsilon(0)}{ln\ \gamma}, \frac{\varepsilon(1)}{ln\ \gamma}, \ldots\right\}$$

is a sequence of iid variables distributed Poisson with parameter $\varphi(\hat{n})$. Thus (3.8) can be rewritten as:

$$ln\ y(\tau + 1) = ln\ y(\tau) + \lambda\hat{n}\ ln\ \gamma + e(\tau); \quad \tau = 0, 1, \ldots \tag{3.9}$$

with

$$e(\tau) \text{ idd., } E(e(\tau)) = 0, \text{ var } e(\tau) = \lambda\hat{n}(ln\ \gamma)^2 \tag{3.10}$$

where $e(\tau) \equiv \varepsilon(\tau) - \lambda\hat{n}\ ln\ \gamma$. From (3.9) and (3.10), the discrete sequence of observations on log output follows a random walk with constant positive drift. It also follows that:

$$AGR = \lambda\hat{n}\ ln\ \gamma, \quad VGR = \lambda\hat{n}(ln\ \gamma)^2, \tag{3.11}$$

where VGR is the variance of the economy's growth rate.

Combining (3.11) with Proposition 1 allows one to sign the impact of parameter changes on AGR and VGR. Increases in the arrival parameter, the size of innovations, the size of skilled labor endowment and the degree of monopoly power all raise AGR. Increases in the rate of interest lower it. These parameter changes have the same qualitative effect on VGR as on AGR. The effects are intuitive and straightforward. The effect of monopoly power, combined with the earlier finding (in the Cobb-Douglas case) that a minimal degree of monopoly power is needed before growth is even possible, underline the importance of imperfect competition for the growth process.

The effects also highlight another Schumpeterian theme: the tradeoff between current output $y_t = A_tF(\hat{x})$ and growth. Any parameter change (other than N) that raises AGR also reduces y_t (taking A_t as predetermined), by drawing skilled labor out of manufacturing. (In the case of α that effect is amplified by the attendant shift of the production function in the consumption-good sector.) However, the loss of current output is not, as Schumpeter argued, a static efficiency loss of monopoly. Our assumption of inelastic labor supply prevents imperfect competition from putting the economy inside its current frontier of manufacturing and research. Whether the economy picks a good point on that frontier will be examined in the next section.

4. Welfare properties of the balanced growth equilibrium

This section compares the *laissez-faire* AGR derived above with the AGR that would be chosen by a social planner whose objective was to maximize the expected present value of consumption $y(\tau)$. Since every innovation raises $y(\tau)$ by the same factor γ, the optimal policy will consist of fixed values of x and n, denoted by x^* and n^* respectively. Expected utility from a policy (x, n) is:

$$U = \int_0^\infty e^{-r\tau} \sum_{t=0}^{\infty} \pi(t, \tau) A_t F(x) d\tau, \tag{4.1}$$

where $\pi(t, \tau)$ equals the probability that there will be exactly t innovations up to time τ. Given that the innovation process is distributed Poisson with parameter $\lambda \cdot \varphi(n)$, we have:

$$\pi(t, \tau) = (\lambda\varphi(n)\tau)^t e^{-\lambda\varphi(n)\tau}/t! \tag{4.2}$$

From (4.1) and (4.2):

$$U = \frac{A_0F(x)}{r + \lambda\varphi(n)(1 - \gamma)} \tag{4.3}$$

One recognizes U as the initial flow of output $A_0F(x)$ discounted at the social rate $[r + \lambda\varphi(n)(1 - \gamma)]$ (see section 2d above).

The social planner will chose (x^*, n^*) to maximize (4.3) subject to: $n + x = N$. This produces the first-order condition:

$$F'(x^*)[r + \lambda\varphi(N - x^*)(1 - \gamma)] + F(x^*)\lambda\varphi'(N - x^*)(1 - \gamma) = 0, \tag{4.4}$$

which can be reexpressed as:

$$\lambda\varphi'(N - x^*)(\gamma - 1)\frac{F(x^*)}{F'(x^*)} = r + \lambda\varphi(N - x^*)(1 - \gamma). \tag{4.5}$$

The left side can be interpreted as the marginal social rate of return to skilled labor in research; the incremental flow probability of an innovation $\lambda\varphi'$ times the resulting increase in output $(\gamma - 1)A_tF(x^*)$, all divided by the social opportunity cost of skilled research labor. Thus (4.5) requires the marginal social return to equal the social rate of discount (right side).

The social optimum is a balanced growth path, with real output growing according to the same stochastic process as before, but with innovations arriving at the rate $\lambda\varphi(N - x^*) = \lambda\varphi(n^*)$ instead of $\lambda\varphi(N - \hat{x}) = \lambda\varphi(\hat{n})$. Accordingly *laissez-faire* produces an AGR more (less) than optimal if $\hat{x} < (>) x^*$. Which way these inequalities go can be checked by comparing (3.3) and (4.5).

The former can be rewritten as:

$$\lambda\varphi'(N - \hat{x})\gamma\frac{\tilde{\pi}(\tilde{x}^{-1}(\hat{x}))}{\tilde{x}^{-1}(\hat{x})} = r + \lambda\varphi(N - \hat{x}). \tag{4.6}$$

The left side is the marginal private rate of return to skilled labor in research; the incremental flow probability times the gain to the successful innovator $\gamma A_t\tilde{\pi}(\hat{\omega})$, divided by the private cost of the labor $A_t\hat{\omega}$. Thus (4.6) requires the marginal private return to equal the private discount.

Again, the linear Cobb-Douglas case $(\varphi(n) = n,\ F(x) = x^{\alpha})$ is easy to analyze. This case produces the following equations respectively for x^* and $\hat{x}$:

$$\lambda\gamma(1 - \gamma^{-1})x^*/\alpha = r + \lambda(N - x^*)(1 - \gamma) \tag{4.7}$$

and

$$\lambda\gamma(1 - \alpha)\hat{x}/\alpha = r + \lambda(N - \hat{x}). \tag{4.8}$$

There are three differences between (4.7) and (4.8): The first difference is the presence of the term $(1 - \gamma)$ on the right side of (4.7). This corresponds to the intertemporal spillover effect discussed in section 2. A private innovator will capture rent from an innovation for one interval only, whereas the social rent continues forever. Hence the difference pointed out in section 2d between the private rate of discount $(r + \lambda n)$ and the social rate $r + \lambda n(1 - \gamma)$: the private rate is always larger. This, in turn, leads the *laissez-faire* economy to underinvest in research $(\hat{x} > x^*)$, which leads to a less than optimal AGR.

The second difference is the presence of the term $(1 - \alpha)$ on the left side of (4.8). This corresponds to a consumer-surplus effect. The private innovator does not internalize the increase in consumer surplus induced by his innovation. This also leads the *laissez-faire* economy to underinvest in research.[15]

The third difference is the presence of the term $(1 - \gamma^{-1})$ on the left side of (4.7). This corresponds to a 'business-stealing' effect. The private innovator does not internalize the loss of surplus to the monopolist whose rents he destroys. In contrast the social planner is aware that by innovating he partially replaces himself. The business-stealing effect leads the *laissez-faire* economy to overinvest in research.

The intertemporal spillover effect will dominate when the size of innovations γ is large – in this case the private reward to research will be small relative to the big social reward – and *laissez-faire* will generate an AGR less than optimal.[16] On the other hand when there is much monopoly power (α close to zero) and innovations are not too large the consumer-surplus effect will be small, and the business-stealing effect will dominate, leading to a greater than optimal AGR under *laissez-faire*.[17]

The same three effects obtain in the general case by comparison between equations (4.5) and (4.6), except that the consumer-surplus effect is now part of a more general *monopolistic distortion effect* which can work in either direction, i.e. toward $\hat{x} \gtrless x^*$. This effect corresponds to the presence, in the left side of (4.6), of the term

$$\frac{\tilde{\pi}(\hat{w})}{\hat{w}},$$

the ratio of profits to the private cost of skilled workers in research, instead of

$$\frac{F(x)}{F'(x)},$$

the ratio of consumption to the social cost of skilled workers in research.

5. Endogenous size of innovations

This section generalizes the analysis of balanced growth developed in sections 3 and 4 by allowing research firms to choose not only the frequency but also the size of innovations. It shows that under *laissez-faire* innovations will be too small. This new effect reinforces the intertemporal spillover effect in tending to make the economy grow too slowly. On the other hand, private firms will tend to overinvest in the speed of innovations which reinforces the business-stealing effect in making the economy grow too fast.

Assume that the arrival rate of innovations to a firm employing the factor combination (z, s) and aiming for innovations of size γ is $\lambda\Phi(z, s)v(\gamma)$, where

$v'(\gamma) < 0$; the bigger the innovation the harder it is to discover. Assume $v''(\gamma) < 0$; the marginal cost (in terms of lower arrival rate) of aiming for larger innovations increases with the size of innovations. Then the product $\gamma v(\gamma)$ is a concave function of γ.

The analysis focuses again on stationary equilibria with positive growth. By the same logic as before, the payoff to the t + 1st innovator will be:

$$V_{t+1} = \frac{A_{t+1}\tilde{\pi}(\hat{\omega})}{r + \lambda\varphi(\hat{n})v(\hat{\gamma})} \tag{5.1}$$

where $\hat{\gamma}$ is the stationary-equilibrium value of γ. If the t + 1st innovation has size γ, not necessarily equal to $\hat{\gamma}$, then $A_{t+1} = \gamma A_t$ and $V_{t+1} = \gamma V_t$. Therefore the expected flow of profits to the research firm in interval t will be:

$$\lambda\varphi(z, s)v(\gamma)\gamma V_t - w_t z - w_t^s s. \tag{5.2}$$

The firm takes V_t as given. Thus its profit-maximizing choice of γ will also maximize the product $v(\gamma)\gamma$. Because this product is a concave function of γ, therefore $\hat{\gamma}$ is defined by the condition:

$$v(\hat{\gamma}) + \hat{\gamma}v'(\hat{\gamma}') = 0 \tag{5.3}$$

The first-order condition for profit-maximization with respect to skilled labor, together with (5.1) produces an expression analogous to (4.6):

$$\lambda\varphi'(N - \hat{x})v(\hat{\gamma})\hat{\gamma}\frac{\tilde{\pi}(\tilde{x}^{-1}(\hat{x}))}{\tilde{x}^{-1}(\hat{x})} = r + \lambda\varphi(N - \hat{x})v(\hat{\gamma}). \tag{5.4}$$

The comparative statics analysis of section 3 carries through unchanged, since $\hat{\gamma}$ is determined by (5.3) independently of all parameters that do not enter the function v, with the obvious exception that it is no longer permissible to investigate the effects of changing γ.

Consider now the social optimal. As in section 4, the expected present value of consumption equals:

$$U = \frac{A_0 F(x)}{r + \lambda\varphi(N - x)v(\gamma)(1 - \gamma)}, \tag{5.5}$$

where the denominator is the social discount rate. Therefore, independently of the choice of x, the social planner will choose γ so as to minimize the expression $v(\gamma)(1 - \gamma)$. The socially optimal value of γ^* is then defined by:

$$-v(\gamma^*) - \gamma^* v'(\gamma^*) + v'(\gamma^*) = 0 \tag{5.6}$$

By concavity of $\gamma v(\gamma)$, we obtain:[18] $\hat{\gamma} < \gamma^*$. Innovations are too small under *laissez-faire*.

This result is another manifestation of the business-stealing effect. The smaller the innovation the larger the losses through obsolescence relative to the gains of innovation. The private firm ignores this cost.

The fact that $\hat{\gamma} < \gamma^*$ is another reason why the *laissez-faire* economy might invest too little in research. The socially optimal level of research employment n^* satisfies the condition:

$$\lambda\varphi'(n^*)v(\gamma^*)(\gamma^* - 1)\frac{F(N - n^*)}{F'(N - n^*)} = r + \lambda\varphi(n^*)v(\gamma^*)(1 - \gamma^*). \tag{5.7}$$

Comparison between (5.4) and (5.7) reveals the same three welfare effects as in the analysis of section 4. In addition, the fact that $\hat{\gamma} \neq \gamma^*$ in itself makes $\hat{n} < n^*$. This is because, as we have seen, $v(\hat{\gamma})(\hat{\gamma} - 1) < v(\gamma^*)(\gamma^* - 1)$. So if the other three effects were absent, then according to (5.7) the effect on research employment would be the same as if the *laissez-faire* economy had a smaller arrival parameter λ, which would reduce $\hat{n}$ below n^*.

The economy's average growth rate: $\lambda\varphi(\hat{n})v(\hat{\gamma})\ ln\ \hat{\gamma}$ is affected by the fact that innovations are too small under *laissez-faire*, although the direction of the overall effect is ambiguous. The direct effect on $ln\ \hat{\gamma}$ is to decrease AGR. The direct effect on the arrival rate $(\lambda\varphi(n)v(\gamma))$ is to increase AGR. But the indirect effect on the arrival rate working through $\varphi(n)$ is to reduce AGR.

6. Strategic monopsony effect

In this section the intermediate monopolist is assumed to take into account its influence on the amount of current research and thereby on the probability of its replacement. In particular, by increasing its demand x_t for skilled labor, beyond the level defined by (2.8) above, the monopolist would raise the wage rate that must also be paid to skilled workers in research. The effect is to reduce the equilibrium amount of research n_t and consequently the arrival rate of the $(t + 1)^{st}$ innovation: $\lambda\varphi(n_t)$. The monopolist will trade this gain off against the higher wages it must pay its own skilled labor.

The analysis focuses on positive balanced growth equilibrium. The monopolist during interval t will choose x_t to maximize the expected present value of profits:

$$x_t = \underset{x}{\operatorname{argmax}} \frac{(p_t - w_t)x_t}{r + \lambda\varphi(N - x)} = V_t$$

$$\text{s.t.: } (*)\ p_t = A_t F'(x)$$

$$(**)\ w_t = \lambda\varphi'(N - x)V_{t+1}$$

where (**) follows from (2.5). The magnitudes x_t, V_t/A_t and w_t/A_t are constant, at the equilibrium values $\bar{x}$, $\bar{V}$ and $\bar{w}$.

Consider first the linear case: $\varphi(n) = n$. Equation (**) becomes:

$$w_t = A_t\bar{w} = \lambda\gamma A_t\bar{V}$$

Therefore:

$$\bar{V} = \max_{x} \frac{[F'(x) - \bar{w}]x}{r + \lambda[N - x]},$$

and $\bar{x}$ must satisfy the first-order condition:

$$F'(\bar{x}) + \bar{x}F''(\bar{x}) = \bar{w} - \lambda\bar{V} = \bar{w}(1 - \gamma^{-1}).$$

Hence:

$$\bar{x} = \tilde{x}((1 - \gamma^{-1})\bar{w})$$

where the function $\tilde{x}$ was defined in section 2. Not surprisingly, for a given productivity–adjusted wage rate $\bar{w}$ the demand for manufacturing labor $\bar{x}$ is larger than when the strategic monopsony effect is not taken into account.

This additional effect will tend to reduce the equilibrium amount of research, thereby reinforcing the intertemporal spillover effect. More formally, the social optimum x^* is still defined by (4.5), whereas $\bar{x}$ satisfies the analogue to (4.6):

$$\lambda\gamma\left[\frac{(1 - \gamma^{-1})F'(\bar{x})}{F'(\bar{x}) + \bar{x}F''(\bar{x})} - 1\right]\bar{x} = r + \lambda(N - \bar{x}).$$

In the Cobb-Douglas case:

$$\lambda\gamma\left[\frac{1 - \alpha - \gamma^{-1}}{\alpha}\right]\bar{x} = r + \lambda(N - \bar{x}), \qquad (6.1)$$

and eq. 4.7:

$$\lambda\gamma\left[\frac{1 - \gamma^{-1}}{\alpha}\right]x^* = r + \lambda(N - x^*)(1 - \gamma).$$

In addition to the three welfare effects pointed out in section 4, the strategic monopsony effect appears through the term $(-\gamma^{-1}/\alpha)$ on the left side of (6.1).

In this linear Cobb-Douglas case the strategic monopsony effect will always dominate: the economy will grow too slowly. But this is not generally true. For example, suppose that $\varphi(n)$ is also Cobb-Douglas, with:

$$\varphi(n) = \frac{1}{1 - \sigma}n^{1-\sigma}, \quad 0 \leqslant \sigma \leqslant 1.$$

When $\sigma > 0$, the wage rate becomes strictly increasing in x:

$$w_t = \lambda(N - x)^{-\sigma}V_{t+1}.$$

This, in turn, will have a double effect on the incumbent firm's present value of profits, which disappears when $\sigma = 0$: the first (direct) effect is to reduce the monopolist's flow of profits since the wage rate increases; the second effect, which goes in the opposite direction, it is to discourage current research and thereby increase the average duration of the incumbent's monopoly power. The first effect turns out to be dominant when the degree of monopoly power is sufficiently large: i.e. when α is sufficiently small. It can be shown that whenever $\sigma > 0$ and α is small enough compared to γ, the business-stealing effect still dominates and the economy grows too fast.

7. Random arrival parameter: creative destruction can destroy growth

This section allows the arrival parameter λ to vary randomly. This generalization illustrates the force of creative destruction by allowing an increase in λ in some states of the world to deter research in other states. In fact, an example is shown where in the limit as the value of λ in one state becomes infinite, the economy's average growth rate falls to zero.

Let $\{\lambda_1, \ldots, \lambda_m\}$ be the finite set of possible values of λ. At the moment of any innovation a new λ is drawn, according to the transition matrix B, and all research firms learn this value. Transition into a high-λ state could represent a fundamental breakthrough that leads to a Schumpeterian wave of innovations, whereas transition to a low state could represent the exhaustion of a line of research.

The stochastic equivalent of balanced growth equilibrium is an equilibrium in which the productivity-adjusted wage rate depends only on the state of the world, not on time. Let A_tV_j be the value of making the tth innovation and moving into state j. Assume the linear case of $\varphi(n) = n$. In any state i, the marginal expected return to research in interval t is

$$\lambda_i \sum_{j=1}^{m} b_{ij} A_{t+1} V_j .$$

This will equal the wage if positive research occurs in state i. If research occurs in all states, the V_i's must satisfy the analogue to (2.6):

$$V_i = \frac{\tilde{\pi}(\lambda_i \gamma \sum_j b_{ij} V_j)}{r + \lambda_i [N - \tilde{x}(\lambda_i \gamma \sum_j b_{ij} V_j)]}, \quad i = 1, \ldots, m.$$

The AGR equals f ln γ, where f is the asymptotic frequency of innovations. Define $n_i \equiv N - \tilde{x}(\lambda_i \gamma \Sigma b_{ij} V_j)$. Then:

$$f = \sum_{i=1}^{m} \lambda_i n_i q_i$$

where q_i is the asymptotic fraction of time spent in state i.

Our example has $m = 2$ and $b_{ij} = 1/2 \ \forall_{ij}$. It is easily verified that in this case:

$$q_1 = 1 - q_2 = \frac{\lambda_2 n_2}{\lambda_1 n_1 + \lambda_2 n_2}$$

$$f = \frac{2\, \lambda_1 \lambda_2\, n_1 n_2}{\lambda_1 n_1 + \lambda_2 n_2}$$

To complete the example, take the Cobb-Douglas case ($F(x) = x^\alpha$) and suppose $\alpha = \gamma^{-1} = 1/2$, and $r = N = \lambda_1 = 1$. Since

$$\bar{x}(\omega) = (\omega/\alpha)^{1/(\alpha-1)}$$

and

$$\tilde{\pi}(\omega) = \left[\frac{1-\alpha}{\alpha}\right]\omega\tilde{x}(\omega),$$

the formula for each V_i can be rewritten as:

$$\begin{cases} V_1 = [16(V_1 + V_2)]^{-1} - \{1 - [4(V_1 + V_2)]^{-2}\}V_1 \\ V_2 = [16\lambda_2(V_1 + V_2)]^{-1} - \lambda_2\{1 - [4\lambda_2(V_1 + V_2)]^{-2}\}V_2 \end{cases}$$

When $\lambda_2 = 1$, the solution to these equations is

$$V_1 = V_2 = \sqrt{\frac{3}{2}}\Big/8,$$

which implies $n_1 = n_2 = 1/3$, and $f = 1/3$. When $\lambda_2 = \infty$, the solution is $V_1 = 1/4$, $V_2 = 3/4$, which implies $n_1 = 0$, $n_2 = 1$, and

$$f = \frac{2n_1n_2}{n_1/\lambda_2 + n_2} = 0.$$

Thus raising the productivity of research in one state can cause research to be discouraged in the other state, through the threat of creative destruction, to such an extent that growth is eliminated.

8. Cycles

This section briefly explores the other perfect foresight equilibria (PFE) to the model; that is, the sequences $\{\omega_t\}_0^\infty$ satisfying the difference equation (3.2) in each interval. The fact that ψ is a decreasing function (Lemma 1 above) rules out complicated periodic PFE such as k-cycles with $k \geqslant 3$ or chaotic PFE. The only possible asymptotic trajectories are fixed points of the second-iterate mapping Ψ^2, which consist of stationary equilibria and 2-cycles. Existence and uniqueness of stationary equilibria have been established. The following discussion focuses on 2-cycles.

A 2-cycle is a PFE that alternates each period between two values ω^0, ω^1, such that:

$$\omega^0 = \Psi(\omega^1),\ \omega^1 = \Psi(\omega^0) \text{ and } \omega^0 \neq \omega^1.$$

If both ω^0 and ω^1 are strictly greater than ω it is called a 'real' 2-cycle. Otherwise it is called a 'no-growth trap' (NGT).

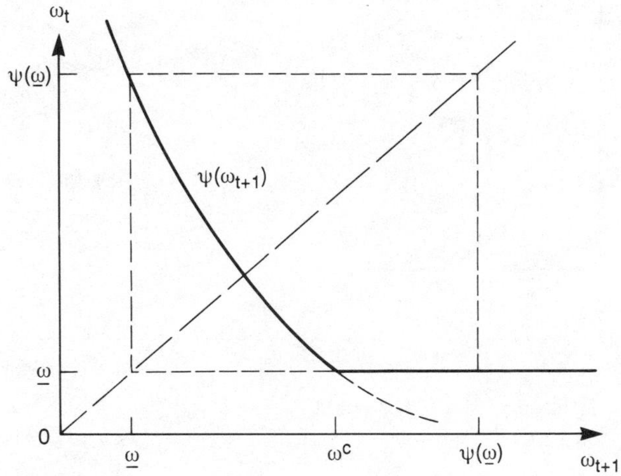

Figure 7.2 No-growth trap

A real 2-cycle is a PFE in which research oscillates between a high and a low value with each innovation. The prospect of high research in odd intervals discourages research in even intervals by (a) increasing likelihood of creative distribution, and (b) raising the wage that must be paid by the monopolist who buys the next patent. Likewise the prospect of low research in even intervals stimulates research in odd intervals. A no-growth trap is the extreme case in which the prospect of high research in odd periods shuts down research entirely in even intervals. Even though NGT defines an infinite sequence $\{\omega_t\}_0^\infty$, the oscillation of the economy will stop after one innovation. From then on the economy will behave as if in stationary equilibrium with no growth.

The existence of NGT is assured if $\underline{\omega} < \omega^c < \psi(\underline{\omega})$, as shown in Figure 7.2, where $\omega^c = \psi^{-1}(\underline{\omega})$.

Provided that skilled labor is necessary in research ($\varphi(0) = 0$), NGT will exist when the rate of interest r is small enough. This is because with a low enough rate of interest and no threat of creative destruction in even intervals the present value of the return to research will be indefinitely large in odd intervals, the prospect of which will shut down research in even intervals.[19]

Note that the smaller is r the easier it is to satisfy conditions (3.5) above; in other words, a no-growth trap is more likely to exist in an economy whose stationary equilibrium exhibits positive growth.

Figure 7.3 shows that a real 2-cycle will exist if there is an NGT and

$$\frac{d}{d\omega}\psi^2(\hat{\omega}) < 1.$$

In the linear case where $\varphi(n) = n$, this will occur if r is small enough and the demand for intermediate input (η)[20] is sufficiently close to 1. The assumption that r is small guarantees the existence of NGT as we have seen. Also:

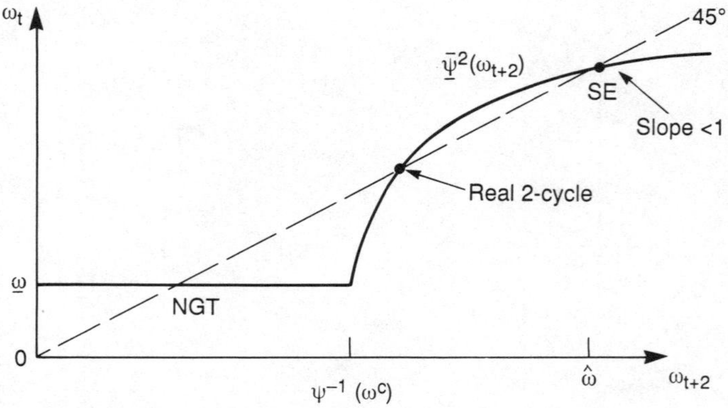

Figure 7.3 Existence of a real 2-cycle

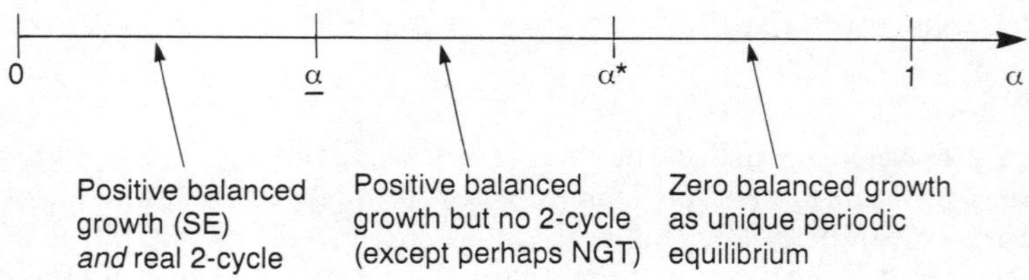

Figure 7.4

$$\psi'(\hat{\omega}) = (\eta - 1)\left[\frac{(1-\eta)/\gamma}{1 - \frac{1}{\eta} - \frac{x\eta}{\eta}} - 1\right] \tag{8.1}$$

(see Appendix 3). This expression approaches zero as η approaches 1. Existence then follows from the fact that:

$$\frac{d\psi^2}{dx}(\hat{\omega}) = \psi'(\omega)\psi'(\psi(\hat{\omega})) = (\psi'(\hat{\omega}))^2.$$

In the Cobb-Douglas case:

$$\eta = \frac{1}{1-\alpha}$$

so there exists a real 2-cycle whenever α and r are both small. We can summarize the conclusions obtained in the Cobb-Douglas case by Figure 7.4. Again α can be interpreted as an inverse measure of the degree of monopoly power in the intermediate industry.

There are several questions that would need to be addressed before judging the likely empirical relevance of these 2-cycles. One is their observability under realistic expectational assumptions; equilibria that are stable under perfect foresight (as in Figure 7.3 above) are often unstable under learning (Grandmont and Laroque, 1986). Another is the size of the resulting output fluctuations; modern economies typically allocate no more than 2 or 3% of their labor force to research. However, even if the movement of labor between manufacturing and research would cause small proportional fluctuations in the level of output it could produce large proportional fluctuations in research, and hence in the growth rate of output.[21]

9. Conclusion

The chapter has presented a model of economic growth based on Schumpeter's process of creative destruction. Growth results exclusively from technological progress, which in turn results from competition among research firms that generate innovations. Each innovation consists of a new intermediate good that can be used to produce final output more efficiently than before. Research firms are motivated by the prospect of monopoly rents that can be captured when a successful innovation is patented. But those rents in turn will be destroyed by the next innovation, which will render obsolete the existing intermediate good.

It would be useful to generalize and extend the analysis in several directions, such as allowing the size of innovation eventually to fall, so that technology is ultimately bounded. Also, a richer intersectoral structure would be useful to study the positive dynamics of diffusion. The model would gain richness and realism if capital were introduced, either physical or human capital embodying technical change, or R&D capital that affects the arrival rate of innovations. Allowing unemployment, by introducing search into the labor market, would facilitate study of the reciprocal interaction between technological change and the business cycle. All these extensions seem feasible because of the simplicity and transparency of the basic model.

Appendix 1: Assumption A.1 is satisfied in the CES case

Let $F(x) = (x^\rho + M^\rho)^{1/\rho}$, where $0 < \rho < 1$.

We have:

$$F' = 1/\rho(x^\rho + M^\rho)^{1/\rho-1}\rho x^{\rho-1} = x^{\rho-1}(x^\rho + M^\rho)^{1/\rho-1}$$

Hence:

$$\begin{aligned} F'' &= (\rho - 1)x^{\rho-2}(x^\rho + M^\rho)^{1/\rho-1} + x^{\rho-1} \cdot (1/\rho - 1)(x^\rho + M^\rho)^{1/\rho-2}\rho x^{\rho-1} \\ &= (\rho - 1)(x^\rho + M^\rho)^{1/\rho-2}[x^{\rho-2}x^\rho + x^{\rho-2}M^\rho - x^{2\rho-2}] \\ &= (\rho - 1)(x^\rho + M^\rho)^{1/\rho-2}x^{\rho-2}M^\rho < 0 \end{aligned}$$

and:

$$\begin{aligned}\frac{1}{M^{\rho}}F''' &= (p - 1)(1/\rho - 2)(x^{\rho} + M^{\rho})^{1/\rho-3}\rho x^{\rho-1}x^{\rho-2} \\ &\quad + (\rho - 1)(x^{\rho} + M^{\rho})^{1/\rho-2}(\rho - 2)x^{\rho-3} \\ &= (\rho - 1)(x^{\rho} + M^{\rho})^{1/\rho-3}[(1 - 2\rho)x^{2\rho-3} + (\rho - 2)x^{2\rho-3} + \\ &\quad M^{\rho}x^{\rho-3}(\rho - 2)] \\ &= (\rho - 1)x^{\rho-3}(x^{\rho} + M^{\rho})^{1/\rho-3}[(\rho - 2)M^{\rho} - (1 + \rho)x^{\rho}].\end{aligned}$$

Therefore:

$$\begin{aligned}2F'' + xF''' &= (x^{\rho} + M^{\rho})^{1/\rho-3}M^{\rho}(\rho - 1)[2x^{\rho-2}(x^{\rho} + M^{\rho}) \\ &\quad + x^{\rho-2}[(\rho - 2)M^{\rho} - (1 + \rho)x^{\rho}]] \\ &= (\rho - 1)M^{\rho}(x^{\rho} + M^{\rho})^{1/\rho-3}[x^{\rho-2}][x^{\rho}(1 - \rho) + \rho M^{\rho}] < 0\end{aligned}$$

Appendix 2: Aggregate demand spillovers

This appendix adopts the alternative interpretation of the model, where innovations improve upon the quality of consumption goods (see section 2e above), to consider the case of a multiplicity of goods subject to quality improvements, and to analyze the additional welfare effects that might result from the existence of demand spillovers between goods.

Assume a continuum of consumption goods x_i where $i \in [0, 1]$. Instantaneous utility in interval t is:

$$u_t = A_t\, G\left(\int_0^1 F(x_{it})di\right)$$

where F and G are respectively homogeneous of degree k and α/k, with $0 < k < \alpha < 1$. The quality parameter A_t applies to all consumption goods: this implicitly assumes that a quality improvement in some final sector i_0 spills over to all other sectors $i \neq i_0$. The polar case, where quality improvements in different sectors are not correlated, is analyzed in Grossman-Helpman (1989). For notational simplicity assume the linear case: $\varphi(n) = n$.

Again, choose instantaneous utility as the numéraire. Then given competitive markets the price of any bundle of goods must equal the marginal utility of that bundle; that is, during interval t the price of each good i is:

$$p_{it} = A_t G'(G^{-1}(u_t/A_t))F'(x_{it}).$$

Given homogeneity this can be expressed as:

$$p_{it} = \alpha A_t x_t^{\alpha-k} x_{it}^{k-1}$$

when all other firms are producing at the level x_t. Thus an increase in all other firms' output shifts firm i's demand curve to the right, given that $\alpha - k > 0$. This is the aggregate demand spillover. (Under the alternative intermediate-goods interpretation it would be a technological complementarity.)

As in the Cobb-Douglas case analyzed above, each consumption-good monopolist faces a demand curve with constant elasticity. In this case the elasticity equals $(1 - k)^{-1}$. Profit maximization implies that the share of profits in each sector will equal $1 - k$. Thus in a symmetric equilibrium profits in each sector will equal:

$$\pi_t = \left[\frac{1 - k}{k}\right] w_t x_t$$

Assume that a positive balanced growth equilibrium exists, with $w_{t+1} = \gamma w_t$ and $x_t = \hat{x}$. As before, the necessary condition for equilibrium in research is:

$$w_t = \frac{\lambda \pi_{t+1}}{r + \lambda[N - x_{t+1}]}$$

Combining the last two equations with the conditions defining stationarity yields:

$$\lambda\gamma\left[\frac{1 - k}{k}\right]\hat{x} = r + \lambda(N - \hat{x}). \tag{1*}$$

This equation determines the level of research $(N - \hat{x})$ in balanced growth, and it is the same as the earlier equation (4.8) for the Cobb-Douglas case with α replaced by k. All the previous positive results then carry through unchanged.

By reasoning analogous to that of section 4 above, a social planner would choose a common level of output x^* in all sectors at all times, so as to maximize:

$$\frac{A_0 G(F(x))}{r + \lambda(N - x)(1 - \gamma)} = \frac{A_0 x^\alpha}{r + \lambda(N - x)(1 - \gamma)}.$$

This is the same maximand as in the Cobb-Douglas case section 4, so again the optimal x^* must solve (4.7):

$$\lambda\gamma(1 - \gamma^{-1})x^*/\alpha = r + \lambda(N - x^*)(1 - \gamma).$$

If k were instead equal to α, then (1*) would be the same equation as (4.8), and the welfare analysis would be the same as in section 4, with the same three effects (intertemporal spillover, consumer's surplus and business-stealing). The fact that $k < \alpha$ thus introduces a fourth effect by making $\hat{x}$ smaller than it would otherwise be, without affecting x^*. This is similar to the effect of aggregate-demand externalities in recent macroeconomic theory (for example, Hart, 1982). Private manufacturing firms do not internalize the positive externalities they confer on each other by producing more. By reducing $\hat{x}$ this effect tends to lead a *laissez-faire* economy to overinvest in research. It thus reinforces the business-stealing effect in tending to produce a less than optimal AGR under *laissez-faire*.

This is contrary to the effect of aggregate-demand externalities in the model of Shleifer and Vishny (1988), which opposes the business-stealing effect. In that model innovation is a means of creating more monopolies in the economy. In this model innovation is an alternative to exploiting monopoly rents.

Appendix 3: Derivation of (8.1)

We have:

$$\psi'(\hat{\omega}) = \frac{d}{d\omega}\left\{\frac{\lambda\gamma\tilde{\pi}(\omega)}{r + \lambda[N - \tilde{x}(\omega)]}\right\}\Bigg|_{\omega=\hat{\omega}}.$$

Since $\hat{\omega} = \psi(\hat{\omega})$:

$$\psi'(\hat{\omega}) = \hat{\omega}\left\{\frac{\tilde{\pi}'(\hat{\omega})}{\tilde{\pi}(\hat{\omega})} + \frac{\lambda\tilde{x}'(\hat{\omega})}{r + \lambda[N - \tilde{x}(\hat{\omega})]}\right\}$$

$$= \frac{\hat{\omega}}{\tilde{\pi}(\hat{\omega})}\{\tilde{\pi}'(\hat{\omega}) + \tilde{x}'(\hat{\omega})\hat{\omega}/\gamma\}.$$

Since $\tilde{\pi}' = -\tilde{x}$:

$$\psi'(\hat{\omega}) = \frac{\hat{\omega}\tilde{x}(\hat{\omega})}{\tilde{\pi}(\hat{\omega})}\{-1 - \eta^x_\omega/\gamma\},$$

where $\eta^x_\omega = -\hat{\omega}\tilde{x}'(\hat{\omega})/\tilde{x}(\hat{\omega})$. It follows from the first-order condition (2.8) that $\omega\tilde{x}(\omega)/\tilde{\pi}(\omega) = \eta - 1$. Also, (2.8) can be rewritten as:

$$F'(x)\left[1 - \frac{1}{\eta(x)}\right] = \omega$$

Differentiate this w.r.t. ω, to reexpress η^x_ω as:

$$\eta^x_\omega = \frac{\eta - 1}{1 - 1/\eta - \dfrac{x\eta'}{\eta}}.$$

Hence:

$$\psi'(\hat{\omega}) = (\eta - 1)\left[\frac{(1 - \eta)/\gamma}{1 - 1/\eta - \dfrac{\hat{x}\eta'}{\eta}} - 1\right].$$

Notes

1. See, for example, Romer (1986). Shell (1973) emphasized the importance of imperfect competition, and derived an endogenous growth model with no obsolescence or uncertainty but with capital accumulation.
2. Stokey's (1988) interesting analysis generates obsolescence but because it assumes

perfect competition it has no research. Segerstrom, Anant and Dinopoulos have a multisectoral model of obsolescence but with no uncertainty in the length of patent races.

3. King and Rebelo (1988) have a similar analysis, in which the payoff to acquiring human capital is random, in a perfectly competitive model with no obsolescence.
4. Mention should also be made of the preliminary work of Corriveau (1988), who is developing a similar model of growth and cycles in a discrete time setting. The discreteness of time in Corriveau's model introduces complications from the possibility of simultaneous innovations.
5. Scherer (1984) combines process- and input-oriented R&D into a measure of 'used' R&D, which he distinguishes from pure product R&D. He estimates that during the period 1973–8 in US industry the social rate of return to 'used' R&D lay between 71% and 104%, whereas the return to pure product R&D was insignificant.
6. Gradual diffusion could be introduced by allowing the cost parameter after each innovation to follow a predetermined but gradual path asympotically approaching the limit A_t, and then to jump to A_t upon the next innovation and follow a gradual path approaching A_{t+1}. This would produce a cycle in research within each interval, as the gradual rise in productivity induces manufacturing firms to hire more and more workers out of research until the next innovation occurs.
7. It is straightforward to relax the assumption of complete spillover by allowing the tth innovator to have an advantage in finding the $t + 1^{st}$ innovation; to have an arrival parameter $\lambda' > \lambda$.
8. This analysis of research is derived from the patent-race literature surveyed by Reinganum (1987) and Tirole (1988, ch. 10). Within that literature the paper closest to this chapter is Reinganum (1985), which emphasizes the affinity to creative destruction. Most other papers do not have the sequence of races necessary for creative destruction. The present chapter goes beyond that literature by embedding the sequence of races in a general-equilibrium setting. Thus (a) the flow of profits π_{t+1} resulting from an innovation will depend upon the amount of research during the next interval, (b) the welfare criterion is the expected lifetime utility of the representative consumer, (c) a stochastic process is derived for GNP, and (d) appendix 2 analyzes an aggregate-demand spillover. In addition, the analysis in section 5 of the endogenous determination of the size of innovations, and the analysis of cycles in section 8, go beyond anything found in the patent-race literature.
9. That is, the strategic influence that an intermediate monopolist can have on current research through raising its demand for skilled labor is provisionally ignored. In section 6 this influence is referred to as the 'strategic monopsony effect' of the incumbent intermediate firm: this firm can indeed use its size in the market for skilled labor in order to raise its supply price in research and thereby reduce the probability of being replaced with a new innovation.

   Unlike the other effects to be analyzed in the paper, this 'strategic effect' becomes negligible once a multiplicity of intermediate sectors is introduced. (See Grossman-Helpman (1989) for a multisectorial extension of the present model). In other words, the monopsony effect appears to be relatively unimportant from a macroeconomic point of view. Furthermore, as section 6 will show, taking this additional effect into account does not invalidate our main results except in one particular case.
10. This means, however, that in the case of non-drastic innovations the successful innovator and previous incumbent would be in a situation of bilateral monopoly. Thus the analysis of research in the previous section would implicitly give all the bargaining power to the innovator. Neither of these implications would follow under the alternative assumption of a fixed reentry cost.

11. The existence and uniqueness of such balanced growth equilibrium is established in the next section.
12. From (2.4), (2.10) and the fact that $D_t = V_t$:

$$R_t V_t + \lambda\varphi(\hat{n})V_{t+1} = \pi_t$$

From this, (2.6), and the fact that $\pi_{t+1} = \gamma\pi_t$:

$$R_t\left[\frac{\pi_t}{r + \lambda\varphi(\hat{n})}\right] + \lambda\varphi(\hat{n})\gamma\left[\frac{\pi_t}{r + \lambda\varphi(\hat{n})}\right] = \pi_t.$$

Equation (2.11) follows immediately.
13. The existence of nonstationary PFE is discussed in section 8 below.
14. In the extension of the model considered in section 7 the second effect may dominate.
15. Two additional kinds of spillover could easily be included. First, researchers could benefit from the flow of others' research, so that an individual firm's arrival rate would be a constant returns function of its own and others' research. Second, there could be an exogenous Poisson arrival rate μ of imitations that costlessly circumvent the patent laws and clone the existing intermediate good. Both would have the effect of lowering AGR. Also, as shown in Aghion and Howitt (1988), the inclusion of μ would introduce another source of cycles in the economy, since each imitation would make the intermediate industry perfectly competitive, which would raise manufacturing employment, until the next innovation arrives.
16. From (4.7), as γ rises to the upper limit $1 + r/\lambda N$, x^* approaches zero, whereas from (4.8) $\hat{x}$ approaches a strictly positive limit.
17. If $\frac{1}{\alpha} > 1 + r/\lambda N$, then $\hat{n} > 0$ for all γ, whereas $n^* = 0$ when $\gamma \leqslant 1 + \alpha r/\lambda N$.
18. Since $v' < 0$, (5.6) implies that $v(\gamma)\gamma$ is locally decreasing at γ^*, so that γ^* exceed the point $\hat{\gamma}$ at which $v(\gamma)\gamma$ is maximized.
19. By the definitions of ω^c, ω and ψ:

$$\frac{\omega}{\varphi'(0)} = \frac{\lambda\gamma\tilde{\pi}(\omega^c)}{r + \lambda\varphi[N - \tilde{x}(\omega^c)]}.$$

It follows that when $r = 0$:

$$\varphi[N - \tilde{x}(\omega^c)] = \gamma\tilde{\pi}(\omega^c)\varphi'(0)/\omega > 0,$$

whereas:

$$\varphi[N - \tilde{x}(\omega)] = 0;$$

from these results and the fact that φ is increasing and $\tilde{x}$ decreasing, $\omega^c > \omega$. Also, as $r \rightarrow 0$, $\psi(\omega) \rightarrow \infty$; otherwise

$$\frac{\psi(\omega)}{\varphi'[N - \tilde{x}(\psi(\omega))]} \text{ remains bounded while } \frac{\lambda\gamma\tilde{\pi}(\omega)}{r + \varphi[N - \tilde{x}(\omega)]} = \frac{\lambda\gamma\tilde{\pi}(\omega)}{r} \rightarrow \infty,$$

in contradiction to the definition of ψ.

20.

$$\eta(x) \equiv -\frac{F'(x)}{xF''(x)}.$$

21. The cycles can be distinguished from those of Shleifer (1986) by noting that innovations, the key endogenous variable here, are exogenous in Shleifer, that the aggregate-demand spillovers central to Shleifer's cycle are absent here, and that Shleifer has no creative destruction. Also the cycles of Deneckere and Judd (1986) arise from unique dynamic equilibria because of instability of stationary equilibria, rather than from multiple equilibria.

References

Aghion, Philippe and Howitt, Peter (1988), 'Growth and cycles through creative destruction', Unpublished, University of Western Ontario.

Azariadis, Costas (1981), 'Self-fulfilling prophecies', *Journal of Economic Theory*, 25, December, pp. 380–96.

Campbell, J. and N.G. Mankiw (1987), 'Are output fluctuations transitory?', *Quarterly Journal of Economics*, 102, November, pp. 857–80.

Cochrane, John H. (1988), 'How big is the random walk in GNP?', *Journal of Political Economy*, 96 October, pp. 893–920.

Corriveau, Louis (1988), 'Entrepreneurs, growth, and cycles', unpublished, University of Western Ontario.

Deneckere, Raymond J., and Kenneth L. Judd (1986), 'Cyclical and chaotic behavior in a dynamic equilibrium model, with implications for fiscal policy', unpublished, Northwestern University.

Grandmont, Jean-Michel and Guy Laroque (1986), 'Stability of cycles and expectations', *Journal of Economic Theory*, 40, October, pp. 138–51.

Griliches, Zvi (1979), 'Issues in assessing the contribution of research and development in productivity growth', *Bell Journal of Economics*, 10, Spring, pp. 92–116.

Grossman, Gene and Elhanan Helpman (1989), 'Quality ladders in the theory of growth', unpublished.

Hart, Oliver (1982), 'A model of imperfect competition with Keynesian features,' *Quarterly Journal of Economics*, 97, pp. 109–38.

Hausman, Jerry A., Bronwyn Hall and Zvi Griliches (1984), 'Econometric models for count data with an application to the patents–R and D relationship', *Econometrica*, 52, July, pp. 909–38.

King, Robert G., and Sergio T. Rebelo (1988), 'Business Cycles with Endogenous Growth,' unpublished, University of Rochester.

Lerner, Abba P. (1934), 'The Concept of Monopoly and the Measurement of Monopoly Power,' *Review of Economic Studies* 1, June, pp. 157–175.

Nelson, Charles and Charles Plosser (1982), 'Trends and Random Walks in Macroeconomics,' *Journal of Monetary Economics* 10, Sept., pp. 139–62.

Pakes, Ariel (1986), 'Patents as Options: Some Estimates of the Value of Holding European Patent Stocks,' *Econometrica* 54, July, pp. 755–84.

Prescott, Edward (1986), 'Theory Ahead of Business Cycle Measurement,' Federal Reserve Bank of Minneapolis Research Department Staff Report 102, February.

Reinganum, Jennifer (1985), 'Innovation and Industry Evolution,' *Quarterly Journal of Economics* 100, February, pp. 81–99.

Reinganum, Jennifer (1987), 'The Timing of Innovation: Research, Development and Diffusion,' unpublished, University of Iowa.

Romer, Paul M. (1986), 'Increasing Returns and Long-Run Growth.' *Journal of Political Economy* 94, October, pp. 1002–1037.

Romer, Paul M. (1990), 'Endogenous Technological Change,' *Journal of Political Economy* forthcoming.

Scherer, F.M. (1980), *Industrial Market Structure and Economic Performance*. 2nd ed. Chicago: Rand McNally.

Scherer, F.M. (1984), *Innovation and Growth: Schumpeterian Perspectives*. Cambridge, MA: MIT Press.

Schumpeter, Joseph A. (1934), *The Theory of Economic Development*. New York: Oxford University Press.

Schumpeter, Joseph A. (1942), *Capitalism, Socialism and Democracy*. New York: Harper and Brothers.

Segerstrom, Paul S., T.C.A. Anant and Elias Dinopoulos (1987), 'A Schumpterian model of the product life cycle', Econometrics and Economic Theory Paper no. 8606, Michigan State University.

Shell, Karl (1973), 'Inventive activity, industrial organization and economic activity', in *Models of Economic Growth*, J.A. Mirrlees and N.H. Stern (eds), London, Macmillan, pp. 77–96.

Shleifer, Andrei (1986), 'Implementation cycles', *Journal of Political Economy*, 94, December, pp. 1163–90.

Shleifer, Andrei and Robert Vishny (1988), 'The efficiency of investment in the presence of aggregate demand spillovers', *Journal of Political Economy*, pp. 1121–31.

Stokey, Nancy (1988), 'Learning by doing and the introduction of new goods,' *Journal of Political Economy*, 96, August, pp. 701–17.

Tirole, Jean (1988), *The Theory of Industrial Organization*. Cambridge, Mass., MIT Press.

Woodford, Michael (1986), 'Stationary sunspot equilibria,' unpublished, University of Chicago.

8 National effects of learning, international specialization and growth paths

Bruno Amable

1. Introduction

With the recognition of phenomena such as imperfect competition and increasing returns to scale, renewed interest has recently begun to be shown in the determinants of international specialization and the latter's impact on a country's economic growth. The latest theories in international economics to some extent challenge the belief that there are necessarily advantages to be gained from opening up to international trade and from free trade (Krugman, 1986). In particular, the specific nature of certain 'strategic' sectors can in some circumstances, and contrary to the most elementary canons of conventional international economics, justify the use of protective measures against international competition. The type of international specialization involved is liable to have a long-term impact on a country's pattern of growth (Lucas, 1988), which means that the choice of international specialization cannot be confined to a static choice in favour of those sectors in which the country concerned is most competitive at a particular point in time. It is essential to take account of dynamic criteria relating to developments with respect to foreign demand, technical change or market structures.

A country's technological specialization raises the same problems because of the specific characteristics of R&D activities and high-tech sectors: high fixed costs, market imperfections, increasing returns, strategic nature, the role of a small number of activities in spreading technical change to all sectors of the economy, the strong growth of world demand for these products, beneficial externalities linked to technology, etc. While static resource allocation criteria may suggest that specialization based on the sectors be rejected, long-term growth considerations may prompt the reverse conclusion (Ros, 1986). So a country anxious to become established on a strong growth path in the long term must direct its specialization towards this type of activity. The link between technology, competitiveness and growth is in this case simple: specialization in activities with a high technology content and with strong potential demand guarantees overall industrial competitiveness and rapid long-term growth. In the context of the information technologies paradigm, it would seem a priori preferable to have production and export structures which favour the few sectors concerned with the technologies which generate that paradigm (electronics, data processing, etc.), sectors which appear to satisfy the dynamic specialization criteria referred to above (high income elasticity, considerable scope for technical progress, etc.[1]).

A number of objections regarding this view of international specialization have been voiced however, particularly in the context of the debate about France's need for industrial restructuring and a shift in the focus of foreign trade. Analyses of France's international specialization have revealed two things. First, French industry is 'unbalanced' (Mathis, Mazier and Danset, 1988) *vis-à-vis* that of its main competitors, particularly Germany, in the sense that industrial transformation is moving towards foreign trade and production structures which are too much like those of the main industrial countries, with France at a competitive disadvantage. Here, it is the overly 'Manichean' nature of industrial restructuring which is challenged, leading as it does to losses of market shares in the neglected traditional sectors without any significant offsetting gains in the spearhead sectors. Second, it has become apparent that France is not sufficiently involved in activities where world demand is growing rapidly, or in trade with high-growth areas (CEPII, 1984; Lafay and Herzog, 1989). Its exports are still concentrated on not very dynamic sectors and partner areas.

Conflicting conclusions as regards the right sort of industrial policy may be drawn from these two findings. A country may, for example, consider not specializing in leading-edge sectors with a high technology content and strong growth of world demand, and decide instead to aim for market segments in more traditional sectors where competition from the industrial countries is less keen. By gaining experience in these sectors and modernizing their equipment, it is possible to improve relative competitiveness and win market shares. It may, on the other hand, be thought that the only appropriate specialization for an industrial country must be based on long-term criteria which involve choosing activities with high income elasticity and significant potential for technical progress – witness the example of Japan (Freeman, 1987; Komiya, Okuno and Suzumura, 1988). Any specialization which neglected these sectors would tend to steer the country towards slow growth in the long term, there being a limit to the market shares that can be won in sectors in relative decline.

The French situation provides an illustration of this problem via the various attempts made to outline an industrial policy: should specialization focus on activities benefiting from strong growth of world demand and with a high technology content, so as to enjoy the favourable effects of a buoyant market (which would be consistent with the long-term criteria of international specialization), or is it preferable to consolidate one's position on more traditional markets and benefit from market share gains deriving from greater competitiveness (i.e. in line with the short-term specialization criteria)? While Japan is an example of a country which has moved towards rapid growth of high-tech activities, Italy has chosen the opposite course, i.e. it has opted in favour of the traditional sectors (clothing and textiles) where it enjoys a competitive advantage. Both these countries recorded substantial industrial surpluses over the period 1967–87. France, which is broadly unspecialized and which began to have an industrial deficit towards the late 1980s, falls between these two examples (Amable and Mouhoud, 1990). However, the choice of activities in which a country specializes is but one aspect of the broader question of structural competitiveness (Chesnais, 1986). Specific national factors, institutional in particular, also decisively influence the national growth path by virtue of the way they determine the interraction between technology and competitiveness. So it is all the factors comprising the 'national system of innovation' (Freeman, 1988; Lundvall, 1988; Nelson, 1988) that have to be taken into consideration. There are, for example, several national ways of adapting to a given technological paradigm, and they will not all have the same consequences in terms of growth paths (Cohen and Zysman, 1989).

The aim of this chapter is to present a simple model by which to define the relations between the different determinants of competitiveness and the growth/technological change relationship. The model, which is of post-Keynesian inspiration, is presented in section 2. The properties of equilibrium and the conditions for stability are discussed in section 3, while section 4 sets out industrial and technology policy conclusions and recommendations for improving overall competitiveness.

2. A growth and international specialization model with endogenous technical change

The following model is a post-Keynesian model of growth in an open economy, based on the balance-of-payments constraint (Dixon and Thirlwall, 1975; Thirlwall, 1979). As McCombie (1989) emphasizes, it is implicit in this approach that an economy's long-term growth rate is determined not by the growth (exogenous) of production inputs, but by the growth of demand (foreign in particular) for the country's output, the supply of inputs being a consequence of the level of activity. The factors relating to supply are apparent in the relative competitiveness of the country considered, which determines its export and import volumes. The need to comply with the external constraint dictates what level of activity the country can sustain over the long term. The country's relative competitiveness is determined endogenously by the conditions governing factor payments, the law of technical progress and how the 'quality' of the country's international specialization evolves. The situation of the country concerned here is that of a small, open economy, while the variables for the rest of the world are assumed to be exogenous. The fact that the behaviour of the rest of the world is not built into the model means that the reverse effect of the country's growth on foreign demand for its goods is not taken into account, that effect being assumed to be negligible, but it does avoid the symmetry of a two-country world economy.

The export function of this model takes account of factors representing non-price competitiveness. The latter is considered as a function of the quality of exported products. This quality is a specific factor which plays a role in the trade flows and is a major influence on the non-price competitiveness of a country. Quality is distinct from other characteristics of goods, particularly from the growth of world demand for these goods. Traditionally, the income elasticity is meant to represent the bulk of 'non-price' factors of a country's competitiveness. As we wish to isolate the specifically national factors of non-price competitiveness, we include an extra factor in the export function. The function featured in this model distinguishes between two types of factors of non-price competitiveness. This competitiveness depends first on the sort of products that are being exported, the sectoral specialization of trade. This aspect is reflected in the value of the income elasticity. To a certain sectoral specialization corresponds a certain rate of growth of the world demand addressed to the country, function of the growth of world income and the income elasticity of the country's exports. The higher this elasticity, the more specialized in products for which world demand is strong, that is for products at the beginning of their life cycle, is the country.

Other things being equal, two countries having the same sectoral specialization will benefit to the same extent from world growth. One can then expect equal values for the income elasticity of their exports. To the sectoral specialization, which does not account for specifically national factors, one can add the quality of products, which is a function of national determinants. Two countries that would be identically specialized for their exports could differ as regards the level of quality of fabricated goods, the national pattern of production exerting an influence on it. If we take a particular product, a computer for instance, a French and an American computer will not be equivalent. Each product will have its own characteristics, which are represented in the quality variable in this model.

The quality of products changes in an endogenous way, as a function of the efforts made to improve it, which are represented as learning effects. These effects encompass simple learning by doing, by which experience in the making of a product allows an improvement of its characteristics, as well as more elaborate types of learning referring to various interactions in the production process that lead to an internal transformation of this process and an improvement in the level of competence of workers. We are close to the preoccupations of Arrow (1962) and Kaldor (1957), who wanted to take into account the endogenous part of technological change. The improvements in quality appear then as by-products of the production of goods, the strength of these effects depending on the characteristics of the national pattern of production. Learning effects related to the production process play an important role, but other elements intervene, such as the notoriety in the commercializing of goods (the image and reputation of products and sellers).

Quality is thus expressed as a function of cumulated production, in order to take account of all the effects of experience and learning related to producing and selling goods. Quality is built up slowly, as experience is amassed and learning is accomplished on techniques and goods, within the sectoral specialization that the country has adopted. National parameters that characterize the determination of the quality of goods and its effects on the pattern of external growth are first the quality-elasticity of exports and second the elasticity of quality with respect to accumulated experience.

The former type of elasticity reflects the conditions in which a country's products – of a given quality – are marketed abroad. It shows what impact an actual improvement in quality has on exports, and is therefore a function of efforts on the sales side such, for instance, as advertising, and of the reputation of the country's products and suppliers and, to some degree, the sectoral focus of production. The second type of elasticity is an expression of the link between the experience gained with domestic output and the improvement of product quality. It thus reflects conditions which are part of the national system of production and of the integration of the new technological paradigm. It is a function of the characteristics of domestic production, the conditions of technical learning, the level of skill of the labour force, etc. It is therefore a parameter specific to the mode of domestic production. Sectoral specialization is also partially represented via the possibilities of product quality improving. It might be assumed that products in the initial stages of their life cycle have greater scope for quality improvement than products at the end of the cycle.

We adopt the following notations for the model. Capital letters refer to levels, small letters to rates of growth:

X-exports M-imports Y*-GNP of the rest of the world P-level of prices in the country P*-level of prices in the rest of the world Ω-quality of exports Q-GNP of the country K-cumulative output W-wage rate A-productivity l-Kaldor-Verdoorn coefficient ($0 < 1 < 1$) n-autonomous trend of technical progress (>0) η-price-elasticity of exports (<0) ν-quality-elasticity to cumulative production (>0) β-price-elasticity of imports (>0) α-income-elasticity of imports (>0) λ-quality-elasticity of exports (>0) ε-income-elasticity of exports (>0)

The model's basic equations are thus:

$$X = Y^{*\varepsilon}\left(\frac{P}{P^*}\right)^{\eta}\Omega^{\lambda} \tag{1}$$

$$M = Q^{\alpha}\left(\frac{P}{P^*}\right)^{\beta} \tag{2}$$

The dynamic equilibrium of the balance of payments is expressed as the rate of growth of exports being equal to the rate of growth of imports.

$$x = m \tag{3}$$

i.e.

$$\alpha q + (\beta_{-1})(p - p^*) = \varepsilon y^* + \eta(p - p^*) + \lambda\omega \tag{4}$$

Prices are assumed to be calculated on the basis of a mark-up added to unit costs and, for the sake of simplicity, the mark-up is assumed to remain constant. The rate of increase in domestic prices is thus the difference between the rate of increase in wages and the rate of growth of productivity.

$$p = w - a \tag{5}$$

a is the rate of growth of productivity.

Regarding the growth of the wage rate, a simple rule is adopted in which wages are indexed on productivity, allowing for a growth of the nominal wage.[2]

$$w = \mu a \tag{6}$$

The evolution of the real wage takes account of changes in the level of domestic prices. Expressed with regard to domestic prices only, the real wage is deduced from equation (5):

$$w - p = a \tag{7}$$

the real wage increases at the same rate as productivity.

Technical change expressed through changes in productivity is determined endogenously. Returns to scale are assumed to be increasing and productivity growth is specified as a function of the growth of production, according to a Kaldor-Verdoorn type linear formula. We have here a form sufficiently simple and general for representing a large range of learning effects and increasing returns usually referred to as 'dynamic economies of scale'.

$$a = n + lq \tag{8}$$

Export quality is a function, assuming constant elasticity, of accumulated experience represented by aggregate production.

$$\Omega = K^{\nu} = \left[\int_0^t Q(\tau)d\tau\right]^{\nu} \tag{9}$$

The balance-of-payments constraint can thus be expressed as:

$$g = \frac{\lambda\nu}{\alpha + (\beta - \eta - 1)(\mu - 1)l}k + \frac{ey^* + (\beta - \eta - 1)[p - n(\mu - 1)]}{\alpha + (\beta - \eta - 1)(\mu - 1)l} \tag{10}$$

Since q is the rate of growth if instantaneous production and k the rate of growth of cumulative production, this gives:

$$q = k + \frac{\dot{k}}{k} \tag{11}$$

Equation (9) can thus be expressed as

$$\dot{k} = Ck^2 + Bk \tag{12}$$

if $k \neq 0$
with:

$$C = \frac{\lambda\nu}{\alpha + (\beta - \eta - 1)(\mu - 1)l} - 1 \tag{13}$$

$$B = \frac{(\beta - \eta - 1)[p^* - n(\mu - 1)] + ey^*}{\alpha + (\beta - \eta - 1)(\mu - 1)l} \tag{14}$$

C and B can be positive or negative.

There are therefore four possible cases (see Figures 8.1a–d):

(1) $C < 0$ and $B > 0$
(2) $C > 0$ and $B < 0$

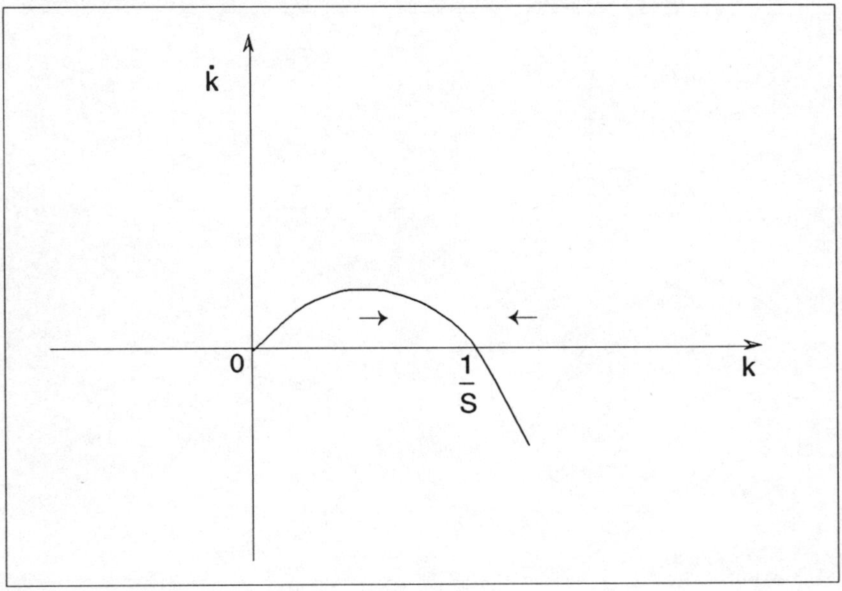

Figure 8.1a C < 0 and B > 0

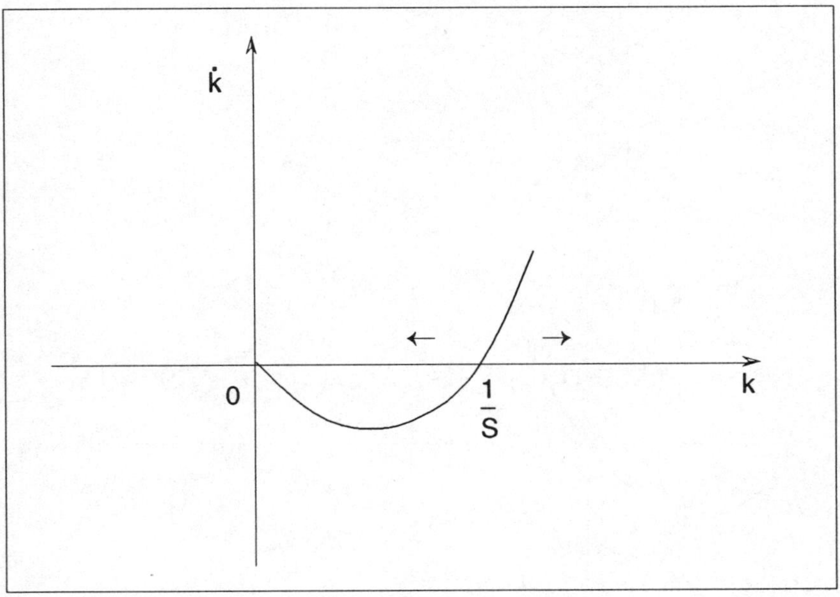

Figure 8.1b C > 0 and B < 0

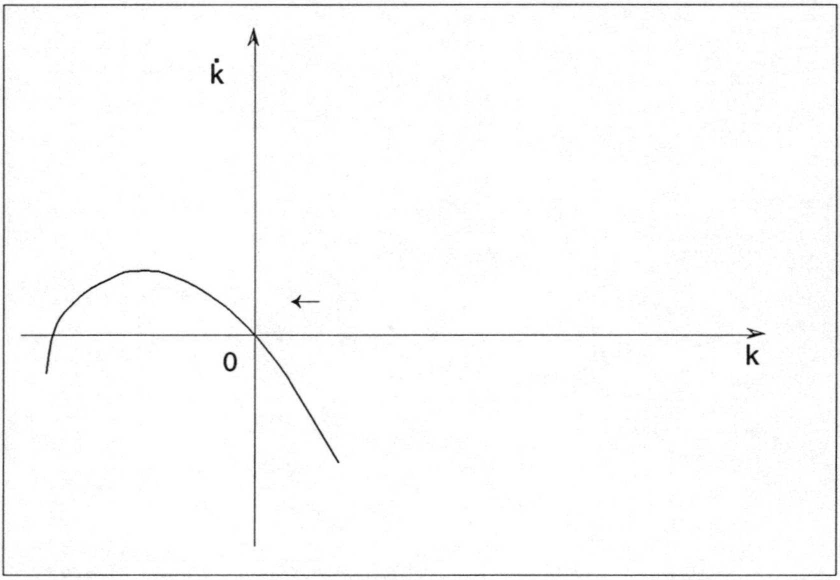

Figure 8.1c C < 0 and B < 0

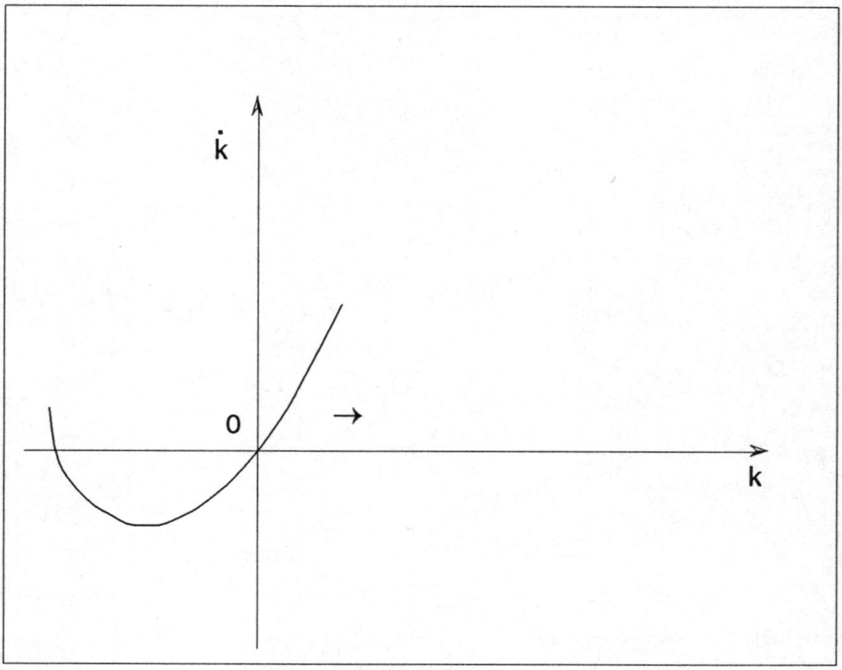

Figure 8.1d C > 0 and B > 0

(3) $C > 0$ and $B > 0$
(4) $C < 0$ and $B < 0$

Equation (11) recognizes two equilibria at the most, including the origin. Only the positive values of k are of economic significance because instantaneous production cannot be negative. In case (1), the strictly positive equilibrium is stable, whereas in case (2) it is unstable. If the origin value of k falls short of equilibrium, production tends to zero: it tends to infinity if the origin value is greater than equilibrium. In cases (3) and (4), the separate equilibrium of the origin is negative and is not therefore economically significant. If the origin value of k is positive, the level of production tends to infinity in case (3) and to zero in case (4). In the latter two cases the instability of growth may be disregarded. They correspond to either very powerful or very weak cases of endogenous mechanisms of technical change and accumulation of experience, accompanied by parameters expressing, respectively, very strong or very weak exogenous growth in the rest of the world. In these cases there is no limit to the combination of endogenous and exogenous mechanisms. Cases (1) and (2) are intermediate cases.[4]

From the economic standpoint the more interesting case is case (1) which is that of an asymptotically stable balanced growth rate. It is this that we shall be concerned with in the bulk of what follows. For the conditions in case (1) to be met, growth in the rest of the world and the income elasticity of the exports of the country considered have to be sufficiently strong. Which means that world economic growth and demand for the country's goods have to be sufficiently buoyant. Also, the endogenous mechanisms of technical progress (the Kaldor-Verdoorn law) and accumulation of experience (quality of exports) should not be too strong, so that this growth of external origin is not infinite. This case may be recognized as being realistic enough to represent the situation of the bulk of industrial countries. The case of unstable growth will be discussed in more detail below.

3. The properties of the growth path

Equation (11) is a first-order, non-linear, Bernouilly-type differential equation. It is solved as follows:

$$k = \frac{1}{\left[\frac{1}{k_0} + \frac{C}{B}\right] \exp(-B \cdot t) - \frac{C}{B}} \tag{15}$$

Since q is deduced from equation (10), this gives:

$$q = U + \frac{R}{S + T \cdot \exp(-U \cdot t)} \tag{16}$$

with:

$$U = \frac{(\beta - \eta - 1)[p^* - n(\mu - 1)] + \varepsilon y^*}{\alpha + (\beta - \eta)(\mu - 1)l} \tag{17}$$

$$R = \frac{\lambda\nu}{\alpha + (\beta - \eta - 1)(\mu - 1)l} \tag{18}$$

$$S = \frac{\alpha + (\beta - \eta)(\mu - 1)l - \lambda\nu}{(\beta - \eta - 1)[p^* - n(\mu - 1)] + \varepsilon y^*} \tag{19}$$

$$T = \frac{1}{k_0} + \frac{\lambda\nu - \alpha - (\beta - \eta)(\mu - 1)l}{(\beta - \eta - 1)[p^* - n(\mu - 1)] + \varepsilon y^*} \tag{20}$$

In case (1), ie $C < 0$ and $B > 0$:

$U > 0$,
$R > 0$,
$S > 0$ and
$T > 0$ or < 0, depending on the value of k at the outset.

The rate of growth *G* describes a logistic curve, increasing from $Rk_0 + U$ to $1/S$ when $k_0 < 1/S$ (Figure 8.2), or a convex curve decreasing from $Rk_0 + U$ to

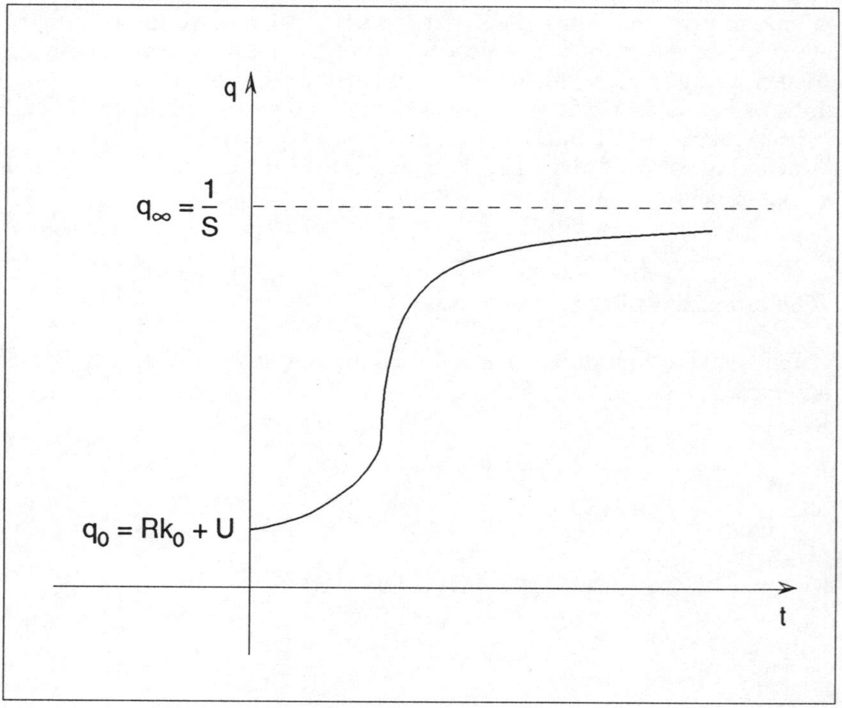

Figure 8.2 An increasing growth path

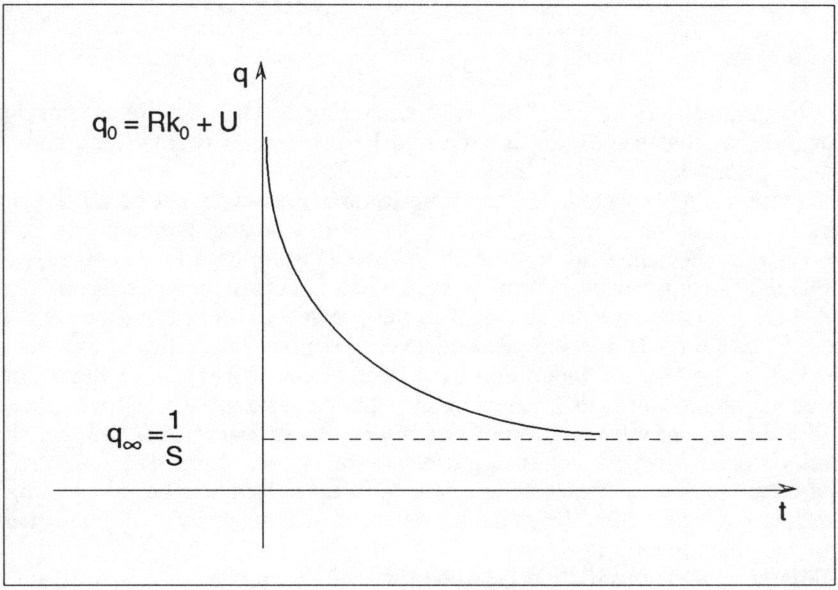

Figure 8.3 A decreasing growth path

1/S when $k_0 < 1/S$ (Figure 8.3). The country's rate of growth rises then over time to its maximum – the asymptotic growth rate.

The asymptotic growth rate is:

$$q_\infty = \frac{ey^* + (\beta - \eta - 1)\,[p^* - n(\mu - 1)]}{\alpha + (\beta - \eta - 1)(\mu - 1)l - \lambda\nu} \tag{21}$$

with:

$$\frac{\partial q_\infty}{\partial \varepsilon} = y^* \geqslant 0 \tag{22}$$

$$\frac{\partial q^\infty}{\partial \lambda} = \frac{\nu[\varepsilon y^* + (\beta - \eta - 1)(p^* - n(\mu - 1))]}{[\alpha + (\beta - \eta - 1)(\mu - 1)l - \lambda\nu]^2} \geqslant 0 \tag{23}$$

$$\frac{\partial q^\infty}{\partial \nu} = \frac{\lambda[\varepsilon y^* + (\beta - \eta - 1)(p^* - n(\mu - 1))]}{[\alpha + (\beta - \eta - 1)(\mu - 1)l - \lambda\nu]^2} \geqslant 0 \tag{24}$$

$$\frac{\partial q^\infty}{\partial l} = \frac{-\{[\varepsilon y^* + (\beta - \eta - 1)(p^* - n(\mu - 1))](\beta - \eta)(\mu - 1)\}}{[\alpha + (\beta - \eta - 1)(\mu - 1)l - \lambda\nu]^2} \geqslant 0 \tag{25}$$

$$\frac{\partial q^\infty}{\partial k_0} = 0 \tag{26}$$

and

$$\frac{\partial q^{\infty}}{\partial(\beta - \eta)} = \frac{[(\alpha - \lambda\nu)(p^* - n(\mu - 1)) - \varepsilon y^*(\mu - 1)l]}{[\alpha + (\beta_{-1} - \eta)(\mu - 1)l - \lambda\nu]^2} \tag{27}$$

will be greater than or equal to zero, depending on the stability assumptions, on condition that wages are not too tightly indexed on productivity and that foreign prices do not fall too much.

As was to be expected, the asymptotic rate of growth will be all the more rapid the higher the income elasticity of exports. So the dynamic criterion of choosing to specialize internationally in products for which growth of world demand is strong certainly applies here. The effects of price competitiveness and market share gains are reflected in the parameters indicating the sensitiveness of foreign trade flows to price changes (the price elasticities of exports and imports). The rule of index-linking wages to productivity and the form of technical change are also determinants of the rate of asymptotic growth. Unlike export growth via income elasticity or quality parameters, market share gains via price competitiveness impose constraints on income distribution. The long-term criteria of choice of specialization may therefore be said to impose fewer constraints, from the point of view of outlining macroeconomic policy, than the short-term criteria.

The possible contradiction between the (static) short-term criterion and the (dynamic) long-term criterion may be illustrated by looking at two different growth paths for two countries. Supposing that country I applies the static criterion and decides to specialize in relatively traditional sectors, with moderate income elasticity, in the hope of winning world market shares through price competitiveness thanks to the experience gained in these areas. Country F, on the other hand, chooses to specialize in high demand-growth

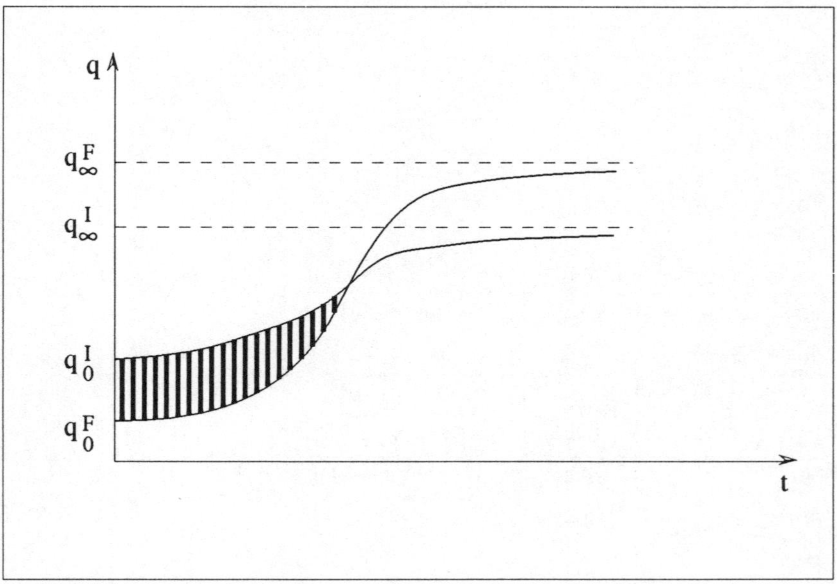

Figure 8.4 Two growth paths (first case)

activities with strong income elasticity and very cumulative technical progress, i.e. it adopts the dynamic criterion. In the long term, the asymptotic growth rate of country F will be higher than the asymptotic growth rate of country I. Country I's initial rate of growth may, however, be faster than that of country F, and may remain so for a certain lapse of time until F achieves its catch-up (Figure 8.4). Until that happens, country F will appear to have been mistaken in its choice of specialization, the result being a lower growth path.

It is even possible to imagine a set of parameters in such a configuration that the two initial growth rates are equal:

$$q_o{}^F = q_o{}^I = R^F K_o{}^F + U^F = R^I K_o{}^I + U^I \tag{28}$$

$$q_{oo}{}^F > q_{oo}{}^I \tag{29}$$

Since, when t = 0, the derivative of the rate of growth with respect to time is:

$$q = \frac{R.T.U.}{(S + T)^2} = K_0{}^2(R.T.U) \tag{30}$$

It is perfectly possible that there are parameters such that the derivative of the rate of growth with respect to time when t = 0 is greater than I, i.e. that there is a period of time during which the growth path of country I is higher than that of country F (Figure 8.5).

$$(K_o{}^F)^2 R^F T^F U^F < (K_o{}^I)^2 R^J T^I U^I \tag{31}$$

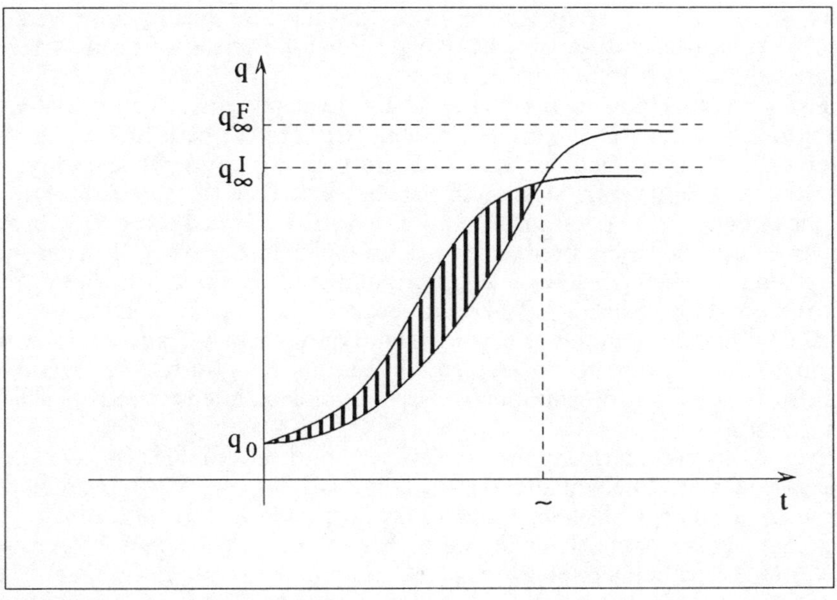

Figure 8.5 Two growth paths (second case)

knowing, also, that:

$$S^F U^F + R^F = S^I U^I + R^I = 1 \tag{32}$$

The less the initial experience in the type of production chosen, the more slowly the growth will pick up initially because it takes time for learning to become established and to pass through to increased product quality. A country specializing in the manufacture of goods in which it already has experience or a reputation for quality will therefore, *ceteris paribus*, achieve a faster initial growth rate than a country which is restructuring more radically and has to build up its experience and reputation for quality on a new range of products.

4. Conclusion

The asymptotic growth rate of an economy will be all higher when specialization is targeted on sectors in which world demand is strong, technical progress is tending to increase and accumulate and the potential for learning and improving on quality is substantial. Growth based on winning market shares in less buoyant sectors, principally thanks to price competitiveness, is more risky. Quite apart from the fact that growth deriving from price competitiveness involves constraints with respect to the method of income distribution, it cannot go on indefinitely. There is a limit to the share that a country can have of a given market and, once that maximum share[5] has been acquired, the rate of export growth will be the same as the rate of growth of demand in the sector, which is assumed to be low. The same reasoning holds when market share gains in these sectors stem from learning and quality-improvement effects which, it may reasonably be thought, tend to wane at an earlier stage in traditional sectors than in new sectors.

The risks attaching to the choice of the dynamic specialization criterion relate to the possible instability of the long-term growth path (case 2). It was observed above that when the combined effects of export growth and technical progress were sufficiently strong, and at the same time the external effects – i.e. the growth of the rest of the world – were sufficiently weak, the equilibrium growth rate of aggregate production was unstable. In the event of the growth rate when t = 0 being below the equilibrium rate, production will tend to zero and the economy will fail to 'take off' because of the unfavourable international environment limiting its expansion and preventing it benefiting from the learning effects and increasing returns. In this event, industrial restructuring and efforts to switch specialization towards technology-intensive sectors will be unsuccessful.

Even when the external economic environment is favourable, specializing in strong-growth, technology-intensive activities do not solve all the problems. There are a number of national modes of integration in a given technological paradigm. These national modes were characterized in the model described above by the parameters of learning and product quality improvement. The values of these parameters are linked to specific national sets of circumstances which include the method of production, manpower skill levels, the effectiveness of R&D expenditure and, more generally, the organizational aspects of the technological changes affecting the method of production. Also to be taken into account are aspects such as the method of marketing and the reputation of the products exported or, more generally, the interplay between producers and consumers. These are the main elements of what many authors call the national system of innovation. Such specific national characteristics qualify the universal normative conclusions that may be drawn from the long-term specialization criterion.

An economy's long-term competitiveness rests on these parameters. Improving a country's structural competitiveness means improving the way the national system of innovation functions, and this has to be accompanied by an increase in the value of the learning parameters. Such industrial policy recommendations reach well beyond the instruments used in macroeconomic policy, since what is involved is changing the actual structure of the growth model by changing the national system of innovation. Also, therefore, it is a question of going further than the traditional industrial policy of sectoral restructuring. It is not enough to invest in a number of sectors which are known to be promising; it is also important to be able to guarantee the prerequisites of long-term competitiveness which remaining in these sectors necessitates. These prerequisites include a 'quality' of production effect. The parameters affecting quality are the factors which enter into structural competitiveness, or else the components of the national system of innovation. In any industrial and technological policy, account has therefore to be taken of specific national features before deciding to apply overall technological criteria.

Notes

1. See OECD (1986) and OECD (1989).
2. Here again, in order not to have more cumbersome notations, we do not take account of wage increases not linked to productivity, that would be integrated in a constant term in eq. (6).
3. We will consider that the Marshall-Lerner-Robinson condition is fulfilled: $1 + \eta - \beta < 0$.
4. Vaglio (1988), in a learning-effect model, encounters the same type of equilibria.
5. In any case, smaller than 1.

References

Amable, B. and Mouhoud, E.M. (1990), 'Changement technique et compétitivité internationale: une comparaison des six grands pays industriels', *Revue d'Economie Industrielle*, no. 54, pp. 22–43.

Arrow, K. (1962), 'The economic implications of learning by doing', *Review of Economic Studies* XXIX (2), pp. 155–73.

Chesnais, F. (1986), 'Science, technologie et compétitivité', *STI Revue* 1, pp. 97–148.

CEPII (1984), *Economie Mondiale 1980–1990: La Fracture?* Economica, Paris.

Cohen, S. and Zysman, J. (1989), 'Diverging trajectories, manufacturing innovation and American industrial competitiveness', in Johnson, Tyson and Zysman, 1989.

Dixon, R. and Thirlwall, A. (1975), 'A model of regional growth-rate differences on Kaldorian lines', *Oxford Economic Papers*, pp. 201–14.

Dosi, G., Freeman, C., Nelson, R., Silverberg, G. and Soete, L. (eds) (1988), *Technical Change and Economic Theory*, Pinter, London.

Freeman, C. (1987), *Technology Policy and Economic Performance: Lessons from Japan*, Frances Pinter, London.

Freeman, C. (1988), 'Japan: a new national system of innovation?', in Dosi *et al.*, *Technical Change and Economic Theory*, Pinter, London.

Johnson, C., Tyson, L. and Zysman, J. (eds) (1989), *Politics and Productivity, How Japan's Development Strategy Works*, Ballinger.

Kaldor, N. (1957), 'A model of Economic Growth', *Economics Journal*, 67, pp. 591–624.

Komiya, R., Okuno, M. and Suzumura, K. (eds) (1988), *Industrial Policy of Japan*, Academic Press.

Krugman, P. (ed.) (1986), *Strategic Trade Policy and The New International Economics*, MIT Press.

Lafay, G. and Herzog, C. (1989), *Commerce International: La Fin des Avantages Acquis*, Economica, Paris.

Lucas, R. (1988), 'On the mechanics of economic development', *Journal of Monetary Economics*, 22, pp. 3–42.

Lundvall, B.A. (1988), 'Innovation as an interactive process: from user-producer interaction to the national system of innovation', in Dosi *et al.*, *Technical Change and Economic Theory*, Pinter, London.

McCombie, J.S.L. (1989), '"Thirlwall's Law" and balance of payments constrained growth – a comment on the debate', *Applied Economics*, 21, pp. 611–29.

Mathis, J., Mazier, J. and Danset, D. (1988), *La compétitivité industrielle*, Dunod.

Nelson, R. (1988), 'Institutions supporting technical change in the United States', in Dosi *et al.*, *Technical Change and Economic Theory*, Pinter, London.

OECD (1986), *Indicateurs de la science et dela technologie*, no. 2, OECD, Paris.

OECD (1989), *Technologies de l'information et les nouveaux domaines de croissance*, OECD, Paris.

Ros, J. (1986), 'Trade, growth and the pattern of specialisation', *Political Economy* 2(1), pp. 55–71.

Thirlwall, A. (1979), 'The balance of payments constraint as an explanation of international growth rate differences', *Banca Nazionale del Lavoro Quarterly Review*, pp. 45–53.

Vaglio, A. (1988), 'Static and dynamic economies of scale in export-led growth', *Economic Notes* 2, pp. 61–81.

Part III
Networking and convergence

Introduction to Part III

Dominique Foray

The chapters in this Part are diverse and based on different disciplinary fields (industrial economics, historical economics and sociology of science and techniques). However, they share the same concern for the analysis of the role of interdependence and networks, as driving forces in the dynamics of science and technology.

In the general introduction I already mentioned the importance of this aspect to illuminate the links between technology and competitiveness. Several chapters in Parts I and II suggest also the key role played by the structural interdependence in the dynamics of invention and innovation (in particular: Freeman, Orsenigo, Carlsson *et al.*, Boyer, Amable and Dosi *et al.*). The interdependence between agents constitutes the essential structure of the process of localized positive feedbacks. In fact, structural interdependencies imbue the individual technical choice, which is essentially static and microeconomic, with macroscopic and dynamic dimensions. In this introduction, I would like to state several propositions derived from the Antonelli, David and Callon chapters in order to characterize and demarcate this field and to identify some avenues for future research. While the next Part will illuminate the driving forces of diversity, through the analysis of the variety of incentive and organization systems, this Part is mainly devoted to the driving forces of standardization and technological convergence, through the analysis of the role of structural interdependences in propagating a decentralized choice.

1. Interdependence is the primary concept compared with those of adoption, diffusion, competition. That is, 'the identity of an actor is nothing else than the series of associations in which he is entered' (Callon, this volume). Thus, the formalization of these interdependences (in Antonelli or David or Callon fashion) allows us to explain the principal mechanisms of the dynamics of adoption, diffusion and competition.

2. As suggested by Paul David, interdependence 'is not restricted to those created by interrelatedness of production technologies; real economic actors function within many varieties of networks . . . social and kinship-related, as well as commercially, transactional and technological'. The David chapter is interesting in that it elaborates a framework for analyzing the diffusion process of an individual choice, without need for any technological basis for the transmission of the positive feedback effects. It is thus not surprising that the illustrative problem David has chosen belongs to the class of 'tipping behaviour' models developed by Schelling (1978).

3. However, it is clear that interdependencies supported by technological

interrelatedness strongly constrain the choices made sequentially. It was the 'message' of the development block approach (Dahmen, 1989). It is also the main argument of Antonelli in this volume. Antonelli focuses on the process of dynamic creation of network externalities, technological complementarities and pecuniary externalities: according to the Schumpeterian analysis, the impact of any innovation has to be related to the characteristics of its diffusion process, which depends upon the relative profitability of this innovation to the existing routines. Antonelli proposes an extension of this framework to a more complex world in which product, process and user innovations are closely interrelated, as they are in the telecommunication industry. Consequently, network externalities appear and deliver a genuine dynamics to the joint system of telecommunications, equipments and related services.

4. It is not surprising that this Part, which focuses on interdependence and network, illuminates also the path-dependent nature of a certain class of allocation processes. The processes described in the Antonelli and Callon chapters are actually path-dependent. It is however in the David chapter that this property is the most deeply investigated by analyzing the operation of structural interdependence in propagating the influence of historical events. It is now well known that a dynamic resource allocation process is path-dependent if it is not possible to describe it by a first-order Markov chain. This formal property manifests itself in the extreme sensitivity of the motion and evolution of a system upon intervening events occurring early in the path. One important result of the David chapter consists of the following argument: while individual choice processes are strictly state-dependent (first-order Markovian), the collective behaviour (the sequence of choices) can be viewed as a stochastic process formed from additively interacting Markov processes (i.e. interdependent Markov chains with locally positive feedbacks). It is in this sense path-dependent (see also DeBresson, 1991). This leads us to a clear understanding of the actual content of the research programme of historical evolutionary economics: that is the analysis of collective behaviour, interdependencies and networks, as well as the rationalities of agents who are embedded in those networks.

5. Interdependence must be studied both in a static and a dynamic framework. Static analysis concerns (i) the decomposition of a system of potential adopters into different reference groups (within which agents make decisions), (ii) the characterization of the linkages between those groups, (iii) the characterization of the physical arrangements of the entities in the networks (see also Schelling, 1971). The goal of static analysis is consequently to explore how 'individual units' decisions become linked indirectly by the intersecting networks of local relationships which transmit and communicate their effect' (David, this volume). Dynamic analysis explores the process of evolution of the structural interdependence (the linkages between the groups of reference): can the technology itself during the diffusion process increase the interdependence between the agents? This dynamics is clarified by Callon, who argues that as the adoption progresses, similarities between agents increase. In this dynamic perspective, the major contribution of Antonelli is to extend the neo-Schumpeterian approach of innovation and competition to the case of a cluster of innovations which are providing increasing returns to scale due to network externalities. Consequently, some cumulativeness is brought into the conventional diffusion model.

Interdependence structures represent one of the most important sources of technical progress. One can observe paradoxically a lack of empirical and theoretical investigations regarding the analysis of individual choice and decentralized decision making in a network context (David, this volume). The goal of this Part thus is to provide useful tools and first results in order to explain how the tendency to standardization and uniformization actually operates.

References

Antonelli, C., 'The dynamics of technological interrelatedness', this volume.

Callon, M., 'Variety and irreversibility in the networks of technical conception, and adoption, this volume.

Dahmen, E. (1989), 'Development blocks in industrial economics', in Carlsson (ed.), *Industrial Dynamics*, Kluwer Academic Publishers, pp. 3–14.

David, P., 'Path-dependence and predictability in dynamic systems with local network externalities: a paradigm for historical economics', this volume.

DeBresson, C. (1991), 'Technological innovation and long wave theory: two pieces of the puzzle', *Evolutionary Economics*, 4, pp. 241–72.

Schelling, T. (1971), 'Dynamics models of segregation', *Journal of Mathematical Sociology*, pp. 143–86.

Schelling, T. (1978), *Micromotives and Macrobehavior*, Norton.

9 The dynamics of technological interrelatedness: the case of information and communication technologies*

Cristiano Antonelli

Introduction

According to a growing literature information and communications technologies constitute a clear case of a technical system. Hence they supply clear evidence about the dynamics of technological interrelatedness: i.e. the strong effects on the evolution of each component played by externalities and complementarities among the components of a broader technical system. In such conditions in fact the features and the behavior of each component of the system influence the features and the behavior of the other components with strong qualitative effects that are not fully reflected by price signals.

David (1987) has highlighted the effects of technological interrelatedness[1] on the emergence of a technological system analyzing the role of complementarities and externalities in constraining sequentially the technological choices about each component. Here we want to analyze the effects of technological interrelatedness on the evolution of the behavior of the components of technological systems.

In this approach the dynamics of supply and demand behavior of each component of a technical system appear to be strongly influenced by the changes in the behavior of all the other components. Because of strong complementarities and hence the interplay of pecuniary, technological and demand externalities, in fact, each component interacts so as to exert major reciprocal effects on the general evolution of supply conditions and demand functions that are not (fully) reflected and mediated by market prices.

Technological interrelatedness in the telecommunications network

The key role played by technological interrelatedness in the growth of the telecommunication industries in the last decade becomes especially clear when the analysis is focused upon the 'telecommunications network', i.e. the vertical relations through specialized industries which cooperate in the manufacture of

* The comments of Bo Carlsson on an earlier version of this chapter are acknowledged. Financial support by MURST National Funds is also acknowledged.

the final product, the telecommunications services. In fact it can be argued that the diffusion of technological innovations within the 'telecommunications network' acted as a dynamic factor which modified the given set of interdependent relations between specialized activities. In this respect the diffusion of innovation revealed 'strong dynamic externalities' and increasing returns which constituted and continue to constitute the engine of growth of this industry.

More analytically, it can be said that while given static exogenous conditions exert externality effects on the adoption of new capital goods, the case of interdependent diffusion of different technological innovations needs a different framework where the origin and the effects of dynamic externalities, which go well beyond the limits of given industries and entail a much broader set of interdependencies and changes, can be understood. Such an approach in fact seems to go beyond the static idea of externalities and it draws more from the notion of increasing returns at the industrial level elaborated by Allyn Young and Adam Smith.[2]

The case of interdependent diffusion appears to be a specification of the interaction between the elasticity of derived demand for intermediary inputs (telecommunications services) and the elasticity of supply of manufactured goods (telecommunications equipment) on which Young focuses his attention.

The case of interdependent diffusion seems to make it possible to go even one step further when combining the effects of increasing returns due to interdependent supply and demand curves with the effects of radical technological changes and significant economies of scale enjoyed in the production of telecommunications equipment.

In the production of telecommunications equipment, the introduction of microelectronics has led rapidly to the digitalization of all switching and transmission. This radical innovation dramatically altered the cost conditions of the firms operating in the industry and eventually the whole market structure of the industry was changed. A lot of wide-ranging empirical evidence confirms that the production of telecommunications equipment is strongly affected by increasing returns and economies of size and these factors are especially relevant when the magnitude of R&D costs necessary to introduce new digital products in switching and transmission are high (Charles, Monk and Sciberras, 1989).

When R&D costs are conceptualized as fixed costs and learning economies are taken into account in the right way, the blueprint, on which a new digital switch or transmission equipment is based can be considered as a form of long-lasting capital. Consequently the shape of long-term cost curves in the production of digital switches can exhibit, for significant portions, a strong negative slope (Amendola and Gaffard, 1988).

The introduction of digital technology and the significant advantages in terms of cost conditions related to the size of firms have led to three significant effects:

(i) The drastic reduction in the number of firms in the market with a strong increase in globalization and concentration.
(ii) The rapid decline in the prices of digital telecommunications equipment.
(iii) The rapid diffusion of digital telecommunications equipment.

The diffusion of digital equipment in the telecommunications services industry in the 1980s, in fact was extraordinarily rapid and the levels of penetration in the OECD area are now very high. With respect to electromechanical technology, digital exchange offers enormous advantages in terms of reduced capital costs, reduced maintenance costs and greatly increased facilities.

Capital costs are reduced by means of (1) substituting cheap microelectronic components for more expensive electromechanical ones, (2) reducing needs of office and plant space, (3) increasing the flexibility of the production process. Savings in maintenance costs are dramatic. They are achieved by means of automatic error detection and correction and there is a consequent dramatic reduction in manpower requirements.

In the United States the average purchase cost of electromechanical exchanges installed between 1980 and 1985 was about 600 US dollars while at the end of 1990 the average purchase costs of electronic switching equipment was already 200 US dollars per line. Growing evidence reveals that the introduction of digital switches and transmission equipment radically affects the shape and the position of the cost curve. According to the extensive empirical evidence available the introduction of digital technology in switching and transmission has reduced the last vestiges of advantage enjoyed by large centrally planned networks and consequently it has made it possible to operate smaller networks efficiently. Moreover the introduction of digital technology has enormously increased the *capacity* of telecommunications services in terms of speed, volume, reliability and overall quality.

Hence the introduction of digital technology makes it possible to reduce both operating costs *and* hedonic prices dramatically, i.e. the costs of telecommunications services to customers weighted so as to take into account the increase of the overall levels of performance favoring the diffusion of advanced telecommunications services (Bradley and Haussman, 1989).

The larger the production of telecommunications equipment, the lower the costs and consequently the costs of telecommunication services which add to the profitability of adoption are lower. Economies of scale in the production of innovated equipment which are made possible by increased adoptions in turn fuel the demand for new services which once again add to the growth of demand for telecommunications equipment.

The demand for telecommunication services consequently appears to be increasingly associated with the introduction of technological innovations embodied in information capital, i.e. machines designed to process information (computers, PBX, peripherals, etc.). The use of telecommunication services by information capital in fact seems to be characterized not only by high levels of complementarity, but also by physical and economic rigidity, so that each piece of equipment of information capital which is installed generates a given amount of telecommunication flows, with little variance. Thus the demand for telecommunication services and the use of information technology products are highly complementary and, in turn, the introduction of new information technology products has important effects on telecommunications demand. Moreover, this shift in demand is likely to continue for as long as diffusion and learning about the opportunities offered by information technologies is taking place in downstream sectors. Furthermore, because of complementarities, price-elasticities of demand are likely to be relatively low. Hence, demand is more likely to be determined by the diffusion of new technology and the quality of service than by tariffs. Therefore it is possible to say that the structure of demand is changing from a final demand, i.e. a demand mostly expressed by the final consumer, to a derived demand for sources that are inputs to production processes.

In sum, within a technical system, the behavior of each component is strongly affected by the behavior of all the others. Hence it seems clear that the analysis of the demand and supply conditions can be approached only when the technical interrelatedness between the different stages of production, and between the technical and economic complementarities of innovations and industries are understood.

The basic ingredients of such an analysis can be drawn together as follows:

(i) Diffusion of information technology products is stimulated by a fall in their market prices and by network externalities.
(ii) Long-term cost curves in production of telecommunications equipment are characterized by strong economies of scale and learning processes.
(iii) Demand of telecommunication equipment is 'derived' from the demand of telecommunications services.
(iv) Hedonic prices of telecommunications services are reduced by the diffusion of digital technology.
(v) Demand of telecommunications services is stimulated by the diffusion of information technologies and by the reduction of the hedonic prices of telecommunications services.
(vi) Growth of the industry is the result of this dynamic interaction of the industries considered.

These stylized facts suggest that a cumulative process with strong virtuous features is likely to develop in the telecommunications industry. Growth is fed by technological change introduced in a number of industries which belong to the same telecommunications network, and by the swarming diffusion of information technologies which affect the demand for telecommunication services in depth. Technological change in the manufacturing industries belonging to this network reduced absolute costs and facilitated important economies of scale. The interdependence of demand and supply in this dynamic context is likely to produce a classical case of increasing returns which can be understood only if attention is focused on the whole system of vertically related industries and on the changes which have taken place over time. Such an industrial dynamics can be understood by the notion of *dynamic externalities*, i.e. the process of dynamic creation of network externalities, technological complementarities and pecuniary externalities brought about by the *sequence of diffusion* of new technological innovations in industries which in turn is characterized by technological interrelatedness and vertical complementarity on intermediary markets.[3]

Such conclusions lead to an empirical analysis of the context of the adoption of each innovation within a specific 'network' framework where the causal role played by the dynamic externalities at work can be modeled and the interdependence between vertically related industries is made explicit.

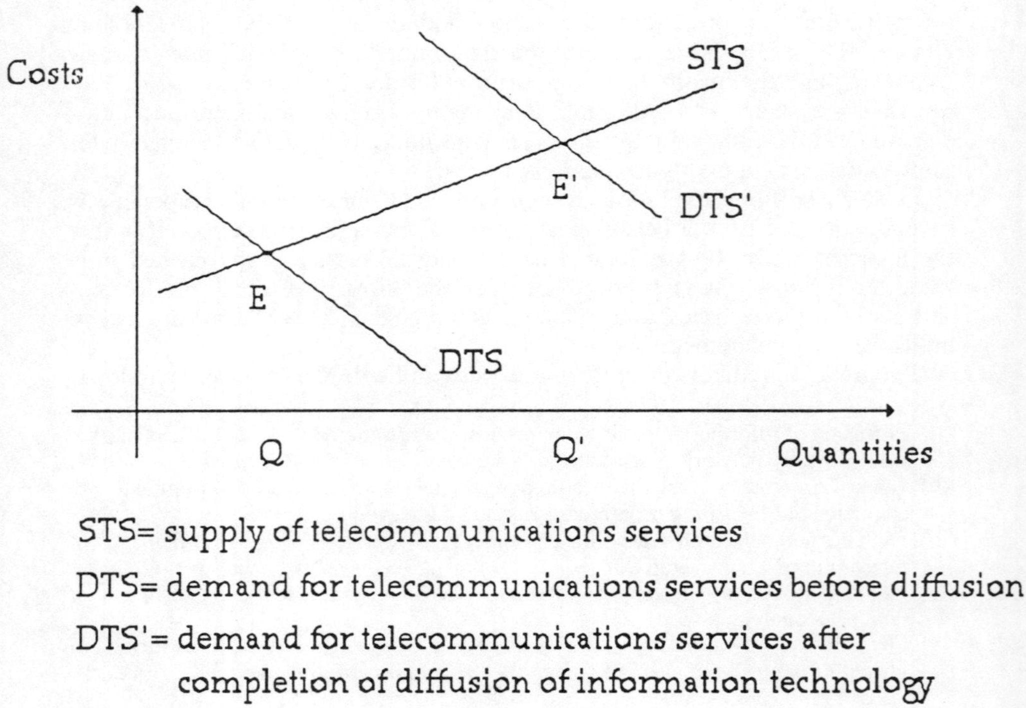

Figure 9.1

The model: graphic exposition and econometric estimates

The dynamics of the model can be summarized in the following steps:

(i) Because of relevant technological complementarities between information capital and telecommunication services the diffusion of information technologies, engendered by *network externalities*, pushes, under conditions of *ceteribus paribus* applying to supply, the demand curve for telecommunications services from E towards E′ and the equilibrium quantity Q to $\bar{Q}$ (see Figure 9.1).

(ii) The increased demand for telecommunications services feeds an increased demand for telecommunications equipment (see Figure 9.2).

(iii) *Long-term supply* of digital telecommunications equipment is characterized by a significant reduction in costs that are determined by the rates of diffusion of digital technology, economies of scale and learning curves (see Figure 9.3).

Because of the introduction and diffusion of digital technology relevant *pecuniary externalities* spill from the telecommunications equipment industry to the telecommunications services industry.

(iv) Plotting Figure 9.3 over Figure 9.2 we can visualize the combined effect of technological change brought by the introduction of information technology and microelectronics (see Figure 9.4). The shift of equilibrium from E towards E′ leads to a parallel decline in prices of digital telecommunications equipment and a further growth in their diffusion levels.

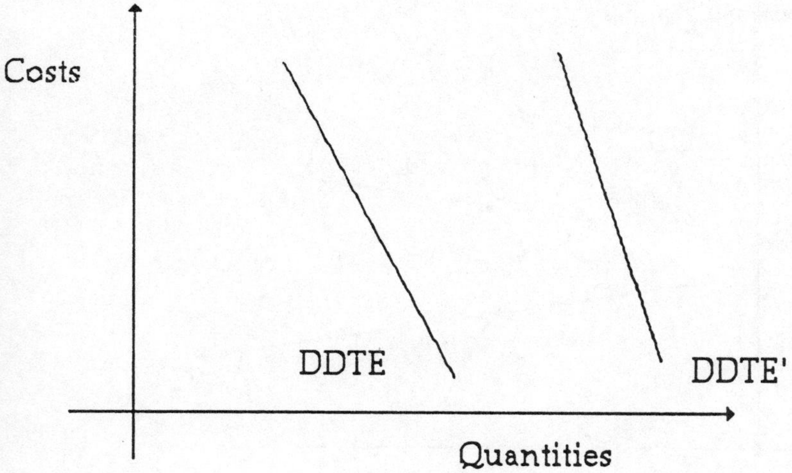

DDTE= derived demand for telecommunications equipment before diffusion

DDTE'= derived demand for telecommunications equipment after completion of diffusion

Figure 9.2

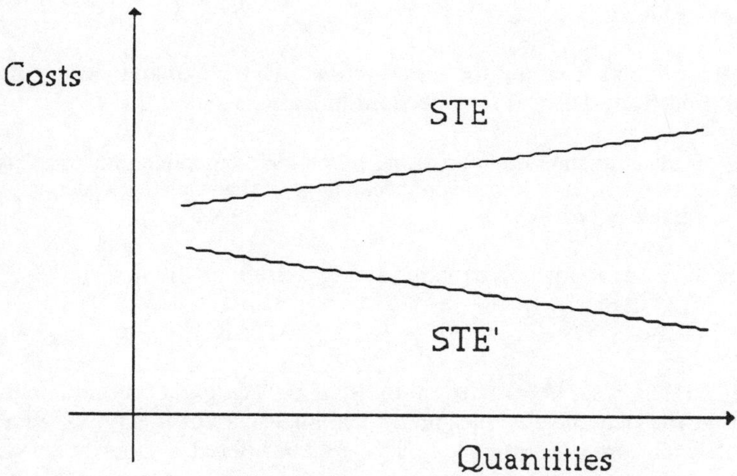

STE= long term supply curve of electromechanical telecommunications equipment

STE'= long term supply curve of digital telecommunications equipment

Figure 9.3

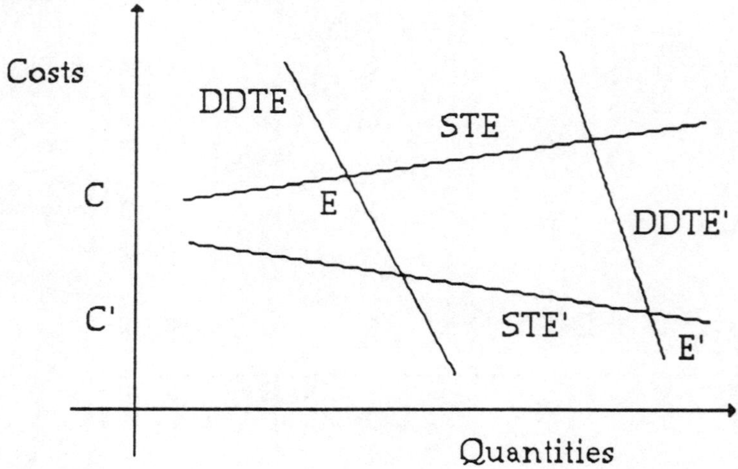

Figure 9.4

(v) The diffusion of digital telecommunications equipment engenders a downward shift of the supply curve of telecommunications services which in turn determines a further trend of increase in the equilibrium quantity of telecommunication services from $\bar{Q}$ to Q′ (see Figure 9.5).

Our model predicts that at the intersection of the demand and supply curves the equilibrium demand for telecommunications services is larger:

- the higher the penetration of digital switches in total telecommunications capacity;
- the higher the penetration of information technology in the economic system;
- the lower the levels of tariffs.

This leads to the econometric equation being written as follows:

$$D = a + b_1 LINES + b_2 DIGITAL + b_3 IT - b_4 TARIFF + e \quad (1)$$

Equation (2) has also been written in such a way to accommodate the hypothesis that the demand for telecommunications services is highly revenue elastic, i.e. that telecommunications services are considered a superior good:

$$D = a + b_1 LINES + b_2 DIGITAL + b_3 IT + b_4 GNP - b_5 TARIFF + e \quad (2)$$

All the variables enter logarithmically and are defined in Table 9.1.

Equation (2) has also been written in the following intensive form:

$$\left(\frac{D}{LINES}\right) = a + b_1\left(\frac{DIGITAL}{LINES}\right) + b_3\left(\frac{IT}{POPULATION}\right) - b_4 TARIFF + e \quad (3)$$

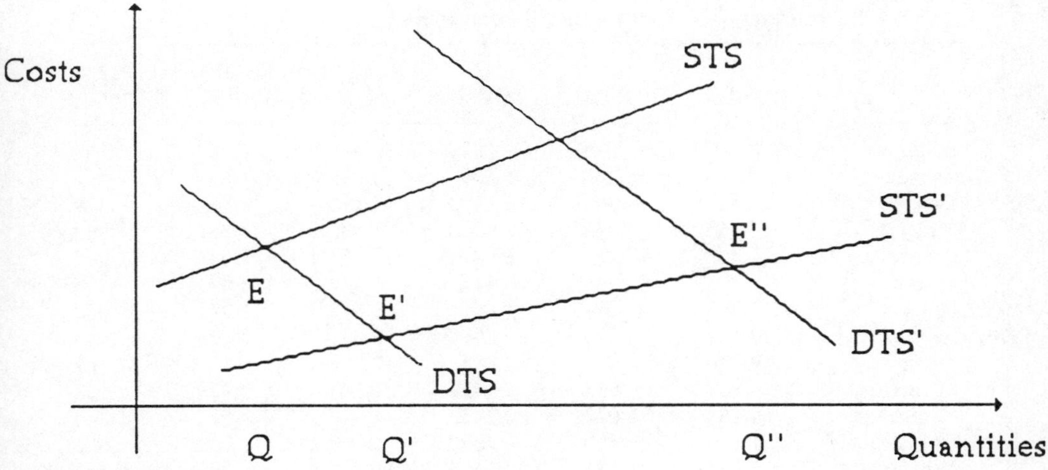

Figure 9.5

Table 9.1 Variables in equations (1), (2) and (3)

LINES	=	total number of telecommunications lines installed in each country in 1987 (ITU, 1988)
DIGITAL	=	total number of digital telecommunications lines installed in each country in 1987 (AT&T, 1988)
IT	=	indicator of the level of diffusion of information technology in the economic system in each country in 1987 (Juliussen and Juliussen, 1989)
TARIFF	=	absolute level of tariffs for three minutes of telecommunications within 100 km (Siemens, 1989)
GNP	=	Gross National Product per capita in US dollars in 1987 (ITU, 1988)
D	=	total sales of telecommunications service in each country in 1987 (ITU, 1988)

In order to understand the specific role of information technology on the demand side, three different indicators have been used for IT:

- MIPS, i.e. the total number of MIPS installed in each country. This value measures the levels of penetration of information technology in terms of capacity of computer hardware;
- D.P., i.e. the dollar value of output of the data processing industry in each country. This value is considered to measure the intensity of the use of the informatic hardware;
- MIPS-DP, i.e. the combined product of the two variables and this should measure both the size and the intensity of use of informatic hardware installed in each country.

Table 9.2 Relevant data for the countries considered

	(a) LINES	(b) DIGITAL	(c) MIPS	(d) DP	(e) TARIFF	(f) D
Belgium	3,367	34	183.4	2,908.5	1.28	1,344
Canada	11,399	6,360	1,008.0	12,931.6	1.69	7,677
Denmark	2,827	305	149.4	2,195.1	0.51	879
Finland	2,365	366	104.8	1,685.8	1.36	1,009
France	24,803	17,287	1,331.0	19,274.4	1.62	12,663
Ireland	798	366	115.3	1,479.2	2.67	316
Israel	1,393	397	81.3	1,042.9	1.37	632
Italy	19,104	2,273	893.5	11,001.0	1.79	9,062
Japan	49,976	18,041	4,088.0	42,713.0	1.85	24,361
Mexico	3,987	622	102.0	1,308.0	0.45	638
Netherlands	6,234	3,073	348.0	4,610.0	0.51	2,800
New Zealand	1,376	400	86.5	1,109.1	1.03	626
Norway	1,984	1,984	146.7	1,981.8	2.23	1,363
Philippines	444	187	53.1	681.2	0.46	300
Saudi Arabia	1,504	1,404	58.2	746.6	0.27	999
S. Africa	2,495	380	76.6	982.7	0.51	977
S. Korea	8,695	6,112	195.5	2,508.1	2.41	2,042
Spain	10,236	1,627	182.4	4,251.7	1.04	2,421
Sweden	5,373	1,612	243.7	3,817.6	0.77	2,222
Switzerland	3,499	1,095	201.8	3,694.8	0.97	2,796
UK	23,183	11,220	2,591.0	18,575.3	1.12	10,073
USA	118,400	90,220	22,812.0	125,870.6	1.40	102,202

Table 9.2 shows the 1987 data gathered for 22 countries with consistent information for total telecommunications lines installed (in thousands) (LINES) (column a); total electronic lines installed (in thousands) (DIGITAL) (column b); total number of MIPS installed in each country (MIPS) (column c); total output of the data processing industry in each country (DP) (column d); tariffs in D-marks for three minutes of telecommunications within 100 km (TARIFF) (column e); total telecommunications revenue (D) (column f).

The results of OLS tests on equations (1), (2) and (3), listed in Tables 9.3 and 9.4 respectively, are highly satisfactory both in terms of total variance explained and for the statistical significance of each of the coefficients considered. The important role of IT variables (MIPS, DP, MIPS-DP) confirms that the variance of intensity of use of telecommunications infrastructure across countries is significantly influenced by the technological *complementarity* of the use of telecommunications services and the size of the stock of information capital as measured by MIPS, DP and the interaction variable MIPS-DP.

The elasticity of telecommunications services to the stock of information capital is very high and it falls between the high of estimated value of 0.588 (in eq. 1.2) and the low of 0.225 in eq. 2.3 where GNP is also considered. Inspecting the matrix of Pearson correlation (Table 9.5) it is evident that there is some collinearity between revenue and the intensity of informatic capital per

Table 9.3 Results of OLS estimates of equations 1.1, 1.2, 1.3, 2.1, 2.2, 2.3

a	−1.695	−1.932	−0.951	−2.224	−2.319	−1.519
	(2.657)	(3.991)	(1.271)	(3.344)	(3.917)	(1.871)
LINES	0.420	0.269	0.341	0.420	0.299	0.358
	(2.866)	(1.770)	(2.299)	(3.055)	(1.948)	(2.496)
DIGITAL	0.199	0.220	0.200	0.197	0.224	0.203
	(1.739)	(2.190)	(1.876)	(1.839)	(2.242)	(1.975)
MIPS	0.400	–	–	0.364	–	–
	(3.095)			(2.969)		
DP	–	0.588	–	–	0.526	–
		(3.892)			(3.287)	
MIPS-DP	–	–	0.251	–	–	0.225
			(3.564)			(3.209)
TARIFF	−0.113	−0.177	−0.144	−0.144	−0.187	−0.164
	(0.944)	(1.586)	(1.258)	(1.271)	(1.677)	(1.482)
GNP	–	–	–	0.161	0.100	0.133
				(1.822)	(1.120)	(1.523)
R^2	0.956	0.964	0.961	0.961	0.964	0.963
F	115.110	140.140	129.159	105.317	114.036	111.806

Table 9.4 Results of OLS estimates of equations 3.1, 3.2 and 3.3

a	2.008	1.817	2.053
	(10.579)	(7.426)	(10.895)
DIGITAL/LINES	0.300	0.323	0.300
	(2.567)	(2.748)	(2.529)
MIPS/P.C.	0.149	–	–
	(2.060)		
DP/P.C.	–	0.144	–
		(1.898)	
MIPS-DP/P.C.	–	–	0.074
			(1.940)
TARIFF	−0.183	−0.177	−0.174
	(1.293)	(1.222)	(1.216)
R^2	0.417	0.400	0.404
F	4.295	3.998	4.072

capita. This result compared to the low explicative value of GNP seems to confirm that revenue effects, which traditionally have always been considered relevant, are much less important and reflect some supply-effects in terms of higher labor productivity in more advanced economic systems where diffusion of information technologies has reached higher levels of penetration.

The levels of diffusion of digital switching capacity exert strong effects in terms of both estimated elasticity and significance. It seems clear at this point that because of information technology the modernization of the telecommunications infrastructure exerts the two-fold effect of a fall in the position

Table 9.5 Pearson correlation matrix of the variables in equations (1), (2), and (3)

	D	*MIPS*	*DP*	*MIPS-DP*	*TARIFF*	*LINE*	*DIGITAL*
MIPS	0.959	–	–	–	–	–	–
DP	0.970	0.988	–	–	–	–	–
MIPS-DP	0.967	0.997	0.997	–	–	–	–
TARIFF	0.337	0.400	0.426	0.414	–	–	–
LINE	0.961	0.926	0.944	0.938	0.337	–	–
DIGITAL	0.935	0.909	0.907	0.911	0.394	0.911	–
GNP	0.529	0.487	0.539	0.513	0.319	0.448	0.451

of the supply curve of telecommunications services *and* a significant reduction in the hedonic price of telecommunications services for innovative customers who use telecommunications services to interconnect information capital goods (computers, peripherals, data banks and so on). The reliability, speed, capacity and privacy of telecommunications services ensured by means of the digital infrastructure are in fact higher with respect to the quality of services supplied by electromechanical equipment.

The same reasoning should also be a valid explanation as to why TARIFF provides such poor results. Tariff levels not only reflect the institutional rigidities and the non-market price-setting policies which exist in most countries but also and more relevantly measure only the monetary element of a more complex definition of a price level that, in fact, should also take into account the 'quality' of the product delivered.[4]

It seems especially interesting to note that in equations (1) and (2) the combined estimated elasticities of MIPS-DP and DIGITAL are far larger than the estimated coefficient of LINE. This seems to confirm the relevance of the diffusion of technological changes from upstream industries and in downstream ones in assessing the patterns of growth of the telecommunications services industry. The demand for telecommunications services responds far more to the effects of interdependent technological changes than to the sheer growth of the economic system.

The results of our empirical analysis in fact confirm the role played in understanding the demand of telecommunications services by:

- *dynamic complementarities*
  between the levels of diffusion of computers and informatics – as measured by MIPS and DP – and the innovative uses of telecommunications;
- *dynamic externalities (both pecuniary and technical)*
  generated by the diffusion of digital switches – as measured by DIGITAL – that enable higher transmission speeds in data communication (thus reducing hedonic prices of telecommunications services) and make possible to spread the use of advanced services (such as those provided by Minitel in France).

Conclusions

When technology changes radically a variety of interdependent innovations are introduced that simultaneously affect *products*, *processes* and *uses*. In such a context the analysis must embrace the main sources of change and take into account the role of dynamic externalities; that is, the process of dynamic creation of network externalities, technological complementarities and pecuniary externalities brought about by the sequence of introduction and diffusion of technological innovations in industries that are characterized by technological interrelatedness and vertical complementarity in intermediary markets.

Such an approach has been elaborated here to analyze the determinants of the growing diffusion in the patterns of growth of the telecommunications services industry in different countries over time. The framework of the analysis elaborated suggests that attention should be focused on the role of the diffusion of process innovations from upstream industries, the diffusion of product innovations in downstream industries and the spread of use-innovations within the industry if the determinants of the growth of the intensity of telecommunications services across countries are to be assessed accurately.

The implications of such an approach are important methodologically and empirically: the results of the analysis in fact show that local equilibria should be considered as points along a growth path that is shaped by the rates and patterns of diffusion of interdependent innovations. Both policy-makers and telecommunications economists should thus extend the scope of their analysis of what controls the evolution of the telecommunications service industry to include the much wider area of that portion of the telecommunications network which is involved in the introduction of new information technology.

Notes

1. More specifically David (1987) uses the notion of technical interrelatedness.
2. See the celebrated quotation:

   Adam Smith's famous theorem amounts to saying that the division of labor depends in large part on the division of labor. This is more than mere tautology. It means, if I read its significance rightly, that the counter forces which are continually defeating the forces which make for economic equilibrium are more pervasive and deeply routed in the constitution of the modern economic system than we commonly realise. Not only new or adventitious elements coming in from the outside, but elements which are permanent characteristics of the ways in which goods are produced make continuously for change. Every important advance in the organization of production regardless of whether it is based upon anything which in a narrow or technical sense, would be called a new 'invention' or involves a fresh application of the fruits of scientific progress to industry, alters the conditions of industrial activity and initiates responses elsewhere in the industrial structure which in turn have a further unsettling effect. Thus change becomes progressive and propagates itself in a cumulative way. (Young, 1928: 533)

3. See David and Greenstein (1990) for an analysis of network externalities, Scitovsky (1954) for an analysis of pecuniary externalities, Foray (1987), Teece (1986) and Milgrom-Roberts (1990) for an analysis of complementarities.
4. 'The reason why these different varieties or models sell at different prices must be due to some differences in their properties, dimensions or other "qualities" real or imaginary' (Griliches, 1971: 57).

References

Amendola, M. and Gaffard, J. L. (1988), *The Innovative Choice. An Economic Analysis of the Dynamics of Technology*. Basil Blackwell, Oxford.

Antonelli, C. (ed.) (1988), *New Information Technology and Industrial Change. The Italian Case*, Kluwer Academic, Boston.

Antonelli, C. (1990), 'Profitability and imitation in the diffusion of process innovations', *Rivista Internazionale di Scienze Economiche e Commerciali*, February, pp. 109–26.

Antonelli, C. (1991), *The Diffusion of Advanced Telecommunications in Developing Countries*, OECD Development Centre, Paris.

Antonelli, C. Petit, P. and Tahar, G. (1990), 'The diffusion of interdependent innovations in the textile industry', *Structural Change and Economic Dynamics* 2, pp. 207–24.

Arena, R. (eds) (1988), *Traite d'Economie Industrielle*, Economica, Paris.

AT&T (1989), *The World Telephones*, Morristown.

Bradley, S.P. and Haussman, J.A. (eds) (1989), *Future Competition in Telecommunications*, Harvard Business School Press, Boston.

Carlsson, B. (1987), 'Reflections on "industrial dynamics": the challenge ahead', *International Journal of Industrial Organization* 2, pp. 135–48.

Carlsson, B. (1989), 'Flexibility and the theory of the firm', *International Journal of Industrial Organization* 2, pp. 179–203.

Carlsson, B. (ed.) (1989), *Industrial Dynamics. Technological Organizational and Structural Changes in Industries and Firms*, Kluwer Academic Publishers, Dordrecht.

Charles, D. Monk, P. and Sciberras, E. (1989), *Technology and Competition in the International Telecommunications Industry*, Pinter Publishers, London.

Dang Nguyen, G. (1985), 'Telecommunications: a challenge to the old order', in M. Sharp (ed.), *Europe and the New Technologies*, Pinter, London.

David, P.A. (1975), *Technical Choice Innovation and Economic Growth*, Cambridge University Press, London.

David, P.A. and Greenstein, S. (1990), 'The economics of compatibility standards: an introduction to recent research', *Economics of Innovation and New Technology* 1, pp. 3–42.

Foray, D. (1990), 'L'Economie des Rendements Croissents et l'Economie de la Firme Innovatrice: Perspective sur le Changement Technique', OECD Mimeo, Technology and Economic Program, June.

Foray, D. (1987), *Innovations Technologiques et Dynamique Industrielle*, Presses Universitaires de Lyon, Lyon.

Gaffard, J.L. (1990), *Economie Industrielle et de l'Innovation*, Dalloz, Paris.

Griliches, Z. (ed.) (1971), *Price Indexes and Quality Change*, Harvard University Press, Cambridge, Mass.

ITU (International Telecommunications Union) (1989), *Yearbook of Common Carrier Statistics*, Geneva.

Juliussen, E. and Juliussen, K. (1989), *Computer Industry Almanac*, Brady, New York.

Mansell, R. (1990), 'Rethinking the telecommunication infrastructure: the new black box', *Research Policy*, December, pp. 501–15.

Milgrom, P. and Roberts, J. (1990), 'The economics of modern manufacturing: technology strategy and organization', *American Economic Review*, June, pp. 511–28.

Rosenberg, N. (1982), *Inside the Black Box: Technology and Economics*, Cambridge University Press, New York.

Scitovsky, T. (1954), 'Two concepts of external economies', *Journal of Political Economy*, April, pp. 143–51.

Siemens, A.G. (1989), *Study on National Telephone Tariffs Worldwide*, Munich.

Sounders, R.J. Warford, J.J. and Wellenius, B. (1983), *Telecommunications and Economic Development*, Johns Hopkins University Press, Baltimore.

Stoneman, P. (1983), *The Economic Analysis of Technological Change*, Oxford University Press, London.

Taylor, L.D. (1980), *Telephone Demand: A Survey and Critique*, Ballinger, Cambridge, Mass.

Teece, D.J. (1986), 'Profiting from technological innovation', *Research Policy*, December, pp. 208–305.

Young, A. (1928), 'Increasing returns and economic progress', *Economic Journal*, December, pp. 527–42.

10 Path-dependence and predictability in dynamic systems with local network externalities: a paradigm for historical economics

Paul A. David

Credo, context, and contents – an introduction

I believe that the future of economics as an intellectually exciting discipline lies in its becoming an *historical* social science. On earlier occasions I have tried to convey a strong sense of how 'history matters' in economic affairs by applied studies, focused on the behavior of stochastic systems whose outcomes are *path-dependent*. A path-dependent process is *non-ergodic*: systems possessing this property cannot shake off the effects of past events, and do not have a limiting, invariant probability distribution that is continuous over the entire state space.

In a paper conceived jointly with this chapter,[1] I observe that those sharing my beliefs should find some encouraging signs for the future of our discipline in the fact that the existence of *local positive feedback* mechanisms – and consequent *multiple equilibria*, arising from a variety of sources – has begun to be more generally acknowledged in current economic modeling exercises. Rather than shunning models with multiple equilibria on account of their indeterminacy, economists increasingly are content to leave open a door through which aleatory and seemingly transient influences, including minor perturbations specific to an historical context, can play an essential role in shaping the eventual outcomes. As the implications of mechanisms of self-reinforcing change in economics come to be more thoroughly and extensively investigated in the context of stochastic processes, using available methods from probability theory, economic theorists and economic historians will be able to begin cooperating on useful programs of research.

Although path-dependence will be found where a dynamic resource allocation process is not described by a first-order Markov chain, it can be encountered also in some stochastic systems where the transition probabilities are strictly state dependent (first-order Markovian) – namely, systems where a multiplicity of absorbing states exists. There, as in the structure to be examined presently, small events of a random character, especially those occurring early on the path, are likely to figure significantly in 'selecting' one or another among the set of stable equilibria for such a system.[2] Under those conditions, *ex ante* predictions of outcomes may not be possible. Yet, there are some stochastic path-dependent processes in which knowledge of the initial conditions enables one to make statements about the probabilities attached (at the very outset) to the alternative limiting states among which the system is bound to end up. The latter property (predictability) manifests a distinct form of path-dependence, one that will be seen to apply in the case of the particular dynamic structure that I have taken as my vehicle for the following excursion.

My main purpose here is to continue attempting to promote cooperation between theorists and economists of an historical persuasion, so that the beliefs of the latter will become more widely shared. I therefore wish to emphasize the variety and richness of the dynamic problems in economics that possess a common structure arising from the interdependence of individual choices under conditions of positive local feedback, and which would thus lend themselves to a research approach that allowed for historical contingency. For reasons to be indicated shortly, in the section immediately after this, I will try to make this case heuristically – by exhibiting one very simple, illustrative path-dependent equilibrium system which could lend itself to a number of quite different economic interpretations. The potential topical applications cited in the later sections of this chapter include macroeconomic phenomena, arising from exchange and investment coordination problems created by supply-side externalities, as well as 'bandwagons' in public opinion formation, and the emergence of collusive oligopoly behavior (cartel cooperation). But, as there is some virtue in having cobblers stick to their lasts, I shall offer some more detailed comments on the application of this type of model to problems of technological change and the emergence of institutions, which must be central and perennial preoccupations of economic historians – indeed, of all those concerned with long-term processes of economic development.

Of many technological brass tacks and occasional sparks

On past occasions, as a rule, I have sought to convey my vision of the importance of remote historical events in shaping the course of subsequent economic changes by giving it a very concrete form, in one practical demonstration or another of the economic historian's craft. Practice, not preach, I told myself; spread the faith through specific 'good works' – by explicitly postulating non-ergodic models that could be applied to concrete problems in economic history, and actually applying them.

To delve deeply into some set of realistic historical contexts is of course essential for any convincing demonstration of the way what we might want to call 'path-dependent equilibrium analysis' would look and feel in practical applications. The amount of specialized knowledge of technologies and institutions required to appreciate the details of this style of empirical inquiry, however, soon strains the patience of even the most tolerant general audience of economists. For better or worse, ours is an intellectual environment in which it is slow and tedious work making sparks with brass tacks.

That is why I have asked myself whether it would do any further good to spend much time here recapitulating results of previous such efforts, either those on my part, or similar research, by others. Will it be generally edifying for me to wheel out that old model of localized technical changes generated by stochastic learning-by-doing in industrial production,[3] which I found useful in explaining nineteenth-century Anglo-American contrasts in the rate and direction of technological progress? To guess the answer is not hard. It is true that the model I have just cited contained all the essential ingredients for strong (non-ergodic) history, and showed how myopic localized learning in a fluctuating factor-input price environment could result in particular events remote in a region's factor-price history causing its manufacturing industries to 'lock-in' to one or another of several available technological trajectories. But, its pitch was tuned to accompany one particular song, and those especially interested in the non-neoclassical reformulation of the so-called Rothbard–Habakkuk thesis probably have seen it already, or can find it without too much difficulty.[4]

The same framework of analysis, while attempting to synthesize Atkinson and Stiglitz's (1969) model of localized technological progress with Rosenberg's (1972) ideas about the way existing production techniques acted as 'focusing devices' for further technical improvements,[5] allowed also for changes in fundamental scientific or engineering knowledge to occasion radical innovations. These could initiate new trajectories which held out prospects for faster 'learning', and wider adoption, and so would threaten to disrupt long-established environments for localized learning – namely those complex systems that previously had been built up incrementally, through the sequential generation and adoption of many small and technically interrelated sub-systems. Nevertheless, the usefulness of this way of looking at the process of technological change remains hard to assess in the abstract; and, again, those prepared to get into specifics, most probably, will have found their way to my various applications of it in studies of agricultural mechanization, and of the parallels between contemporary industrial experience with the application of robotic technologies and historical experience in the development and adoption of farm machinery.[6] Moreover, if they have not seen those particular realizations of the vision, they certainly will have encountered kindred expressions of this approach to technological history in the works of Nathan Rosenberg.[7]

More recently, pursuing a slightly different course towards the same general goals, I have tried to whip up some more easily digestible offerings from the kitchens of 'applied path-dependent equilibrium analysis'. One such effort presented a brief account of a familiar but striking piece of technological and economic history: a *histoire raisonnée* of the standardization of typewriter keyboards that occurred around the beginning of this century.[8] Although my version of the story of QWERTY invoked some powerful theoretical results obtained by Arthur, Ermoliev and Kaniovski[9] about the properties of a certain class of path-dependent stochastic processes (namely, generalized Polya urn processes), at heart it was a tale of how, when historical contingency rules, 'one damn thing follows another'. I meant it to be read by the many as a kind of allegory – and actually concluded it with a moral-like 'message'. I thought that a careful investigation of just how we collectively came to be stuck using the suboptimal QWERTY keyboard layout would provide an effective medium for transmitting my main methodological point about the necessity of taking an historical approach to certain classes of economic phenomena.

Looking on the brighter side of that enterprise, my attempt to launch the science of 'Qwerty-nomics' was a half-success. For, not surprisingly, telling modern economists that a dynamic market process has everyone 'locked-in' to using an inefficient technology is a reliable way to at least get their momentary attention.[10] But, you would be amazed – I, at least, was amazed – by the number of fine economists (economic historians, even) who seemed to have utterly missed the larger point which this particular tale was supposed to underscore.[11] They passed the reprints along to friends as a cute story about the anomalous economics of typewriters, or, at most, about certain kinds of information technologies that featured network externalities. However, with the appearance subsequently of other case studies from which the generic features of 'lock-in' phenomena are also evident,[12] the initial skeptical view of the story of QWERTY as being (at best) a *curiosum* has begun to give way to its acceptance as a metaphor for a class of multiple equilibrium processes – dynamic coordination games – that may yield Pareto inferior outcomes.

Despite the flurry of interest among applied microeconomists in industrial organization issues raised by problems of technological compatibility and standardization,[13] no great rush has yet been seen among economists to generalize broadly from the analysis of 'excess momentum' and 'lock-in' mechanisms in that specific technological context. Perhaps it is overspecialization that inhibits wider recognition of the deeper connections that should be made, say, with Thomas Schelling's (1978) path-breaking work on static models of 'tipping behavior'; and which keeps many economists from leaping ahead to see what bearing QWERTY-like dynamics might have for our understanding of other kinds of decentralized decision situations – where individual agents embedded in social and informational networks are subjected to analogous positive feedback forces that can cause collective choice processes to tip towards one extreme or another.

Whatever the cause for it, I confess I did get a little discouraged when smart people took these detailed stories so literally that they asked: 'Are you suggesting that historical influences upon present day economic life come only through the path-dependent nature of technological change, and not via institutions, or the formation of tastes?' Quite the opposite is the case, of course: I meant to suggest the fundamental homomorphism of the dynamic processes that are at work in all of those domains. Evidently, the problem is how to convey my main point simply and effectively. Although I have no intention to give up brass tacks, resign the life of teaching through exemplary works, and try for a career as a television evangelist crusading for 'generalized path-dependence and the historicization of all economic analysis', a change in the style of argument may be called for at the present juncture.

I am persuaded that there remains something important for economists (myself included) to learn about individual choice and decentralized decision-making in 'network contexts', and about the operation of such structures in propagating the influence of historical events, or what I call 'strong history'.[14] The network contexts I have in mind are not restricted to those created by the interrelatedness of production technologies; real economic actors function within many varieties of networks – social, and kinship-related, as well as commercially transactional and technological. Each of these potential webs of interaction and positive reinforcement into which individual agents may be drawn provides a theatre for the unfolding of historical dramas.

So I shall make another attempt here at getting that general idea across, this time with the aid of an entirely hypothetical example. Given my technical limitations, any economic model that I would design to have others take home to play with is going to be a very simple toy, indeed. But, perhaps this will be an advantage, in that its structure – the starker for having fewer lifelike complications – will be more immediately grasped as one that could be adapted and elaborated by others more adept at rigorous model-building; and eventually put to some good use in studying a wide array of empirical questions. The history of economic analysis surely offers much support for gambling on the lure of artificiality: those models which have succeeded best in becoming paradigms for thinking about everything are just the ones whose specifications make it plain that they are not meant to apply directly to any actual situation.

Networks, interdependence and history: a heuristic model

The illustrative problem I have chosen for this purpose is one that (I'm almost sure) must have been thought through thoroughly by Thomas Schelling while he was writing *Micromotives and Macrobehavior*. But, I haven't yet located the place where he, or anyone else of similar bent has fully worked it out.[15] I call it 'the snow-shoveling problem' and will sketch it here in the following way.

The 'snow-shoveling problem'

(i) There is a city block lined with stores, and there is a snow-storm in progress: the snow is falling continuously but gently.
(ii) The pavement in front of each shop can be kept passable if the shopkeeper goes out from time to time and shovels off the snow. In other words, given the available shovel-technology, it is quite possible for each of them to keep pace with this moderate storm – in removing snow from their own portion of the sidewalk.
(iii) A typical store owner rightly believes there will be a net benefit from shoveling her own bit of sidewalk *if and only if* the pavement outside at least one of the shops immediately adjoining is being kept passable; because then customers would be able to get through to her doorway. To keep things simple, consider only shops with neighbors on both sides; if you like, picture the stores being arranged around a four-sided city block.[16]
(iv) However, the merchant's private net gain isn't big enough to make it worthwhile for her to hire someone to shovel off the next-door pavement if and when it becomes obstructed, or to put up with the hassle (transactions costs) of persuading the neighbor to do so.
(v) Being busy with customers and other tasks, the shopkeepers do not continuously monitor the state of the sidewalk. Instead, each follows a strategy of checking on conditions outside when they have a randomly-timed free moment. But things are so arranged that the interval between those moments is short enough for her to expect to be able to keep up with the snowfall should she want to maintain a clear sidewalk throughout the storm's entire (long) duration.[17]

When a shopkeeper happens to peer outside her door, here's what her response pattern is going to be:

(i) First, suppose that her own pavement is clear because her decision at the previous point was to have it shoveled. If she finds the sidewalks on either side are similarly clear, she will decide to keep her sidewalk clear. And if both her neighbors' pavements are piled high with snow, she will decide to let the snow accumulate on hers. But if she finds the pavement is clear on one side and not on the other, she flips a (fair) coin to decide whether or not to continue a policy of shoveling – on the reasonable supposition of there being equal probability that potential customers could be trying to reach her shop front by coming from either direction, right or left.

(ii) Alternatively, suppose that the time before she had decided not to shovel. The payoffs from instantly clearing her pavement will be the same as those which apply under the condition already considered: they depend upon what her neighbors have done. So, again, she will stick to her previous policy (no-shoveling, in this instance) if she finds that is what both neighbors are doing; and her policy will switch (therefore opting to shovel) if it was found to be out of line with the neighbors'. When she receives a 'mixed signal' from the condition of the neighboring sidewalks, there is a 50% chance she will switch from (or continue) the previous policy.

An obvious problem-question has by now formed in your mind: What can we expect to happen on this block as the storm goes on? Will there be sections of pavement that are being kept clear, whereas in front of other rows of shops the snow has been allowed to pile high? Will the entire length of the sidewalk turn out eventually to be impassable, or can the system find its way to the opposite, far happier macro condition – with the entire pavement being kept clear, and business humming? Is there any way to successfully predict which of these configurations eventually would emerge? What does economic intuition tell you?

Snow-shoveling, Markov random fields and the wider world

Before proceeding towards answers, there are four remarks to be made about the foregoing set-up. Three of these are technical observations, and the fourth concerns the generality of the basic structure – and hence the applicability of the methods suitable for its analysis to a wider class of economic phenomena. I shall take up these points, briefly, in turn.

The first thing to notice is that at each random ('sample-and-decide') point in time, the individual shopkeepers' transition probabilities are strictly state-dependent and not path-dependent; they are described completely when the current 'state' is characterized by a triplet indicating the prevailing snow condition (previously shoveled or not) on her own sidewalk, and on the right- and left-adjoining pavements. Thus the evolution of her snow-shoveling policy can be said to follow a *first-order Markov chain*. Notice that many micro-economic decision-problems involve binary (or discrete multinomial) choices, and where these are subject to revisions conditional on changes in the agent's environment (current state) it is quite conventional for economists to model the microbehavior as a first-order Markov chain.

A second point is that the course of action just described is exactly equivalent to another, even simpler strategy. It wouldn't take one of these shopkeepers long to realize that instead of checking both neighbors and flipping a coin on occasions when they do not agree, she could flip the same (fair) coin in order to decide which one of the adjoining pavements to look at exclusively. When her past policy choice was found to be out of line with the prevailing policy of the selected neighbor, she would switch to match what that neighbor was doing; and if the two of them were found to coincide, she would continue with her previous policy. But, either way, her strategy can be reduced to this: at a random moment adopt whichever policy a neighbor selected at random has been following most recently.

The third point is now immediately apparent. When we move to the macrolevel and ask about the properties of a system composed of all the shopkeepers on the block, their collective behavior can be viewed as a stochastic process formed from *additively interacting Markov processes* or, as we might say, 'interdependent Markov chains with locally positive feedbacks'. This last observation is quite a help, because there already exists a substantial literature in probability theory for economists to consult on the subject of interacting Markov processes and *Markov random fields*.[18]

Some formal definitions may be supplied at this point to make clearer just why the latter measures are of relevance. So: let G = G (A, E) be *a graph* with A = ($a_1, a_2, \ldots, a_n$) the vertices or nodes, and E = ($e_1, e_2, \ldots, e_n$) the edges or shortest line connections between nodes. A (spatial) *configuration* x is an assignment of an element of the finite set S to each point of A. In our example we can denote the policies of 'shoveling' and 'not shoveling' by + and −, respectively and define S = (+, −); a configuration is then an assignment of a + or − to every shop location (node) of the graph (the horizontal base line of Figure 10.1) describing our city block. The *neighbors N(a)* of shop a are the set of all points b of A, such that (ab) is an 'edge'. (My notation follows Kindermann and Snell, 1980.)

Now we can say: a *random field p* is a probability measure p(x) assigned to the set X of all configurations x such that p(x) > 0 for all x. Random field p is called a *Markov random field* if

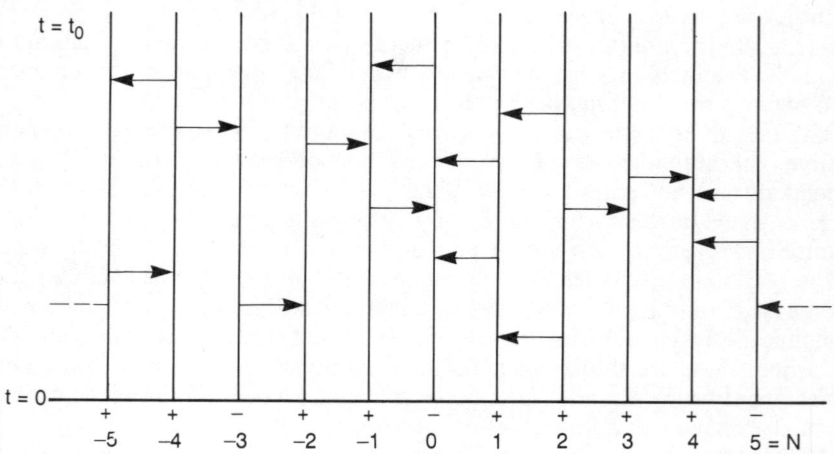

Figure 10.1 A schematic representation of the 'snow-shoveling' model, uses a Harris-type diagram (adapted from Kindermann and Snell, 1980) to show how each store will revise its policy in exponential time – adopting the policy that was being followed by a randomly selected neighbor

$$p[x_a = s|x_{A-a}] = p[x_a = s|x_{N(a)}];$$

which is to say, if in order to predict the value obtaining at node a (policy state, + or −, at shop location a), given the entire configuration of A, we need only know the values (the prevailing policy states) assigned to the neighbors of a.

There is more than the satisfaction of intellectual curiosity to be gained from exercising one's intuition on the particular mental jungle-gym I set up in the previous sub-section. Although artificial and deliberately simplistic, the 'snow-shoveling' construct is meant to bear a resemblance to core features of many natural, and more complicated dynamic processes involving *coordination equilibria*. These appear in macroeconomics, industrial organization economics pertaining to oligopoly behavior, the microeconomics of technological competition and standardization, and the economic analysis of social conventions and institutions.[19] Recognition of this wider sphere of relevance should be sufficient motivation for the moment. It will be easier to grasp some details of those connections after working through the simple illustrative model.

Interacting Markov processes: properties of the model

The structure of the particular interacting Markov process that we have been looking at is one that has been introduced in different contexts,[20] by Clifford and Sudbury (1973) and Holley and Liggett (1975), and has since been very thoroughly investigated. In briefly presenting its dynamic properties I can crib from the work of Ross Kindermann and J. Laurie Snell (1980), who have, in turn, made use of a neat graphical representation due to Harris (1978).

A linear graph of the location of stores (points) situated contiguously around the perimeter of the city block appears in Figure 10.1 – along the horizontal axis. Eleven locations are indicated, 5 on each side of store 0. (In general one can consider the regular graph consisting of the points $-N, -N + 1, \ldots, 0, 1, \ldots, N$, but the Figure shows the case $N = 5$.) The reference set (neighbors) of a shop at point i consists of neighboring points (shops) $i - 1$ and $i + 1$ for $-N < i < N$, i.e., except for the points at the extreme left and right end of the graph. Since we have laid out the block-surrounding sidewalk on a line, the neighbors of N are actually points $N - 1$ and $-N$, and the neighbors of $-N$ are $-N + 1$ and N.

Time is indicated in Figure 10.1 by distance along the vertical lines constructed at each point from the base line (corresponding to $t = 0$). The times *between* a shopkeeper's reassessments of the snow conditions on the adjoining sidewalks is assumed to be distributed exponentially with a mean 1. To represent this, along each time line one can randomly draw horizontal arrows pointing to the right neighbor, letting the distance separating these arrows be exponentially distributed with mean 2; and then do the same for arrows pointing in the other direction.

The dynamic evolution of a shopkeeper's snow-shoveling policy from the initial assignment now is readily determined: move up the shop's time line until you come to an arrow pointing towards it, at which time the shop adopts the policy of the neighbor from which that arrow comes. The whole process can therefore be described as *a finite state continuous time Markov chain* with *states* being configurations of the form x = (+ + − + . . . + + 1), where x(i) is the policy of the i-th shopkeeper. The configurations x^+ and x^-, where everyone is in agreement on shoveling or not-shoveling, represent the *absorbing* states of this process. It is self-evident that when either of those states is reached no further policy changes could occur.

So now we have a pretty good intuition of what we should expect to find as the snowstorm was infinitely extended: complete standardization on either a positive or a negative policy in regard to snow-shoveling represents the stable attractors into whose domain the dynamics of the system will be pushed. In fact, a quite simple proof will be given shortly for the proposition that for any starting state (initial configuration) x the chain will, with certainty, eventually end up in one or the other of these two absorbing states, x^+ or x^-.

Notice that the existence of a multiplicity of absorbing states means this process plainly is *non-ergodic*, inasmuch as it cannot invariably shake loose from all initial configurations. If the block started off with everyone out shoveling the pavement, things would stay that way; and likewise if everyone started off by not-shoveling until they saw what their neighbor(s) would do. Which of these extreme outcomes will emerge when this game is actually played cannot be indicated with certainty at its outset. Quite obviously, if the process remained undisturbed, the identity of the eventual equilibrium will be determined by the initial configuration of policy assignments *and* the subsequent random timing of the visits the shopkeepers make to check on conditions in the immediate neighborhood.

Predictability of the limiting macro state

Even more interesting than the certainty that the outcome will be one of those extremal solutions, is the predictability of the asymptotic state of the system for more general initial configurations – that is, when the initial condition is not a trivial case involving complete unanimity of policy assignments. It turns out that for this continuous time finite state Markov chain prediction *ex ante* is possible, and the following intuitively plausible proposition can be shown to hold: starting in configuration x, the probability that the chain will end in absorbing state x^+ is equal to the proportion of + policies in that initial configuration.

A simple graphical device due to Harris (1978) has been applied by Kindermann and Snell (1980) in a way that makes this proposition quite easy to confirm: Figure 10.2 introduces what is called the dual graph for the percolation substructure represented by the 'snow-shoveling process'. The dual graph is constructed simply by reversing the direction of each of the arrows in Figure 10.1.

To determine the probability that store 0 in Figure 10.1 starting with policy assignment x will arrive at policy + after time t_0, one can go to the dual graph and trace events backwards time from t_0 towards $t = 0$. So, starting on the store 0-line at time t_0, follow the path *downward* in the direction of the arrows until $t = 0$. By construction, store 0 eventually acquires the policy (+) of the store at the end of the path indicated in Figure 10.2.

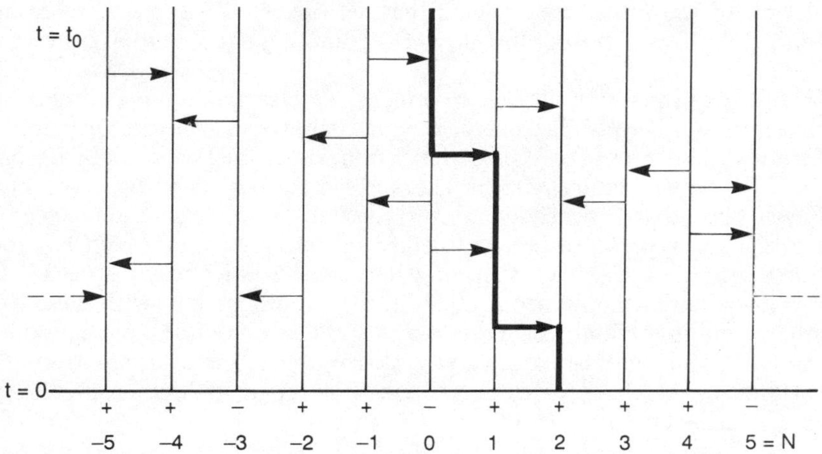

Figure 10.2 The dual representation for the snow-shoveling process reverses the arrows in Figure 10.1. By tracing a random walk backwards through time for store O, it is seen that the probability of the + policy being followed by that store at (large) time t = to in the actual process will approach the proportion of +'s in the original policy assignment

This random path is the outcome of a continuous time random walk with states defined (in general) as the integers −N to N; it moves downward an exponential time whose mean is 1, and then with probability 1/2 leftward one step, and with probability 1/2 rightward one step. It is known that the limiting distribution for this (symmetric) random walk, as t becomes infinitely large, assigns uniform probabilities over all the attainable states −N to N. Therefore the probability that the downward path for store 0 in the dual graph will end at a (location with) + policy is just the proportion of +'s in the configuration at t = 0. But this represents, in the original process, the probability that after a long time storekeeper 0 will be following a + policy.

Now consider shops at two different locations, and ask: what is the probability that at some time t both will have the same policy (say, +)? Going to the dual graph in Figure 10.2, two random walks can be traced downwards, starting from those locations at time t_0. Clearly these will not be independent; if they ever intersect they will continue together as a single random walk. But it is known (from a theorem due to Polya[21]) that in infinite time two symmetric random walks performed in one dimension – and random walks in two dimensions also – will meet with probability one. The answer to the question now before us, therefore, can be given directly from the preceding one: the probability that after some long time both shopkeepers will be following the + policy approaches the proportion of +'s in the initial configuration.

This argument can be generalized directly to arrive at the result that the probability of the shops at all of (the finite numbered) locations after some long time having a + policy, i.e., the probability of x^+ being the limiting configuration, is just the proportion of + assignments in the initial configuration. By analogy, the probability of x being the limiting configuration is the proportion of the initial assignments that are *not* +. Together these results establish that there is zero probability of the limiting configuration containing both +'s and −'s.

The fact that random walks in two dimensions, like those in one dimension, will intersect with probability one in infinite time means that these asymptotic properties will hold also for Markov random fields defined on two-dimensional graphs. The square lattice (or the checkerboard array) is perhaps the most easily visualized two-dimensional network design: the decision units located in such arrays are represented by the nodes (or cells), each of which has four nearest neighbors – east and west, and north and south.[22] Sequential binary choices made under the influence of additive local interactions with these near neighbors, will in the limit converge to one or the other uniform configuration; and the outcome will be predictable, as has been seen, simply from the proportions in which the alternative policy assignments were made in the initial configuration.

Limitations on generalizations of the model

Intuitively, one would not expect these results to be robust to further generalizations of the model in this direction. The reason is that there is no certainty of meetings between random walks performed in three-dimensional, or still higher-dimensional spaces. When agents' decisions are subject to 'network externality effects' of such complexity that they cannot be reduced to a two-dimensional representation, there is no guarantee of their policies becoming *perfectly* correlated in the long run; all possible assignments of policies for any particular finite set of agents will have positive probability in the limit. But, at the level of three dimensions, some strong correlation still can be expected to emerge.[23]

There is another direction in which the foregoing model cannot be generalized without losing the qualitative properties of eventually locking in to an extremal solution, and of having that outcome be predictable on the basis of some 'initial' or 'intermediate' configuration. Those properties will hold strictly only for finite populations of agents. The behavior of very large finite systems, however, may well be better indicated by that of an infinite population – a continuum of shopkeepers. For the case of a continuum, a random walk being performed on all the real integers, it is found that there will not be a time after which the system remains in an absorbing state;[24] in the limit of the snowstorm the process oscillates between the extrema of perfectly correlated policies – with all sidewalks shoveled clear at one moment, and all the shops being snowbound at some other.[25]

What may we surmise, if the evolution of collective behavior in extremely large networks with positive local externalities is likely to resemble that of the infinite population case? Even under stationary structural conditions, such populations should not be expected to become inextricably 'locked in' to one of a number of locally stable equilibria. But, as noted below, they may linger for many periods of time in those neighborhoods.

What can then be said with regard to the qualitative macro behavior of smaller ensembles of interacting decision agents, when these remain isolated from external interference and their internal structural relations remain undisturbed? If there is no inherent tendency for the behavioral orientation of the individual agents to undergo 'spontaneous' changes, that is changes not prompted by interactions or signals received from the macro state of the system, these systems will be most prone to exhibiting the workings of history in its strongest form: convergence to an indefinitely persisting equilibrium (i.e. 'lock-in') characterized by perfectly correlated individual policy choices. Moreover, the prevailing collective policy, or behavioral norm arrived at will have been 'selected' adventitiously, as a result of some particular concatenation of small events on the historical path. Inasmuch as the policy thus settled upon may well be one that is globally inefficient, indeed, collectively disastrous, this may be a reason why small systems of social and economic interaction appear historically to have a greater tendency to become extinct or absorbed into larger systems.

In dynamic models where the orientation (say, + or −) of the individual particles are subject to continuous *spontaneous* random perturbations, as well as to the influence of positive feedback from the macro state of the system (say, the difference between the global proportions of + and − particles), the expected time interval separating the transitions the system makes between the neighborhoods of its local equilibria will lengthen rapidly if the number of particles in the (finite) system is increased.[26] An intuitive explanation can be given for this size effect: it is the spontaneous random perturbations of the system's micro units that serve to push it out of the neighborhood of one potential minimum and beyond the point of instability, whence it can move towards the other point of locally minimum potential. In a larger population it is a lower probability event that enough independently distributed random perturbations will be positively correlated to produce such a 'shock'. This is the same principle that prevents a large herd of horses that are tethered together from moving any great distance in finite time, whereas, as any cowboy knew, a small band of horses tied together for the night could easily be out of sight by the sunrise.

Reinterpretations and immediate applications

The foregoing model lends itself readily enough to several loose reinterpretations that show the general bearing these methods and results can have on the programs of applied research currently being pursued by a wide assortment of economists. Among these I would include some of the following:

(i) Macroeconomists concerned with the implications of so-called 'trading externalities', 'investment spillovers' and 'Keynesian coordination problems'.[27]

(ii) Microeconomists working on the range of classic problems relevant to the 'organic' or spontaneous emergence of social customs and social institutions affecting resource allocation.[28] Included with these are the important historical processes of institutionalization of non-obligatory contractual and organizational forms governing long-term trading and employment relations.[29]

(iii) Those interested in endogenous aspects of consumer taste formation, and in the interplay between narrowly economic psycho-social forces in the formation of public opinion on matters of political economy, particularly processes through which emerge a public consensus in 'collective conservatism'.[30]

(iv) Industrial organization economists working on oligopoly dynamics and strategic behaviors sustaining implicit collusion or cartel cooperation – particularly where monitoring of cartel adherence is imperfect.[31]

(v) Applied economists worrying about standards for technological compatibility in 'network' industries, especially the competitive diffusion of alternative technologies characterized by 'increasing returns to adoption' and the emergence of *de facto* standardization.[32]

The seeming diversity of this collection notwithstanding, an area of common ground remains clear. It lies in the interdependence of choices made sequentially by agents pursuing their private interest within contexts where substantial benefits arise from coordinated or synergetic actions, and where individual decision-agents each are limited (for various reasons) to considering only the actions of the members a particular reference group, or sub-set of 'significant others' within the collectivity. As in the snow-shoveling problem, however, the reference groups of agents in question generally are not disjoint; because the 'neighbors' of one agent are also 'neighbors' of some other agents, individual units' decisions become linked indirectly by the intersecting networks of local relationships which transmit and communicate their effects.

Rather than attempting to elaborate all these contexts of application explicitly, I prefer to invite those more deeply involved in each of the various specialized areas to take up the task. In that spirit I will direct the main thrust of my remaining comments to the microeconomics of competitions among alternative technologies, and among alternative organizational and institutional forms. Having elsewhere avoided dwelling on the applicability of this particular model for analyses of the economics of compatibility standards and the emergence of *de facto* standardization, I feel freer to do so here without inviting the surmise that I associate path-dependence solely as a feature of the gradual evolution of technological systems and social institutions.

When we consider what application the preceding model might have to analysis of the dynamics of technology adoption under conditions of positive network externality, the correspondence between a binary policy choice and a binary technology choice is immediately transparent. The relevance of the analysis of the preceding section – based as it is on the structure described by Holley and Liggett (1975) as a 'voter model' – to Kuran's (1987) theory of the formation of public policy consensuses should also be sufficiently self-evident to require no further comment here. Only slightly less self-evident, in my view, is its relevance to the dynamics of the institutionalization of certain organizational practices, such as the granting of procedural rights to university professors under the tenure system, in which the significance of withholding the right is affected by how may other organizations aspiring to the same status (i.e. the neighborhood set for comparisons) have done so.

Equally transparent is the representation of the 'locally bounded' nature of the technological network externalities which impinge upon users' choices. Telephone service subscribers may care only about the ease of communicating directly with their family, friends, and business associates, not with someone drawn at random from the entire community; and computer users may give weight only to the advantages of being able to share specialized software applications and technical information with identified co-workers. This form of local, or neighborhood 'boundedness' has not been explicitly considered – to my knowledge, anyway – by the theoretical economics literature devoted to the dynamics of technological competition under conditions of positive feedback due to network externalities. Arthur (1989) discusses the properties of a generalized Polya urn process featuring a different kind of 'bounded' increasing returns: the probability of adding a ball of one color rather than another to an urn at each moment is assumed to improve as a function of the *global* proportion of the balls in the urn that currently are of that color, with the improvement function's slope decreasing to zero within the domain of the probability distribution's support. Although Arthur has shown that such processes can reach limiting states in which more than one color will be represented by a finite share of the urn's population, this result implies nothing about the behavior of the one- and two-dimensional systems of locally interacting particles considered here. It does have a rough analogue, however, in the behavior of three-dimensional finite particle systems – where the random perturbations impinging upon each particle are capable of overwhelming the local positive feedback effects.[33]

Extension of the model to two dimensions offers a quite natural representation of a spatial context within which choice-of-technique decisions arise affecting compatibility and 'connectivity'. This has been applied, for example, by Puffert (1987) in a dynamic stochastic simulation model of the process through which extended railway networks of a standardized gauge may emerge from decentralized gauge choices by local lines. By contrast, most entries in the recent spate of studies dealing with the dynamics of competitive diffusion of alternative technologies subject to network externalities of one kind or another, assume micro decisions are directly influenced by positive feedback from the macro state of the system; they suppose that individual equipment buyers of VCRs select between Sony Betamax and VHS formats, and personal computer users choose between PCs and MACs, by balancing their intrinsic preferences against considerations that depend upon the relative share of the installed base which each of the competing technologies holds in the *global* market. The same simplification typically is made in constructing analogous models of 'tipping dynamics' and 'bandwagon momentum' in public voting and crowd behavior – in which individuals' private preferences are seen as being first subjugated in favor of deriving the approbation of the majority, and ultimately modified (to minimize cognitive dissonance) by perceiving virtue in the state that prevails.[34]

Because the snow-shoveling model is focused upon the macro dynamics that emerge from the microlevel process of policy revision, it lends itself immediately to the analysis of technology choices that can be revised when equipment wears out.[35] Most theoretical treatments of competitive diffusion abstract from such complications by assuming the alternatives come embodied in infinitely durable capital goods. In such models the only way the share of each variant in the total installed base can change is if there are additions made to the stock, and such additions have a different distribution than the stock itself. This is the set-up of Arthur's (1989) model of sequential commitments to rival technological systems, for example. But, assuming infinite durability is not realistic and suppresses important aspects of the way that *de facto* standardization historically has been achieved.

Nor is strict irreversibility of technology commitments at the micro level or physical durability a condition necessary to produce the result that a decentralized choice process with positive local externalities eventually will 'lock' to one or another of its extremal solutions – complete standardization on one or another version of the technology. Such 'lock-in' has been shown to be a property of the foregoing model – in which the exponential time process of policy revision can be immediately reinterpreted as an exponential process of replacing a one-horse-shay durable (embodying one or another version of the technology) whose expected service life is unitary.

How the global 'replacement market' first comes into existence is not part of the story described by the snow-shoveling model. A more complete, if stylized account would describe some process generating an initial, possibly jumbled configuration of users of incompatible versions – say, the P(lus)-variant and the M(inus)-variant, which could represent PC and Macintosh computers – of the new technology. By a simple modification, for example, we could add a phase at the outset during which additional discrete points (store-owners, or computer-owners) were being inserted on the horizontal line in Figure 10.1, each of whom came with an initial policy assignment reflecting the owner's inherent preferences.

Imagine, then, such a 'birth' or 'entry' process in which new Ps and Ms (+'s and −'s) take up positions randomly in the circular graph – doing so in exponential time with mean interval shorter than the unitary mean of the policy revision (or technology replacement) process. So long as this rapid entry phase lasts, the population will be tending to increase without limit and, correspondingly, the proportion of Ps and Ms in the growing total would be tending towards random fluctuations between the extrema.

But, if the mean interval between the arrival of new entrants was increased with the passage of time, eventually it would exceed the unit mean interval of the exponential replacement process and cause the dynamic properties of the system to alter so that they asymptotically approached the model already described. When a critical point was reached at which the (finite) population had become sufficiently large and the entry rate had become sufficiently slow so that only the stochastic replacement process was driving further changes in the share of the total assignments that were Ps (or Ms), the latter proportion would accurately predict which of these two variant technologies would ultimately emerge as the universal standard. Naturally, as the perturbation of the existing proportions of Ps (or Ms) by entry died away and that critical point was approached, the accuracy of predictions of the eventual outcome of the competition based upon observations of the existing distribution of installed base would be raised.

At that stage the formation of expectations would become a potent additional positive feedback mechanism, particularly if there was a gain to be achieved by individual agents who joined the emerging consensus sooner rather than later.[36] Were the individual agents able to benefit by aligning themselves sooner with the emerging consensus, and were they to possess the knowledge that the identity of the limiting correlated configuration was predictable from the currently observable configuration, monitoring of the sufficient statistic describing the macro state of the system (namely the proportion of Ps (or Ms) would become a worthwhile activity. Moreover, the formation of expectations on that basis would, under many plausible suppositions regarding individual's behavior, drive the system more rapidly towards 'selection' of the probability-favored equilibrium.[37]

Towards further expansion of the historical framework

It is evident that in the finite population situation some a priori beliefs about what the neighbors are going to do (when the snowstorm is just starting) can govern the eventual outcome. David Lewis (1969, pp. 33ff) has made the more general point that the creation of a consistent structure of mutual expectations about the preferences, rationality, and actions of agents can help in achieving rational (non-arbitary) solutions to coordination problems. Shared historical experiences, and conscious perceptions of a shared past, provides one of the principal means by which groups of people may justifiably form a system of consistent mutual expectations when they are not readily able to arrive at a common course of action by directly discussing the problem that currently faces them. When individuals who cannot communicate are confronted with a pure coordination problem, according to the classic observations of Schelling (1960), they try for a coordination equilibrium that is in some way 'salient' – one that distinguishes itself from the other candidates by possessing some unique and remarkable feature, which need not be held to have any intrinsic value.[38] So, precedent is an important source of salience in such contexts: a solution to a coordination game can acquire 'conspicuous uniqueness' by simply having been part of the players' shared history – because they remember having reached it on a previous, similar occasion.

Thus, through the reinforcement of mutual expectations, 'accidents of history' may acquire a status of surprising durability in human social arrangements. As Lewis (1969: 39, 41–2) puts it:

> It does not matter *why* coordination was achieved at analogous equilibria in the previous cases. Even if it had happened by luck, we could still follow the precedent set. . . . Each new action in conformity to the regularity adds to our experience of general conformity. Our experience of general conformity in the past leads us, by force of precedent, to expect a like conformity in the future. And our expectation of future conformity is a reason to go on conforming, since to conform if others do is to achieve a coordination equilibrium and to satisfy one's own preferences. And so it goes – we're here because we're here because we're here because we're here. Once the process gets started, we have a metastable self-perpetuating system of preferences, expectations, and actions capable of persisting indefinitely.

In the hypothetical coordination problem that we have been examining, there is an obvious functional role for a pervasive ideology of civic mindedness, 'block pride' or whatever, if that created expectations which sufficed to induce every shopkeeper to at least begin by adopting a positive policy on the question of whether or not to shovel her bit of sidewalk. And, of course, such community spirit would be reinforced by the happy material outcome: merchants located on blocks that embraced that ideology unanimously would be rewarded, individually and collectively, by staying in business with certainty throughout the storm season, whereas other communities lacking their unanimity could become snowbound.

But, *ad hoc* appeal to *universal* ideologies as a source of consistent mutual expectations is a rather unsatisfying device for historians to employ in explaining (predicting) the outcome of collective behavior in such situations. How would such ideologies come to have been formed in the first place? Moreover, in keeping with the spirit of economic imperialism, economists surely should be attempting to analyze and explain the dynamics of collective opinion-formation (see Kuran, 1987, 1988 for example). Why should we want to place the origins (and decay) of ideologies firmly outside the system, safely beyond reach of being reinforced or undermined by the microeconomic and macroeconomic results of dynamic resource allocation processes such as the one we have been considering? If there are ways to thus represent the co-evolution of microeconomic behaviors with regard to technology choices (technical standardization) or conformance with social norms (custom and convention) and correlated patterns of ideology or beliefs carrying normative force (subjective conformism), the explanatory apparatus available to economists studying long-term trends in technology and social institutions will surely be much more powerful.[39]

The beauty of a system such as the one that has been set out above is that it lends itself neatly to being nested, or 'embedded' in a larger historical model, in which the macrocosmic outcome or distribution of microcosmic outcomes from one extended, storm-like epoch might be supposed to influence the initial policy assignments adopted by players in a subsequent epoch. We would have a choice, of course, between modeling such linkages between successive 'epochs' in a way that made the transition strictly Markovian (state-dependent), or entertaining a richer historical vision in which societies' evolving common knowledge of their past channeled the course of change.

Acknowledgments

The research underlying this chapter has been supported by grants from the Stanford University Center for Economic Policy Research, and the Stanford Humanities Center, at which I held a Mericos Fellowship during 1990–1. Suggestions and references supplied by Donald Brown, Steven Durlauf and Shane Greenstein were very helpful in the original formulation. Presentations of this material in the following forums have yielded many insightful comments and criticisms, for all of which I remain grateful, even though only some could be assimilated in the revisions made for this version: the Technological Innovation Project Workshop in the Stanford Economics Department, in November 1988; the International Symposium on Non-Linear Dynamics and Evolutionary Economics at the University of Texas, Austin, in April 1989; the University of Rome ('La Sapienza') Economics Faculty Seminar in March, 1991, the SPRU Seminar at the University of Sussex, in April 1990; the University of Turin-Einaudi Foundation Conference on the Economics of Information Networks, in April, 1991.

Notes

1. 'Path-dependence: putting the past into the future of economics', Institute for Mathematical Studies in the Social Sciences Technical Report 533, Stanford University, November 1988. The latter is a revision of my presentation to the seminar series held at the IMSSS Summer Workshop (August 1988) on the subject 'Path-dependence – historical processes in economic resource allocation'.
2. For further discussion of these general points, see David, 'Path-dependence' (November 1988: section 2).
3. See P.A. David, *Technical Choice, Innovation and Economic Growth: Essays on American and British Experience in the Nineteenth Century* (Cambridge: Cambridge University Press, 1975, ch. 1.)
4. See also J.A. James and J.S. Skinner, 'The resolution of the labor-scarcity paradox', *Journal of Economic History*, 45(3), September 1985: 513–40, for a more recent discussion, and (happily) econometric evidence supporting the argument I advanced in 1975.
5. See A.B. Atkinson and J.E. Stiglitz, 'A new view of technological change', *Economic Journal* 79 (September) 1969, which has been elaborated upon recently in J.E. Stiglitz, 'Learning to learn, localized learning and technological progress', ch. 5 in *Economic Policy and Technological Performance*, P. Dasgupta and P. Stoneman (eds) (Cambridge: Cambridge University Press, 1987); N. Rosenberg, 'The direction of technological change: inducement mechanisms and focusing devices', *Economic Development and Cultural Change* 18(1) October 1969, is reprinted as ch. 6 of his *Perspectives on Technology* (Cambridge: Cambridge University Press, 1976).
6. See *Technical Choice* (1975), chs 4, 5; more recently, P.A. David, 'La moissonneuse et le robot: La diffusion des innovations fondées sur la micro-électronique', ch. 5 in *Les Enjeux du Changement Technologique*, J.-J. Salomon and G. Schmeder (eds) (Paris: Economica, 1986); and 'The reaper and the robot: the diffusion of labour-saving machinery in the past and present', forthcoming in *Essays in Honor of Sir John Habakkuk*, F.M.L. Thompson (ed.) (Oxford: Oxford University Press).
7. See, e.g., Rosenberg's *Perspectives on Technology* (New York: Cambridge University Press, 1976) and the recent formulation (and references to earlier studies, therein) provided by N. Rosenberg, 'Qualitative aspects of technological change: some historical perspectives'. Institute for Mathematical Studies in the Social Sciences Summer Workshop Paper, Stanford University, July 1988.
8. See my 'Clio and the economics of QWERTY', *American Economic Review* 75, 2 (May): 332–37; also, 'Understanding the economics of QWERTY: the necessity of history', ch. 4 in W.N. Parker (ed.), *Economic History and the Modern Economist* (Oxford: Basil Blackwell, 1986).

   Another case study focused on the early rivalry between alternating and direct current electricity supply systems: P.A. David, 'The hero and the herd in technological history: reflections on Thomas Edison and the battle of the systems', ch. 2 in *Favorites of Fortune: Technology, Growth, and Economic Development Since the Industrial Revolution*, P. Higgonet, D.S. Landes and H. Rosovsky (eds) Cambridge, Mass.: Harvard U.P.; and, relatedly, P.A. David and J.A. Bunn, 'The economics of gateway technologies and network evolution: lessons from electricity supply history', *Information Economics and Policy* 3 (Fall), 1988: 165–202.

   For more formal extensions and broader applications to problems concerning the emergence of *de facto* technical interface standards, see Paul A. David, 'Some new standards for the economics of standardization in the information age', ch. 8 in *Technology Policy and Economic Performance*, P. Dasgupta and P. Stoneman (eds) (Cambridge: Cambridge U.P., 1987).

9. See W.B. Arthur, Y.M. Ermoliev and Y.M. Kaniovski, 'Strong laws for a class of path-dependent urn processes', in *Proceedings of the International Conference on Stochastic Optimization, Kiev, 1984*, Arkin., Shiryayev and Wets (eds); Lecture Notes in Control and Information Sciences, 81. (Berlin: Springer-Verlag, 1986).
10. And, in due course, I did receive in the mail a charming luncheon invitation from the president of the International Dvorak (Keyboard) Society. As was to be expected, not everyone was persuaded: S.J. Liebowitz and S.E. Margolis, 'The Fable of the Keys', *Journal of Law and Economics* 33 (April) 1990: 1–25, dispute the claim that the Dvorak Simplified Keyboard developed in the mid 1930s was an ergonomically superior alternative; and, presumably, also that the Ideal Keyboard (a late nineteenth-century contemporary of QWERTY) also would have been preferable. In Liebowitz and Margolis' view, this casts doubt on my 1985 characterization of the historical episode. But, their discussion begins from the premise that if there were a more cost effective keyboard than QWERTY, some firm would find it possible to make a profit by introducing it, a position that fails to distinguish between efficiency *ex ante* and *ex post*, and denies that network externality considerations and 'installed base effects' could have a bearing on an innovating firm's profitability in this context.
11. For an example of one economic historian's misapprehensions see, e.g., Alexander Field, 'Let us now praise exogenous variables: a review of "Economic History and the Modern Economist"', *Historical Methods* (Spring) 1988.
12. See, e.g., R. Cowan, 'Nuclear power reactors: a study in technological lock-in', *The Journal of Economic History* L(3), September 1990: 541–68; R. Cowan, 'Sprayed to death: on the lock-in of an inferior pest control strategy', C.V. Starr Center for Applied Economics Research Report, RR no. 91-23 (April), 1991; M.A. Cushmano, Y. Mylonadis and R.S. Rosenbloom, 'Strategic maneuvering and mass-market dynamics: the triumph of VHS over Beta', Consortium on Competitiveness and Cooperation Working Paper no. 90-5, Center for Research in Management, University of California, Berkeley, 1990.
13. Among the influential recent theoretical publications in this genre, see J. Farrell and G. Saloner, 'Installed based and compatibility: innovation, product preannouncements, and predation', *American Economic Review* 76(5), December 1986: 940–55; M.L. Katz and C. Shapiro, 'Technology adoption in the presence of network externalities', *Journal of Political Economy* 94(4), 1986: 822–41. More extensive references will be found in P.A. David and S. Greenstein, 'The economics of compatibility standards: an introduction to recent research', *Economics of Innovation and New Technologies*, I(1-2), Fall–Winter, 1990: 3–41.
14. See David, 'Path-dependence' (August 1988: section 2) for a taxonomy of models differing in the strength of the role played by history.
15. In *Micromotives and Macrobehavior* (New York: W.W. Norton, 1978, p. 214) 'shoveling the sidewalk in front of your house' is mentioned among a large collection of decision problems involving a binary choice, the influence of which is external to the agent-actor's own accounting of benefits and costs, but is internal to some other agent's accounting, purview, or interest. The other examples given included getting vaccinated; carrying a gun, or liability insurance; wearing hockey helmets. Schelling's essentially static formal analysis of these problems emphasized positive feedback from the macro state to the micro decision, although, as noted below, his discussion recognizes the implications of local positive (externalities) interaction effects.
16. The condition given for there to be a private net benefit from shoveling one's own sidewalk could hold true for every shop in the system if one supposes there were potential customers already inside the shops when the storm began, as well as others who could reach the pavement from the street. Note that the physical arrangement of the shops also can make a difference. If there are boundaries to the

system, conditions at the boundaries may exercise a special influence. This point is made by Schelling (1978: 214), who gives an example of people who are able to read if they and both their neighbors keep their lights on, and will turn their own light off if either neighbor does so. There will be two equilibria when the group is arranged in a circle; if they are on a line; however, the people at the ends would not keep their light on under any condition and the whole system would go dark.

17. We may assume some special conditions to ensure that this sampling procedure is compatible with a sensible individual decision-strategy. For example, suppose there is – relative to the rate and variability of snowfall – a powerful shoveling technology at each shop's disposal (a large gang of children?) which imposes a fixed cost per 'job', independent of the actual volume of snow which has to be removed per job. So, no matter how long a time elapsed since the last shoveling, the cost (and anticipated net benefit to the owner) of clearing would be the same on every occasion.
18. See e.g., Frank Spitzer, 'Interaction of Markov processes', *Advances in Mathematics* 5 (1970): 246–90, and D. Griffeath, *Additive and Cancellative Interacting Particle Systems*, Lecture Notes in Mathematics, no. 24 (New York: Springer-Verlag, 1979), which are technically rather beyond me. Thomas M. Liggett, *Interacting Particle Systems* (New York: Springer Verlag, 1985), provides a comprehensive recent treatise. Less formidable recent surveys containing numerous references to earlier work also are available: see Thomas H. Liggett, 'Interacting Markov processes', in W. Jager, H. Rost and P. Tautu (eds), *Biological Growth and Spread*, vol. 38 of *Lecture Notes in Biomathematics* (New York: Springer-Verlag, 1980): 145–56; R.P. Kindermann and J.L. Snell, 'On the relation between Markov random fields and social networks', *Journal of Mathematical Sociology* 7 (May) 1980: 1–13, and R.P. Kindermann and J.L. Snell, *Markov Random Fields and their Applications*, Contemporary Mathematics, vol. 1 (Providence, RI: American Mathematical Society, 1980).
19. An early and important connection between games involving coordination equilibria and social conventions, and the role of expectations grounded on historical precedent in supporting conventions, was made by D.K. Lewis, *Convention: A Philosophical Study* (Cambridge, Mass: Harvard U.P., 1969). In this work (p. 42) 'convention' is defined as a regularity in the behavior of agents faced with a recurrent coordination problem, about which it was a matter of 'common knowledge' (in the technical sense of something known to be known, and known to be known to be known, *ad infinitum*) that every agent would conform, and was expected by every other agent to conform, because of it no one would be better not conforming given that the others conformed. For a recent discussion of these and related ideas, see R. Sugden, 'Spontaneous order', and J. Elster, 'Social norms and economic theory', *Journal of Economic Perspectives* 3(4) Fall, 1989: 85–98 and 99–118, respectively.
20. See P. Clifford and A. Sudbury, 'A model for spatial conflict', *Biometrika* 60, (1973): 581–8; R. Holley and T. Liggett, 'Ergodic theorems for weakly interacting systems and the voter model', *Annals of Probability* 3, (1975): 643–63; T.E. Harris, 'Additive set-valued Markov processes and percolation methods', *Annals of Probability* 6 (1978): 355–78; R.P. Kindermann and J.L. Snell, 'On the relation between Markov random fields and social networks', *Journal of Mathematical Sociology* 7 (1980): 1–13.
21. See William Feller, *An Introduction to Probability Theory and Its Applications*, 3rd edition (New York: John Wiley and Sons, 1969): 360–2, for alternative formulations and a proof of Polya's theorem.
22. In the one-dimensional array of Figure 10.1, each shop belongs to three three-party local networks; in the square lattice arrangement just described, each actor would be a member of five 'local networks' – the one in which she occupies the central

position, and those in which her four near neighbors, respectively, are centrally positioned; in a hexagonal arrangement on the two dimensional lattice, with agents positioned at the vertices, every agent participates in seven overlapping local networks.

23. Although there is thus no certainty in the three-dimensional case that the system eventually will 'lock in' to one of the extremal (uniform) configurations, in the limit there will be strong correlation among the policy assignments arrived at by members of the network. See Kindermann and Snell, *Markov Random Fields* (1980: 79–80), and M. Bramson and D. Griffeath, 'Renormalizing the 3-dimensional voter model', *Annals of Probability* 7, 1979: 418–32, for further discussion.
24. See Kindermann and Snell, 'Markov random fields and social networks' (1980: 11).
25. Qualitative results of this sort are familiar in the framework of general dynamic models of systems in which particles are subject to spontaneous random reorientations (policy switches, in the present metaphor), which are influenced by additive local interaction effects. This framework has been studied extensively in connection with dynamic two-dimensional 'Ising models' of magnetization (following E. Ising, 'Beitrag zur Theorie des Ferrogmagnetismus', *Z. Phys.* 31, 1924: 253–8; R.E. Peierls, 'On Ising's ferromagnet model', *Proceedings of the Cambridge Philosophical Society* 32, 1936: 477–81; R.B. Griffiths, 'Peierl's proof of spontaneous magnetization in a two-dimensional Ising ferromagnet', *Physics Review* 136, 1964: A437–9).
26. W. Weidlich and C. Haag (*Concepts and Models of a Quantitative Sociology*. New York: Springer-Verlag, 1983, pp. 37–50) examine explicit solutions for such a model, which exhibits such 'fluctuation dominated motions' between two points of minimum potential, and show that the transition time is an *exponentially* increasing function of the number of units in the system.
27. See e.g., P.A. Diamond, 'Aggregate demand management in search equilibrium', *Journal of Political Economy*, October 1982, *90*: 881–9; P.A. Diamond, *A Search-Equilibrium Approach to the Micro Foundations of Macroeconomics, The Wicksell Lectures 1982*, Cambridge, Mass.: MIT Press, 1984; P. Howitt, 'Transactions costs in the theory of unemployment', *American Economic Review*, March 1985, *75*: 88–100; W. Heller, 'Coordination failure with complete markets in a simple model of effective demand', *Discussion Paper No. 84–16*, Department of Economics, U.C. San Diego, September 1986; S.N. Durlauf, 'Locally interacting systems, coordination failure, and the behavior of aggregate activity', Center for Economic Policy Research Technical Paper, no. 194, Stanford University (January 1990), and S.N. Durlauf, 'Non-ergodic economic growth and path-dependence in aggregate output fluctuations', *American Economic Association Papers and Proceedings* 81 (May), 1991, goes far beyond earlier work in this vein, by examining the time-series properties of a fully specified dynamic stochastic process.
28. My reference to these as 'classic' alludes to the following famous passage from Carl Menger (*Problems of Economics and Sociology* (*Untersuchungen ueber die Methode der Socialwissenschaften und der Politischen Oekonomie insbesondere* Leipzig, 1883), transl. F.J. Nock, Urbana: University of Illinois Press, 1963, p. 130):

    The normal function and development of the unit of an organism are thus conditioned by those of its parts; the latter in turn are conditioned by the connection of the parts to form a higher unit. . . . We can make an observation similar in many respects in reference to a series of social phenomena in general and human economy in particular. . . . Similarly we can observe in numerous social institutions a strikingly apparent functionality with respect to the whole. But with closer consideration they still do not prove to be the result of an *intention aimed at this purpose*, i.e., the result of an agreement of members of society or of positive legislation. They, too, present themselves to us rather as 'natural' products (in a certain sense), as *unintended results of historical development*.

For further discussion of the relevance of information network paradigms to the development of social organization and cultural institutions, see P.A. David, 'Information technology, social communication norms and the state: a public goods conundrum', paper presented to the CREA Conference on Standards, Norms and Conventions, held 27–9 March, 1989 at L'Ecole Polytechnique, Paris, France.

29. See, e.g., G. Wright, 'Historical processes in land tenure and labor market development', Institute for Mathematical Studies in the Social Sciences Summer Workshop Paper, Stanford University (June 1988); P. Murrell, 'The economics of sharing: a transactions cost analysis of contractual choice in farming', *Bell Journal* 14, Spring 1983: 283–93; W.A. Sundstrom, 'Institutional isomorphism: the standardization of rules and contracts in business firms and other institutions', Economics Department Working Paper, Santa Clara University, May 1988. See also D. Romer, 'The theory of social custom: a modification and some extensions', *Quarterly Journal of Economics* 98, November 1984: 717–27, for a different but somewhat related analysis of the macro consequences and reinforcement of micro beliefs affecting adherence to 'social custom' in employment relations.
30. See, e.g., M. Granovetter, 'Threshold models of collective behavior', *American Journal of Sociology* 83, 1978: 1420–43; M. Granovetter, 'Economic action and social structure: the problem of embeddedness', *American Journal of Sociology* 91, 1985: 481–510; T. Kuran, 'Preference falsification, policy continuity and collective conservatism', *Economic Journal* 97, September 1987: 642–55. A critical survey of the literature, with many references, will be found in T. Kuran, 'The tenacious past: theories of personal and collective conservatism', *Journal of Economic Behavior and Organization* 10(2), September 1988: 143–71.
31. See, e.g., D. Kreps, P. Milgrom, J. Roberts and R. Wilson, 'Rational cooperation in the finitely repeated prisoners' dilemma', *Journal of Economic Theory* 27, 1982: 245–52; D.M. Kreps and A.M. Spence, 'Modelling the role of history in industrial organization and competition', ch. 10 in *Issues in Contemporary Microeconomics and Welfare*, ed. G.R. Feiwel (London: Macmillan Press, 1985); E.J. Green and R.H. Porter, 'Noncooperative collusion under imperfect price information', *Econometrica* 52, 1984: 87–100; D. Abreu, D. Pearce and E. Stacchetti, 'Optimal cartel equilibria with imperfect monitoring', *Journal of Economic Theory* 39(1), 1986: 251–69. Kreps and Spence (1985: 364) remark that the technique of setting up games of incomplete information, and seeking a non-cooperative equilibrium of the cartel game where each participant is assumed to condition its actions on the private information it has received, 'captures the idea that history matters. At intermediate points in the game, participants use the actions of their rivals to make inferences concerning those things about which they were initially uncertain. . . . Since rivals tomorrow will be looking back [if they are able] to see what you did today, today you should look forward to gauge the consequences of your actions on tomorrow's competition.' The snow-shoveling process makes history *really* matter, by restricting the informational capabilities of the agents to what they can discover just by observing their neighbors. This is brought out explicitly in David, 'Path-dependence' (August 1988: section 5).
32. The sample of works cited here all give formal and explicitly dynamical treatments of these subjects: W.B. Arthur, 'Competing technologies and lock-in by historical small events', *Economic Journal*, 99(394) March, 1989: 116–31; J. Farrell and G. Saloner, 'Installed based and compatibility: innovation, product preannouncements, and predation', *American Economic Review* 76(5), December 1986: 940–55; M.L. Katz and C. Shapiro, 'Technology adoption in the presence of network externalities', *Journal of Political Economy* 94(4), 1986: 822–41; P.A. David, 'Some

new standards for the economics of standardization in the information age', in *Economic Policy and Technology Performance*, eds. P. Dasgupta and P.L. Stoneman (Cambridge: Cambridge U.P., 1987); D.J. Puffert, 'Spatial network externalities: a model with application to the standardization of railway gauge', Technology and Productivity Workshop Paper (Department of Economics), Stanford University, December 1987.

33. D.J. Puffert ('Spatial network externalities – a model with application to the historical standardization of railway gauges', Social Science History Workshop Paper, Department of Economics, Stanford University, Fall 1987) tackles the problem in a spatial network model that is far more complex than the one suggested here – which he studies using stochastic simulation methods. See text below.
34. Like S. Jones (*The Economics of Conformism*, Oxford: Basil Blackwell, 1984), Kuran (1988) invokes L. Festinger's *A Theory of Cognitive Dissonance* (Stanford, California: Stanford U.P., 1957) in suggesting a micro level (taste formation) mechanism that would serve to rapidly 'lock in' a social (or in Jones' context, a work group) consensus reached by bandwagon dynamics. Taking a longer real-time perspective Kuran (1988: 32–3) also suggests that educational and informal socialization of new cohorts may serve as mechanisms strengthening the tenacity of the past's hold. But, as one so often observes in revolutionary regimes, re-education and 're-socialization' are equally instruments employed in freeing societies from their history – for better or worse.
35. The assumption that the previously installed technology wears out – and so must be completely replaced – abstracts from 'switching costs' or 'conversion costs' that arise from the technical interrelatedness among the producer's portfolio of real assets. The necessity of rewriting code for specialized applications software, for example, creates substantial 'switching costs' for mainframe computer users considering changing to equipment using an operating system different from that of the one that previously had been installed. See, e.g., S. Greenstein, 'Computer systems, switching costs and organization responses: the federal government experience', Technology and Productivity Workshop Paper (Economics Department, Stanford University), October 1988.
36. In the illustrative example of the shopkeepers, since the intervals between policy reconsiderations are exponentially distributed, it is quite possible that a given shopfront would remain blockaded by snow for some extended period of time (thereby missing opportunities for business) even though the sidewalks for some distance to either side had all been cleared. An ability to assign a probabilities to the alternative consensuses that could emerge would thus permit the development of a superior choice strategy than the one based on myopic consideration of randomly arriving information about the state of the neighborhood.
37. On the role of self-fulfilling expectations in dynamic stochastic systems of this kind, see David, 'New standards for the economics of standardization' (1987), and W.B. Arthur, 'Self-reinforcing mechanism in economics', in *The Economy as an Evolving Complex System*, P. Anderson, K. Arrow, and D. Pines (eds) (Reading, Mass.: Addison-Wesley, 1988).
38. See Thomas C. Schelling, *The Strategy of Conflict* (Cambridge, Mass.: Harvard U.P., 1960), pp. 83–118, 291–303.
39. On the importance of taking account of induced and self-reinforcing normative messages, rationalizing and sanctioning behavioral trends among some segments of a population, see Victor R. Fuchs, 'A note on prices, preferences, and behavior', National Bureau of Economic Research Memorandum, Stanford, CA (September 1985). The co-evolution of contraceptive technologies and sexual ideologies during the nineteenth century is examined by P.A. David and W.C. Sanderson,

'Rudimentary contraceptive methods and the American transition to marital fertility control', ch. 7 in *Long-Term Factors in American Economic Growth*, S.L. Engerman and R.E. Gallman (eds) (Chicago: University of Chicago Press for N.B.E.R., 1986), pp. 307–90.

11 Variety and irreversibility in networks of technique conception and adoption

Michel Callon

Introduction

Faced with the proliferation of technical objects – manufacturing processes and consumption goods – modern man may be led to ask himself the question already formulated by Leibniz: Why this world rather than another? The reply given by this famous Leipzig philosopher is well known. God, a moral and rational being, wished 'among the infinite combinations and possible series' the world in which we live to correspond to 'that where the maximum of essence or possibility is able to exist'. The parallel between Leibnizian theodicy and the neo-classical economist's concept of the market has often been observed (Elster, 1975). The definition of divine acts[1] can be easily applied to the attainment of the social optimum within the constraints imposed by resource scarcity. The market is the institution which among all the productive configurations and possible states of the world selects the most rational and most moral combination. Pareto inscribed this elegant Leibnizian metaphysics in social physics.

However the task of Leibniz's god and Pareto's market is devilishly complicated. University research laboratories and company R&D departments produce new beings and previously unknown technical devices daily. The universe of essences is expanding without discontinuity: it condemns Leibniz's moral god and mathematician to begin its complicated calculations anew each morning. It compels the market to anticipate trends on which no information exists. Which world should be selected from all those which can be imagined, and even those which, through our ignorance, cannot? Which procedures should be used to make these choices, in particular so as not to hinder our future margins of manoeuvre? These questions have left the realm of metaphysics and are now firmly at the heart of political philosophy. They address themselves to social sciences, in particular those which concern the study of technical change and its dynamic.

These issues can be summed up in three crucial questions. The first concerns the *accessibility* of goods, knowledge and know-how. The moral viability of a given social order rests not only on an actor's capacity to mobilize science and techniques but also to procure certain products and goods. Once the problem of accessibility has been resolved, a second question concerns the reality of choice itself. The greater the *variety* of goods and knowledge, the more options there are, and the agent or actor has more scope to exercise the decision-making faculty (Bruno *et al.*, 1991; Saviotti, 1991); on the other hand, the less the variety, the more difficult it becomes for the actor's preferences to come into play. The third issue focuses the debate on the *irreversibility* constraint which results from the influence of past decisions on future ones. There are cases when an agent's choice is to a large measure predetermined by those previously made by a wide range of other actors. They impose behaviour which actors have not expressly chosen and often without their knowledge or will (Dosi and Metcalfe, 1991). It is not difficult to imagine a world where accessibility is unevenly distributed and where variety is menaced by extinction so that the trajectories followed become more and more irreversible.

The danger of such trends should not be underestimated, as the various institutions which have been designed over the centuries to counterbalance them have lost some of their potency. This situation is due to an unexpected change. At first the development of science and techniques enriched the variety of goods available and was also viewed as a liberating movement which weakened traditions and customs which were considered to be constraints imposed by history. For some, it even promised more justice by making goods, formerly the preserve of an elite, more accessible. This optimism is a thing of the past. Today science and techniques are accused of increasing inequalities and making a uniform world as they destroy cultural variety and impose the relentless logic of their development with no possibility of return. Even the market, an institution which should guarantee flexibility and openness, finds it difficult to contain this momentum and does not escape from criticism. By treating technology as an endogenous variable, economists of technical change have highlighted the existence of mechanisms which emphasize some of its serious imperfections. On the supply side, analyses have rightly stressed the creation of trajectories which concentrate investment: the concept of learning is vital in accounting for the unjustified preferences shown for particular techniques and products, accompanied by the progressive and often irreversible discarding of other options (David, 1975; David, 1982; Dosi *et al.*, 1988; Dosi, 1988; Freeman, 1990; Foray, 1991; Rosenberg, 1982; Stiglitz, 1987). On the demand side, diffusion and technological competitiveness models stress the importance of mechanisms which, by imposing certain options, sometimes lead to the elimination of particular products or techniques. In consequence, economic actors are left with limited room for manoeuvre in the exercise of choice (David, 1987; Arthur, 1989). Similar analyses have been undertaken by sociologists and historians of science and techniques. T. Hughes, to take the most brilliant case, has shown how socio-technical systems are historically constructed and how their weight increases over time so that they become more and more difficult to counteract (Hughes, 1983). Akrich, Callon, Latour and Law have shown that the production of scientific facts and technical mechanisms cannot be detached from the establishment of networks which determine their circulation space and contribute to irreversibilities, so difficult to undo (Akrich, 1989a, 1989b, 1991; Callon *et al.*, 1986; Callon, 1986, 1987, 1989, 1991; Latour, 1979, 1986, 1989; Law, 1986, 1987, 1988). The conclusions of all these studies confirm the urgent necessity for a renewed debate on the role of science and techniques. This should contribute to our understanding of how they affect the accessibility of goods, the variety of options and the resulting irreversibility of decisions.

In this chapter only variety and irreversibility will be dealt with. These two concepts have been the subject of numerous studies which need to be synthesized as they were carried out using different theoretical and conceptual frameworks. More concretely, the aim is to show that it is possible to unify evolutionary analyses, whether sociological or economic, stressing the importance of technological creation and usually from a supply-side point of view, with those which are more exclusively concerned with adoption processes, whether or not in a technologically competitive situation.

To achieve this, the analysis is organized around the network concept.* Two cases are distinguished. The first (model 1) analyses the adoption process of fixed techniques (end products or production processes) which do not undergo transformation during adoption. In the second (model 2), phases of (re)conception and adoption follow on from one another and, drawing on model 1, the interactions between conception and adoptions network which give birth to trajectories are shown. These models illustrate the mechanisms by which individual decisions are progressively aggregated to create an overall dynamic. While in some cases this encourages the maintenance of variety without any irreversibility, in others variety is progressively lost and trends become more and more irreversible. For each model this chapter is structured around three questions concerning:

- The entities composing the network.
- The law of recruitment, i.e. the mechanisms through which entities join a network.
- The recruitment dynamic, i.e. the extension of the network.

The analysis is intentionally formal.[2] The ultimate aim is to develop simulations, certain of which are already underway (Courtial and Cahlik, 1992). These will allow those variables which influence network evolution and give rise to variety and irreversibility to be identified.

Social adoption networks (model 1)

This model is the simplest both regarding the entities in the network and the prevailing law of recruitment. These networks favour human actors over techniques which play a passive role which is why they are referred to here as social networks.

This model reflects certain sociological and economic analyses on the diffusion of technical objects. The work is based on the central hypothesis that a population of adopters corresponds to a given state of technical objects T, and that it is possible to describe the extension and characteristics of this population.[3] As a result, the analysis of diffusion (even of the more sophisticated models) presupposes deep similarities among the actors considered to be potential adopters. These similarities explain why, under certain condi-

*In this chapter the concept of network does not refer to an intermediate form of organization between hierarchy and the market but to a set of relations between actors and techniques. The advice and encouragement of P. Cohendet, F. Eymard Duvernay, D. Foray, P. Mustar were invaluable in drafting the chapter and I wish to thank them here.

tions, actors adopt. They are considered to share common situations and aspirations. Thus, 'The underlying basis of the study of diffusion is to rationalize why, if a new technology is superior, it is not taken up by all potential users' (Stoneman, 1983), and the response is invariably the same. Actors are sufficiently similar so as sooner or later to be condemned to adopt, but sufficiently different for them to take the decision sooner or later than others.

Over the last few years, the dissimilarities have been examined with a fine toothcomb. Stoneman, at the end of a presentation of the various diffusion models suggests the following (non-exhaustive) list, 'Lack of knowledge, misconception of returns available, positive quasi-rents on existing technologies, expectation of future advances, insufficient expected return and the nature of the oligopoly game' (Stoneman, 1983: 148). In all these models, whether Bayesian or not, the time lags in adoption are explained primarily in terms of differences among actors. All potential adopters are basically similar, distinguished only by a few criteria (size, willingness to take risks, profitability threshold, information needs, level of performance required for the technique, etc.) which account for lags in decisions, or in some cases non-decision. The adopter population is therefore based on an overriding similarity (farmers developing cereal crops, doctors, middle-class households, firms active in the same sector, etc.), the probability of adoption depending on dissimilarities among adopters which can be considered secondary.

A second hypothesis on the effects of the adoption process on similarities and dissimilarities can be added to the first. Once the innovation is diffused the similarities among the effective adopters become more marked as they have made the same decision and now have the same technique in common. Before the introduction of hybrid corn, farmers in Arizona, Iowa and Missouri shared many characteristics. Once the hybrid was accepted they became even closer as they cultivated the same strain and shared the same technical and commercial problems. The corn just by its existence reinforced and extended social links. Such homogenization also occurs in the simpler epidemiological models and Bayesian models. Once a decision has been taken, the variable which explained lateness in adoption reverts to being passive and loses relevance: the early and late adopter are indistinguishable. This process which feeds on secondary differences between extremely similar actors creates even more similarity by removing those differences which precede adoption.

Diffusion models can be made more realistic by taking account of yet more variables to describe the differences which distinguish actors from one another more finely. This approach though hinders understanding of the phenomenon without adding anything really new. The alternative approach is opted for here, strangely neglected by economists and sociologists alike, i.e. the analysis of similarities and their dynamic. It is odd to note that such a central hypothesis has never been fully discussed.

Similarity networks

The response to what is meant by the singular expression 'potential adopters'[4] touches on the uneasy relationship between the two sister disciplines of economics and sociology. Sociologists have no difficulty in explaining the existence of a target population: adopters resemble one another simply because they are a social group with its own identity which means the individual behaviour of each of its members is predictable. Standard economics based on the independence of individual utility functions can but record the permanent miracle whereby consumers, free to behave as they wish, opt for identical goods.[5] Both positions are equally untenable. The sociological argument is too complaisant, as that which needs explanation is taken as given. An evident limitation is its inability to account for the appearance of new interdependencies shaped by the diffusion of a new technique. The economic viewpoint is no more convincing because utility functions can never be totally independent from one another; when economists speak of common knowledge, beliefs, customs, rules, routines, demonstration effects, trust and, more generally, externalities, they are obviously introducing the hypothesis of a relationship among the choices and decisions implemented by the agents.[6]

To resolve the dilemma while staying in line with the recent findings of sociology and economics of technical change, it is suggested here that the concept of structural equivalence, used by authors such as Burt to describe social networks, might be usefully extended (Burt, 1982). This leads to a definition of similarity which encompasses both the existence of preestablished interdependencies and their reconstruction by the actors concerned.

The concept of structural equivalence predates Burt's work but it was he who demonstrated its usefulness in the study of adoption phenomena. In a famous article he showed that the diffusion of a new medicine was explained less by the existence of direct links between actors (contagion) than by the equivalence of their positions within the social networks in which they were actively engaged (Burt, 1987). A distinction should be made between two types of structural equivalence: strict equivalence and regular equivalence (Doreian, 1988). Strict equivalence covers those entities which have the same relationships with the same other entities. This concept contrasts with that of a cohesive group which brings together individuals with a high density of interrelationships and may even result in the formation of a true clique. Regular equivalence is an extension of strict equivalence. Two elements are equivalent if they have the same types of relationships with elements which are themselves equivalent. Figure 11.1 illustrates this definition.

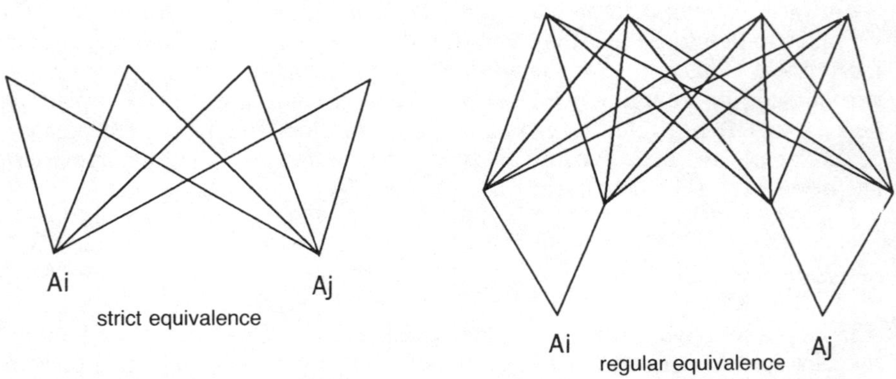

Figure 11.1

Two modifications have been made to these definitions, one to simplify the analysis and the other to extend the application of these concepts. Burt, in common with many social theorists, attempts to qualify the relationships among actors (friendship, commercial transaction, etc.). Here the identification of the existence of a relationship, or what more generally is termed association, will suffice. Thus, the relevant concept is the association profile which for an entity i is defined in terms of the population of entities j with which it is associated. Furthermore, these are not exclusively social relationships. Burt, as all sociologists, was content to plot the links which tied human actors together (individuals and groups). Such a perspective is too restrictive and does not help in understanding the dynamic of technical diffusion nor the resulting transformation of networks. It is considered here that each entity, whatever it is, and each association should be taken into account when sketching the association profile.

For example, the profile of human actor Ai may be drawn by looking at each human and non-human entity with which Ai has links. This set is referred to as S(Ai). Suppose Ai is a doctor, S(Ai) will include patients, certain colleagues, medical representatives, family and relations but also the scanning equipment used, the medicines prescribed, the stethoscope worn, vaccinations given and the car equipped with a radio telephone for visits. Such a list reconstitutes the universe of all elements, human beings, technical mechanisms but also texts and financial resources which form Ai's world. If only direct associations with Ai were taken into account, the set is referred to as being of rank 1 and is called S1(Ai). But, in line with the definition of regular equivalence, indirect links may be considered and each Bi element associated with Ai may be plotted on the graph of its associations S(Bi). The sum of all

$$s\,(Ai, Aj) = \frac{3}{3+4} = 0.43$$

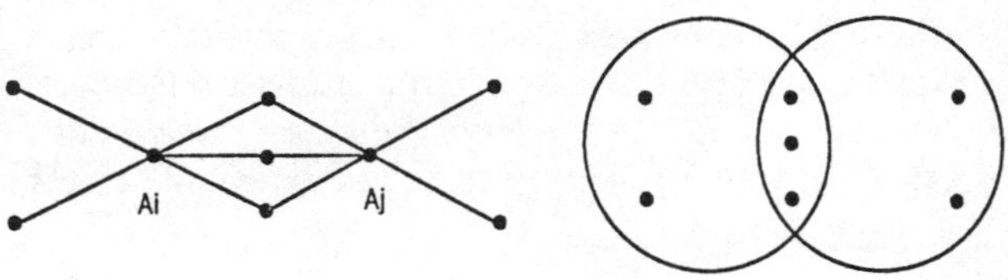

$$s_1(Ai, Aj) = \frac{2}{7+2} = 0.22$$

$$s_2(Ai, Aj) = \frac{\frac{2}{7+2} + \frac{1}{1+4}}{2} = 0.21$$

$$s_3(Ai, Aj) = \frac{\frac{2}{7+2} + \frac{1}{1+4} + \frac{1}{1+2}}{3} = 0.25$$

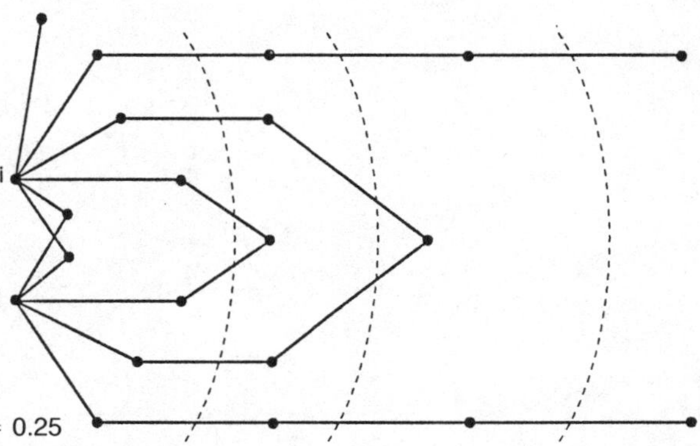

Figure 11.2

these graphs is termed S2(Ai) which also includes S1(Ai). The association profile of rank n and the corresponding Sn(Ai) can thus be defined step by step. The empirical counterpart of this definition is easy to grasp. To take doctors Ai and Aj from the previous example, they might have links with different medical representatives but who are employed by the same pharmaceutical company. They use their own scanners of the same make. In this case it would be erroneous to only look at rank 1 similarities. Although the medical representatives and scanners are quite separate, they are similar because they are themselves associated with the same firms. They thus reinforce the similarity of Ai and Aj.

This results in a similarity index of rank n between the two actors Ai and Aj. It can be said that Ai and Aj are more similar the more the (relative) number of elements shared by Sn(Ai) and Sn(Aj).[7] Sn(Ai, Aj) is the value of this index which will vary between 0 (no similarity) and 1 (perfect similarity) (Figure 11.2).[8]

This concept allows the introduction of similarity and dissimilarity relationships into a given population based on the same premise. An entity's (actor, technical object) identity can only be constituted by the bundle of associations entered into.[9]

The concept of the similarity index gives rise to that of a similarity network. For an Ai population, a similarity network, for threshold z and rank n, is a non-oriented graph linking all the Ai, Aj pairs of which the similarity indices of rank n are superior to z. The similarity network's morphology is given by the matrix of similarity indices between all the Ai and Aj. The matrix is symmetric and all elements on the diagonal are equal to 1. Ai's similarity network, for size n and rank z, is the graph which takes as its starting point Ai. If the graph in Figure 11.3 is a similarity network, the similarity network of A7, for a threshold of 0.3 is shown by the graph in Figure 11.4.

These networks possess a set of properties which will be made use of later. It must above all be stressed that a similarity network and the associated graph *do not illustrate the direct relationships among the Ai* as they build on the concept of extended structural equivalence. Two elements, Ai and Aj, with a similarity

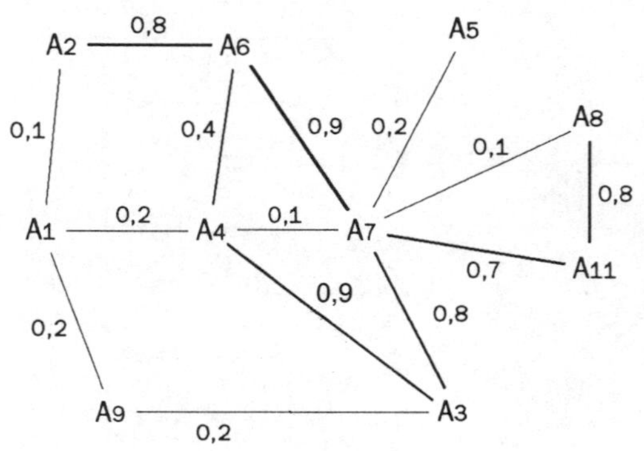

Figure 11.3

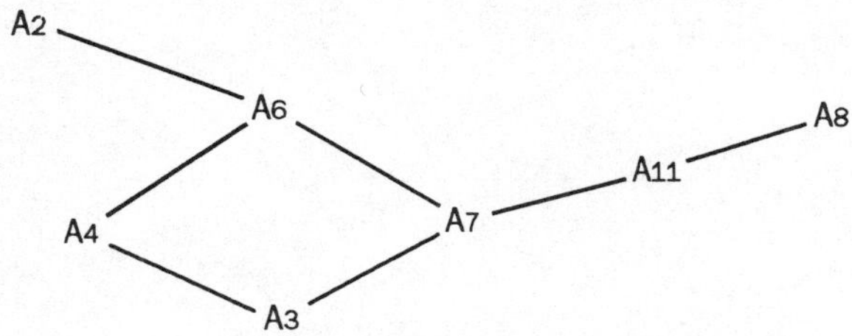

Figure 11.4

index equal to 1 and which are linked on the graph may very well have no direct relationships. Besides similarity indices of rank n are not cumulative and a link between A7 and A11 followed by a link between A11 and A4 does not mean a link exists between A7 and A4. On the other hand, for the same configuration there is a n1 > or = n, so that the similarity index of rank n1 between A7 and A4 is not zero (see Figure 11.5).

A final concept may be useful, that of an aggregated similarity index. If we take a population R(Ai) composed of elements Ai, . . . , Ai − k, Ai and Bi + m, one element does not belong to R(Ai). The aggregated index of Bi + m with reference to R(Ai) (s{(Bi + m, R(Ai)}), is the ratio between the sum of the similarity indices of Bi + m with each of the Ai − k and the number of elements in R(Ai).[10]

Adoption networks without competition

The constituent entities

A priori, two families of entities exist:

1. *Tokens* which could be, for example, any of the following: statements, viruses, tractors, video recorders, hybrid corn, etc. They may be replicated without cost and are transportable without transformation. T represents a family of identical tokens. Ti is the ith copy. Tokens have only one property, that of being capable of being adopted, i.e. forming an associative relationship with any actor. In the following, only tokens which are technical objects are considered, but logically the concept could be extended to any imaginable category of tokens.
2. *Adopters* of tokens who are human beings, individuals or groups of individuals. They are considered to be actors as they are capable of choice. At a given time, t, adopters are divided into two populations: those who have already decided to adopt (effective adopters) and those who are going to choose (perhaps) to adopt during the events which follow (potential adopters). The first population is easy to identify, and by using the concept of a similarity network it will be possible to formulate a precise definition of the second. The last actor to have adopted is referred to as Ai, the object adopted is the ith (Ti). Adoption corresponds to the creation of a new relationship (or association) between Ti and Ai.

The population of adopters at t is called R(Ai). It is made up of A1, . . . , Ai − k, . . . , Ai. The population of candidate adopters at t is called R(Bi + 1)

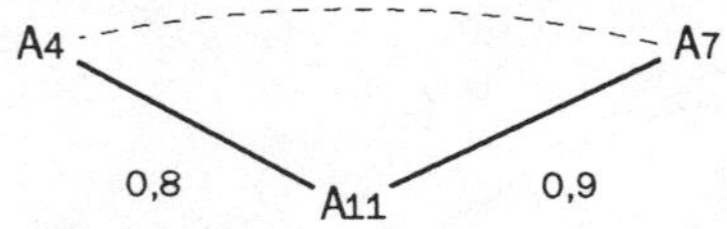

Figure 11.5

and is composed of the actors Bi + 1, . . . , Bi + m, . . . Bi + m may be referred to as a potential adopter [i.e. belongs to R(Bi + 1)] if, and only if, *at least* one Ai − k exists where S(Bi + m) and S(Ai − k) have *at least* one element in common.[11] It can be shown that the *composition* of the population of potential adopters for a similarity threshold of Z and rank n is wholly defined once the first decision to adopt has been made. Similarity indices (which are variable over time) can be calculated between any pair of the type {Ai − k, Bi + m}, {Bi + m, Bi + 1}, {Ai − k, Ai − r}.

For whatever t, it is possible to order all the Bi + m by their descending aggregated similarity indices: Bi + 1, Bi + 2, . . . , Bi + m, . . . with s(Bi + 1, RAi) > s(Bi + 2, RAi) > . . .

Law of recruitment

A definition of the law of recruitment describes the mechanism by which an element of R(Bi + 1) at t adopts Ti + 1, i.e. transforms itself into Ai + 1. This can be stated more concisely as the law of recruitment describes the transformation of R(Ai) into R(Ai + 1). This law should be stated so as to be sufficiently general to integrate the various models of technical diffusion, i.e. it should be based solely on the concept of similarity. The following formulation results: the probability of Bi + m adopting Ti + 1 at t is an increasing function of the aggregated similarity index of Bi + m with R(Ai).

I would like to spend a little time explaining this approach. The definition of similarity implies that two entities are as similar as their past sequence of associations are identical. In social science terms, and if it is accepted that the entities are human actors endowed with the capacity of reflection, it can be said that a strong similarity between Ai and Aj signifies that they have taken the same decisions or even that they have consistently made the same choices. Ai and Aj are involved in the same social interactions, they have adopted the same technical artefacts, they possess the same level of financial resources, etc. Clearly, this observation does not reveal anything about Ai or Aj, and it is not necessary to know why they made these choices or took these decisions, whether they were the result of secret impulses, of vested interest, habitus, empire building or whether they stemmed from a strategic calculation, Machiavellianism, altruism or individualism. It is enough to state that an actor's identity is nothing more than the series of associations entered into; in other words, a web of links which are (painstakingly) developed and consolidated (Latour, 1989; Degenne, 1991; Teil, 1991). This analysis goes no further than the hypothesis that from the similarity of past associations, it is possible to forecast the probability of the similarity of future associations. This hypothesis avoids having to take sides in the determinism and autonomy debate. For example, it can be argued that Ai and Aj are subject to the same deterministic forces and it could equally be argued that they had coincidentally made the same use of their free will. Nor is there a need to hypothesize about a possible mimicry which leads Aj systematically to imitate the behaviour of Ai. Possible explanations are not considered here, it is enough to establish whether or not a similarity exists between the trajectories followed in the building of identities.

The probability function can take very diverse forms as the only, theoretical, constraint is that it be increasing. Strictly speaking, the choice of this function's form can only be empirical. However, from the point of view of this text, this issue is not that important. The only property which will be taken into consideration is whether it is increasing. To facilitate the argument, only simple functions with a fixed threshold will be used here: similarity indices with a value above the s0 threshold have a probability of adoption equal to 1, for values below, probability equals 0.

The law of recruitment can thus be stated as: at t all the Bi + m for which s(Bi + m, RAi) > s0 are recruited.

This law of recruitment underlines the determining role of two variables: the similarity networks and their morphology at t (which will be shown to be dependent on their previous evolution) and the form of the adoption probability function (here, the value of the adoption threshold).

The dynamic

To follow the adoption dynamic, it is useful to develop a variation rule for similarity networks. In this model, similarity indices are not calculated once and for all. Their values change as the adoption process of T advances even though there is no change in the population of potential adopters. When at time t, Bi + 1 adopts Ti + 1 and becomes Ai + 1, the similarity relationships between the various effective adopters and the potential adopters are in reality modified. This variation may be expressed simply as follows: the adoption of Ti + 1 by Bi + 1 results in an increase in all the similarity indices of rank n > n0 (n0 depends on the similarity network's morphology and the position of the first adopter).

This transformation is easy to grasp. If Bi + 1 has been recruited, by definition the value of the aggregated similarity index of Bi + 1 with R(Ai) was above the recruitment threshold or, to put it another way, some Ai − k existed with association profiles sufficiently close to that of Bi + 1. Adoption reinforces the similarity of Bi + 1 with those who have already adopted while simultaneously increasing the similarity indices between all the elements of R(Ai). Bi + 1 becomes Ai + 1 and, if all the Ti are taken as being interchangeable, shares a supplementary association of rank 2 with all the Ai − k. In the same way, all the indices of (Ai − k, Ai − r) pairs increase because the actors share an additional association which links each one of them, through Ti + 1, to Ai + 1. R(Ai + 1) is more connected than R(Ai). Finally, it is possible to show that aggregated similarity indices of rank n, which are not zero, linking any Bi + m (m > 1) with R(Ai) also increase (on condition that n > n0, n0 depending on the morphology of the initial similarity network and, more particularly, on the position of Bi + m in relation to the first adopters).

Growth over time (i.e. after numerous adoption sequences) by reinforcing the similarity among potential adopters and the network of those who have already adopted, contributes to the increased probability of adoption. If the hypothesis is made that the adoption threshold is stable (or falling) over time, this property can be simply expressed as follows: the probability of adoption increases in proportion to the number of effective adoptions (a self-reinforcing process). The form and scope of the variation depends on the morphology of the similarity networks established at t − 1, i.e. on the distribution of the s{(Bi + m, R(Ai)} (connectivity, division into blocks, etc.). From this it can be deduced by recurrence that the adoption trajectory is determined by two essential variables: the form of the similarity networks at t = 0 and the form of the probability functions. To give an idea of the variety of trends that can result from this, two contrasting examples are given where the probabilities depend on prefixed thresholds.

Example 1 At t = 0, i.e. before the first adoption, let us take the following (star-shaped) network (Figure 11.6).

The rank 1 similarity indices of the Bi pairs are very weak, other than the s(1, 2), s(1, 3), s(1, 4), s(1, 5) and s(1, 6) indices. The values of these five indices are such that s(1, 2) > s(1, 3) > s(1, 4) > . . . Moreover, the hypothesis is made that B1 is the first adopter and an adoption function at the threshold s0 is selected, so that where the rank n ⩾ 2, sn(1, 2) > s0 > sn(1, 3) > . . . How does the recruitment dynamic develop from these initial conditions?

At t = 1, B1 adopts. Given the network configuration and 'the index values, at t = 2 only B2 adopts. However at t = 3, several developments are possible.

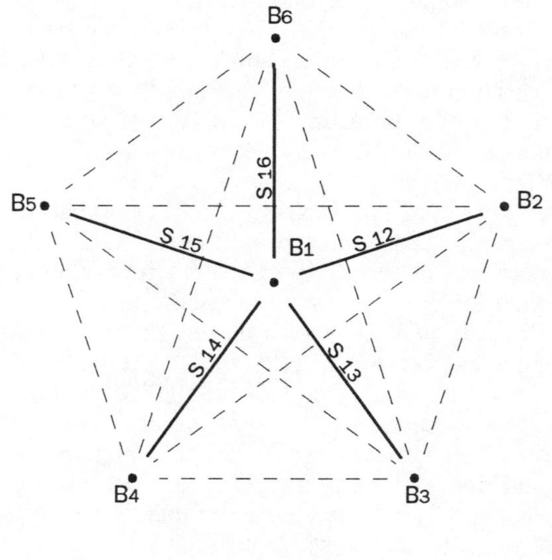

Figure 11.6

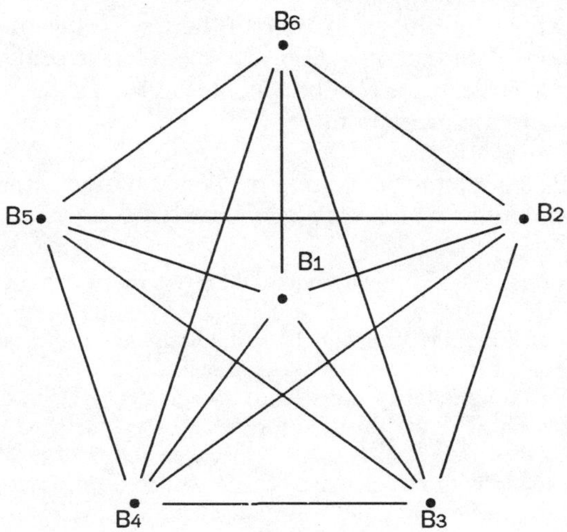

Figure 11.7

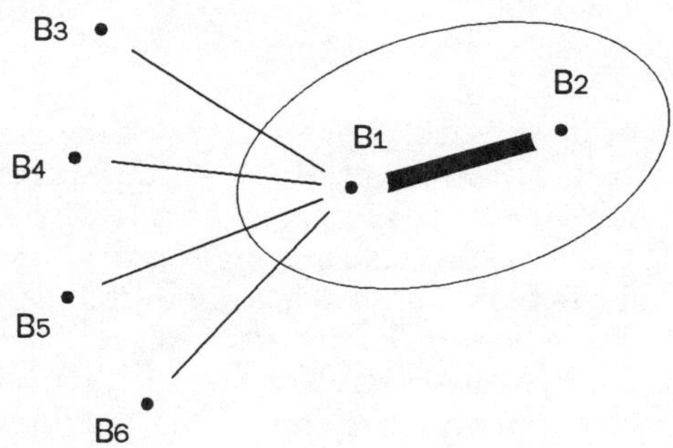

Figure 11.8

It can be shown that the decision to adopt taken by B2 results simultaneously in an increase of s1(1, 2) and of sn(1, 3); sn(2, 3) increases too because it was presupposed to be weak but not zero at t = 0.[12] At t = 3 therefore the aggregated index of B3 with the B1 and B2 adopters increases. As long as the new index sn(B3, B1 + B2) passes the s0 threshold, B3 will adopt. The same logic applies to the ensuing sequences. The morphology of the network at the end of the process will fall between two extreme configurations: either all the Bi (Figure 11.7) have adopted and the end network has reinforced all the initial similarities or only B1 and B2 have adopted and all the others are excluded. In the latter case the adoption dynamic has led to the creation of two sub-sets between which the dissimilarities are greater than prior to adoption (Figure 11.8).

This example underlines the crucial variables. The position of the first adopter (if it were not B1, the dynamic would be different), B1's similarity network and the form of the probability function (value of the threshold). These three variables suffice to describe the recruitment dynamic. This dynamic will result either in a reinforcement of pre-existing networks or, on the contrary, in their fragmentation.

Example 2 This completes the argument. A population P0, formed at t = 0, has 5 SPi sub-populations which may be defined as follows:

- All the elements of a given SP have, when taken two by two, high similarity indices (in the range of 1).
- The elements of an SPi have zero indices with all elements from all the other SPi.
- There are, however, a few exceptions to the above rule. In SP1, four elements stand out: each has a non-0 similarity index with *one* element of *one* SPi (the first with SP2, the second with SP3, the third with SP4 and the fourth with SP5). In SP2, a second element (other than that which is closely linked to the SP1 element) has a non-0 similarity index with *one* element of SP3 (which should not be confused with that linked to SP2). The other links are shown in Figure 11.9. SP5 has a single outside link, that of one of its elements with SP1.
- s(1, 3), s(1, 4) and s(1, 5) are well below the s0 adoption threshold; on the other hand, s(1, 2), s(2, 3) and s(3, 4) are above s0.
- The first adopter is an element of SP1 which has zero similarity links with elements belonging to all the other SPi.
- SP1 only comprises a small number of elements.

The dynamic can be described as follows. At t = 1 the first adoption occurs. At t = 2, all the elements of SP1 adopt. At t = 3, the element of SP2 which is linked with SP1 adopts. At t = 4, all the elements of SP2 adopt. The process spills over to SP3 and SP4 in subsequent rounds according to the same logic. Given the initial configuration, SP5 remains completely outside the adoption process. The similarity indices with SP1 are weak at t = 0 and as the SP1 population is small, the hypothesis can be made that, even if a high n rank is used, the increase in these indices was limited and that the threshold was not attained. On the other hand, at the end of the process the links between SP1, SP2, SP3 and SP4 are considerably reinforced. The gulf between SP5 and the other sub-populations will be definitively widened when SP4, SP3, SP2 and SP1 have integrated still further (Figure 11.9).

Here again three variables show themselves as strategic: the position of the first adopter, the initial state of the similarity network, and the form of the probability function (the value of the adoption threshold).

Consequences

This model is characterized by the extreme simplicity of the hypotheses used. The only data necessary are: the similarity matrix, the position of the first adopter and the form of the probability function which establishes a link between adoption and the similarity index. The vital contribution of the model is to explain how the adoption process itself engenders similarity networks through sequence after sequence. It is the adoption at t which definitively gives birth to the network whose new state will define the adoption process at t + 1.

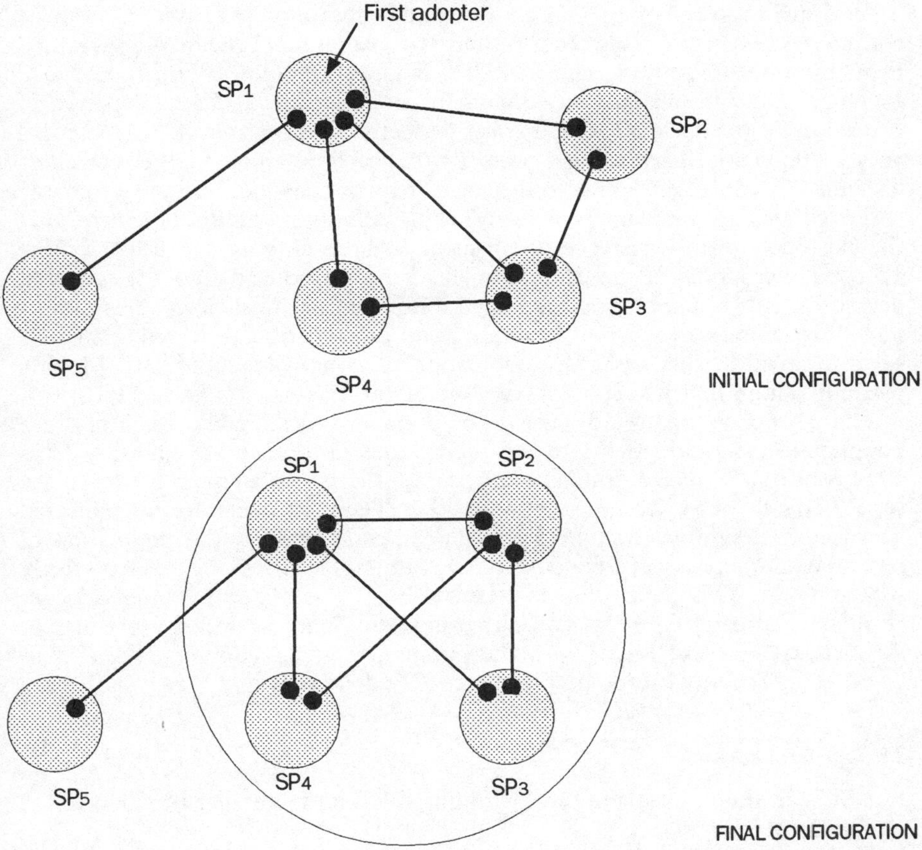

Figure 11.9

R(t) → adoption at t → R(t + 1) → adoption at t + 1

A single appearance of a token in a similarity network of whatever sort can set off a social dynamic which remodels the morphology of the existing relationships.

Competitive adoption networks

It is possible to extend the above model to a situation where techniques compete and can, moreover, give rise to the creation of network externalities. The principal elements are given here.

The entities

A rigorous distinction is made between techniques (technical artefacts) and human actors.

In order to avoid undue complication, two techniques T1 and T2 will be considered to be competing. In the alternative variation of the model presented here, these techniques are considered to be substitutable but incompatible.[14] Incompatibility means that an adopter in choosing T1 (or T2) is deprived of choosing T2 (or T1). T1 and T2 are perfectly substitutable if: (a) potential adopter populations correspond perfectly; (b) each potential adopter considers T1 and T2 equivalent (this equivalence may or may not include economic variables such as price and availability). This definition implies, of course, that the existence or non-existence of substitutability is entirely dependent on the adopters considered. If some realism is to be introduced into the analysis, degrees of substitutability and incompatibility can be considered. An interesting situation arises where techniques are incompatible but only partially substitutable. In this case, the population of potential adopters of T1 only partially obtain that which potential adopters of T2 obtain (Figure 11.10).

Human beings may be divided into suppliers and adopters. Each of these populations can be divided into sub-populations in terms of T1 and T2. Thus at a given time t, the set R1(Ai) comprises all those who adopted a T1 and the set R2(Aj) all those who have adopted T2. These two sets are unconnected because by definition T1 and T2 are incompatible. At t, the population of potential adopters is not divided into two sub-sets as T1 and T2 are perfectly substitutable. This population is called R(Bi + 1) and the potential adopter Bi + m. For each Bi + m, two aggregated similarity indices are calculated, the first s1(Bi + m, R1Ai) with the population of T1 adopters, the second s2(Bi + m, R2Aj) with the population of T2 adopters.

The law of recruitment

For Bi + m the p1 probability of adopting T1 is an increasing function of the s1 similarity index.

The same may be true for p2, for the adoption probability of T2. Besides, it is legitimate to consider that functions supplying adoption probabilities have the same form for both techniques. If for Bi + m, s1(Bi + m, R1Ai) $\gg$ s2(Bi + m, R2Aj), the only probability to be considered is that which applies to the adoption of T1 (or T2 if s2 $\gg$ s1). If both aggregated similarity indices are the same size and the adoption probability is not 0, the choice of one technique over another is random. Thus at any given time, the population of potential adopters can be divided into three sub-populations defined by the relative values of their adoption probabilities of T1 and T2.

It is easy in this model to take into consideration techniques which produce network externalities. If T1 and T2 are network techniques, it is routinely assumed that the choice of Bi + m depends on the number of former adopters of each of these competing techniques. This formulation is extremely simplistic because it does not take into account existing similarity networks. It is more exact to say that Bi + m is interested in interacting (using the network technique) with those with whom there is the most resemblance. In this case the decision will depend more heavily on the choices made by those with whom the similarity index is highest. Bi + m will choose T2, even if N1(Ai) $>$ N2(Aj) for the simple reason that those which he resembles and which form

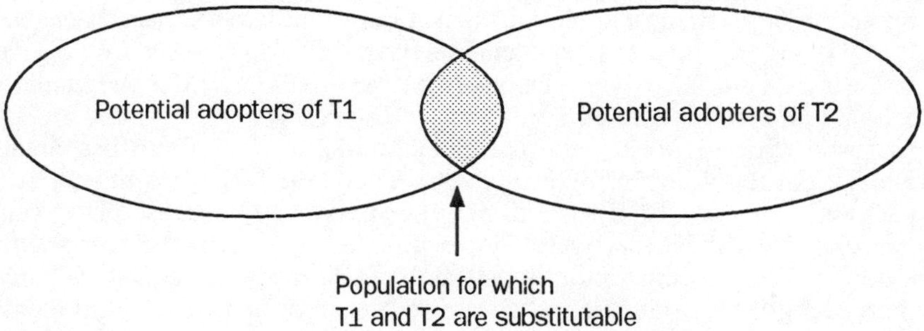

Figure 11.10

the adopter sub-population, have adopted T2. There are, then, two good reasons why the decision of Bi + m is only linked to aggregated similarity indices: the first stems from the logic underlined by the analysis of the adoption process of a single technique; the second is the result of the existence of network externalities.[14]

It should be noted that the law of recruitment set out here is different to that suggested by authors such as Arthur (Arthur, 1989). For them the decision to adopt T1 (or T2) depends on two factors: the first is the natural preferences of the adopters *vis-à-vis* each of the two techniques; the second is an increasing function of N1 (or N2). In the model here, the decision is a probability function which only takes into account the value of aggregated similarity indices. If for example the aggregated similarity index of Bi + m in relation to R1Ai is equal to zero, and the index in relation to R2Aj is higher, he will adopt T2 even if N1 ≫ N2.

The dynamic

In the simplest scenario, that used here, T1 and T2 conserve their characteristics over time. They remain (perfectly) substitutable, non-complementary and incompatible throughout the competition/adoption process.

As adoption advances, the aggregated similarity indices change. The rules which regulate these changes are analogous to those which hold for model 1. If Bi + 1 adopts T1, the similarity indices increase among T1 adopters, but, equally, so do all the s1(Bi + m, R1Ai + 1) similarity indices of all Bi + m which have similarity indices of non-0 n[14] rank with R1(Ai) and/or Bi + 1. In simple terms, this means that self-reinforcement mechanisms are established which depend on the morphology of the various similarity networks. No simple mathematical rule can be used to follow the development of this process and simulations are necessary. Two examples, corresponding to the extreme cases, highlight the main variables which direct this dynamic.

In the first example, the initial configuration (t = 0) is characterised by three sub-populations SP1, SP2 and SP3, the N1 members of SP1 are much more numerous than the N2 of SP2 (N1 ≫ N2). For all pairs of elements belonging to a given sub-population the similarity indices are near 1. Between (SP1, SP3) and (SP3, SP2) some elements have similarity indices well over the adoption threshold. On the other hand, between SP1 and SP2 the similarity indices are all zero (Figure 11.11).

At t = 1 (first sequence), an SP1 element adopts T1 and an SP2 element adopts T2. In the following sequence (t = 2), all the SP1 elements adopt T1 and all SP2 elements T2. In the third sequence (t = 3) because N1 is much bigger than N2, SP3 collectively adopts T1. At the end of the competition/adoption process a new similarity structure has arisen. SP1 and SP3 have further strengthened their links, whereas SP2 is marginalized. Technological competition has played the role of an agent of social change in a more marked way than it would have done in a simple adoption process. The more numerous the competing techniques, the more likely it is that the dissimilarities among the emerging sub-groups will be marked and multiple.

In the second case, the initial configuration is also characterised by three sub-populations (Figure 11.12). The only difference from the preceding case being that the N1 and N2 populations are equal. The first two sequences unfold as previously. The third sequence is different. The probabilities of SP3 elements choosing T1 and T2 are equal. There is therefore a high probability that at the end of the process the SP3 population will divide equally between T1 and T2 adopters. Once again the initial similarity network undergoes a profound transformation, but the final structure is different because it is characterized by the juxtaposition of two strongly unconnected sub-populations of equal size. It should be noted that it would be possible to change the final breakdown if the internal similarity network of SP3 had similarity indices initially distributed in such a way that their values allowed T1 and T2 to advance at different rates within SP3.

In both cases, it is clear that the dynamics would have been different, had the position of the first adopters of T1 and T2 been different. If, for example, in the first sequence SP2 selected T1 and SP3 T2, then in the final sequence SP1 would naturally have selected T2.

In this variant of the adoption model, the explanatory variables are the same as those in the (non-competitive) adoption model of which this model can be seen to be a sort of extension. Similarity networks (or in a more restricted

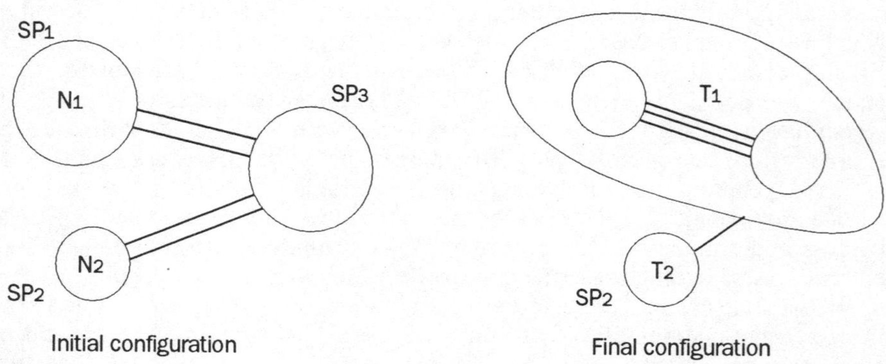

Figure 11.11

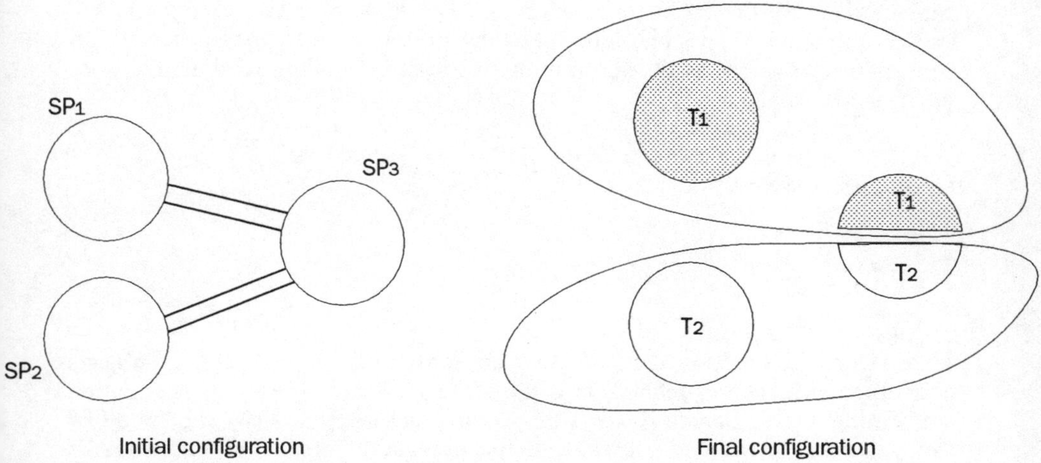

Figure 11.12

sense interconnections: Foray, 1991; David and Foray, 1991) and the position of the first adopters are, together with the form of the probability functions, the parameters which regulate the dynamic of competition.[15]

Conception-adoption-competition networks (model 2)

The main weakness of the preceding model is to consider techniques and networks as given at the beginning of the process, i.e. as exogenous, even if the dynamic during the process is to some degree endogenized.

The model presented in this section, integrates the phases of conception (or reconception) of techniques. Furthermore, it leaves open a certain number of developments which were not possible in model 1. Not only do techniques undergo transformation but with them the suppliers and adopters. The division between passive entities (adoptable techniques) and active entities (suppliers and adopters who decide) may be redrawn during the process as a result of a modification of the conventions in operation.

In this model two categories of network are distinguished: *conception networks* and *adoption networks*. This does not mean that the hypothesis, which has been justly criticized, of a radical separation between invention, innovation and diffusion/adoption has been taken on board. All analyses by economists and sociologists of techniques agree on the importance of transformations during diffusion. Here, it will be sufficient to distinguish between two types of dynamic: conception corresponds to hot situations where actors and techniques co-evolve; adoption corresponds to cold situations where actors and techniques are stable. However, the model allows for successive hot and cold phases, so that all predictable developments can be described. In some cases the dynamic followed will be essentially that of the conception (and reconception) networks, while other trajectories, on the contrary, will be characterized by the rationale of adoption. Between these two extreme cases, any combination is possible. The following presentation focuses primarily on conception networks, as the adoption network and the adoption dynamic were described in model 1.

Conception Networks[16]

The Entities

ACTORS AND TECHNIQUES

Here again a distinction is made between entities which are technical objects and others which are defined as actors. The first are subject to negotiation concerning their characteristics and production process. They are modified over time in function of the state of the prevailing forces and compromises reached among actors. These transformations may be more or less significant. This will depend on the degree of convergence or divergence among the viewpoints of the actors participating in the negotiation. Kidder shows, for example, how within a firm computer scientists, researchers and the sales team discuss and negotiate to define the characteristics of a new mini-computer to be put on the market: degree of compatibility with existing models, bit mode to be selected, micro-codes, machine size, microprocessor features, etc. (Kidder, 1981). The history of an innovation's conception is mixed up with the discussions and compromises to which it gives rise (MacKenzie, 1991). In some cases, at the end of a series of transformations necessary for an agreement acceptable to all the protagonists, the innovation is stabilized and ready to enter the adoption network.

The concept of negotiation leads on to that of actor. An actor is any individual or collective entity who takes part in negotiations and contributes to reach a compromise. Each actor may be simply described in terms of his own view of the technical object to be put into circulation: a member of the marketing team may give more weight to the compatibility of a new model with previous ones, an engineering team may prefer a new microchip promoted by a firm in which they have confidence, the project manager requires that the conception will allow production automatization, etc.

It is possible to say that the techniques and actors who participate in the conception process evolve together. Take two actors Ai and Aj negotiating technique Tk, they will (sooner or later) compel themselves to find a compromise between their preferred options. This will lead to a redefinition of the Tk whose characteristics will be modified as negotiations progress. But Ai and Aj themselves will also develop because of the simple fact that they are involved in a search for a compromise. Their conceptions, interests, and projects will change as they agree to abandon some of their initial demands to take into consideration those of the other actor. Once agreement is reached, the identities of Ai and Aj are transformed. New actors are created while a new Tk is being defined. This is why I use the idea of the co-evolution of techniques and actors.

The nature of the dynamic which ends with a satisfactory compromise. obviously depends on the distinction established at the beginning of the process between the techniques which are the subject of negotiation and the actors who negotiate. It would be natural to consider that 'technical objects' are of necessity material in nature and that actors are human beings, whether individual or collective. This is indeed often the case, but this rule is subject to exceptions more and more. Techniques are often hybrid encompassing material objects and human beings (for example, an automated factory, a nuclear reactor, etc.) and the subject of conception can include in the same phase of development the operators, the turbines, the circulating fluids, and the ever moving electrons, etc. This is reflected by the fact the population of potential actors is not limited to human beings. Distributed artificial intelligence is the best example of the communities in which robots and expert systems, assisted by human operators, negotiate the production schedule of a new product. In some cases, the separation between *what is negotiated* and *what negotiates*, between the intermediary and the actor, has to be accepted as rather ambiguous (Callon, 1991). Usually there are conventions (laws, rules, customs) to clarify the basis of these distinctions. However, nothing may be considered as definitively stable and ambiguous situations are multiplying under pressure from science and technology. For example, when is an embryo recognized as an actor with rights of expression (through an intermediate spokesperson) rather than a thing on which any experiment or transformation can be conducted? Answers to this type of question are far from fixed and change with the networks within which they are formulated.

Other than the conventions which delimit the boundaries between actors and techniques, the list of actors and techniques which are involved at a given time in the conception process is the second variable which affects the dynamic and the outcome of negotiations. According to the identity of the protagonists and the nature of the objects which are subject to creation, the end product and even their diffusion space may be significantly different.

The model presupposes the existence of sequences during which the conventions governing the division between actors and techniques are stable. Moreover, it is considered that at the beginning of a sequence, the list of actors and techniques involved is known (even if, as will be shown, it can evolve during the sequence).

SOCIO-TECHNICAL NETWORKS: THE NEGOTIATION STAKES

Once the identity of the negotiating actors and the techniques to be developed has been defined, it needs to be determined what is at stake in the negotiations. In this context, I would like to recall a crucial finding of innovation sociology (Akrich, 1988; Callon, 1987; Latour, 1988; Law, 1991; Rabeharisoa, 1991).

Take a technical artefact Tk which is the subject of negotiation by various actors in the conception network. It is possible to show that for each Tk there is a corresponding socio-technical network which describes the world that this technical artefact structures and mobilizes. A simple example will illustrate this point. Take an everyday device like an electronic entry barrier to a Parisian metro station. This presupposes a human user in possession of a valid ticket sufficiently competent to understand that he should insert the ticket in a slot designed for this, that the ticket should be recuperated, if it is reusable, and who will not panic if the ticket does not reappear. The most recent versions of this mechanism presuppose a fraudulent and agile user who, even if he can jump over the barrier without paying, will run into the subsequent gate which has been added to make his life more difficult. The device also presupposes transport tickets with a magnetically recorded strip, a central computer capable of reading this information and identifying and classifying the user (valid or not) and subsequently unlocking the gate, . . . the description could be developed *ad infinitum*. The entry barrier at the entrance to metro stations organizes a network of relationships among entities, some human, others technical, of which the roles and links are perfectly defined and are to some degree structured by the device. The consequences of this observation are important for the analysis of the conception process. What is negotiated among actors is not only the characteristics of Tk (e.g. the entry barrier), but also at the same time the socio-technical network linked to Tk. It is easy to imagine that there were deep divergences when defining the project description between engineers at Thomson and those of the subcontractors and between the design office and the RATP (Paris Public Transport Authority) marketing department. These were not solely concerned with the technical characteristics of the barrier (its size, the additional gate, the speed that the electronic strip was read, the casing which protects the barrier from possible vandalism, etc.). They also involved various definitions of complementary techniques (the design of the ticket, the control algorithm, etc.) but also, and above all, the roles given to the passengers and their ultimate division into distinct sub-populations, e.g. would it be necessary to adopt the entry barriers for disabled travellers or those laden down with heavy luggage. The conceivers of techniques became technicians, sociologists, economists, moralists as they traversed the socio-technical network linked to the object being conceived. Controversies arose on all these issues. In discussing the form of the barrier the various departments of the RATP and Thomson discussed the content and morphology of the socio-technical networks to be built. When compromise was reached, in fact they were agreeing on a particular socio-technical network, which is often nothing more than the result of tinkering with networks stoutly defended by the various actors initially (Figure 11.13). The characteristics of the potential actors can thus be determined, and also those of the producers and distributors, etc.

CONVERGENCE AND DIVERGENCE

It is now possible to describe the different states of a conception network. A network is *convergent* if the three following conditions hold: (a) there is agreement on the initial distinction between actors and techniques; (b) there is agreement on the list of actors and techniques involved in the negotiation; (c) there is agreement on the description of the techniques being conceived, i.e. on the description of the related socio-technical networks. When these agreements are not reached, the conception network is divergent.

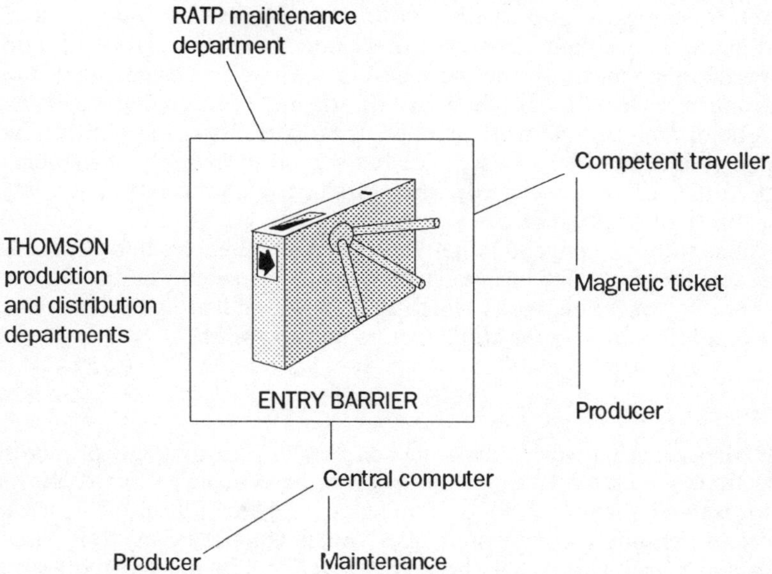

Figure 11.13 Some elements of the socio-technical network of the entry barrier

Recruitment

It has been taken as given that at the beginning of the conception sequence the conventions which allow a distinction to be made between actors and techniques are fixed, together with the list of actors and techniques involved.

The drawing up of this list does not only depend on the contents of the conventions already discussed, but also on the organizational procedures and structures which determine the direction that negotiations concerning techniques should take, for example, concerning the participants, their batting order, their prerogatives, etc. The participating actors may belong to a single organization, may or may not come from different departments or divisions, or be recruited from different organizations (public laboratories, service firms, lead users, ministries, banks, capital risk companies). The form of relationships and their development over time may range from simple bilateral links, moving relentlessly forward, to ongoing multi-directional interactions. All these rules (formal and informal) have been well covered by sociologists and more recently by the economic theory of organizations. They contribute forcefully to the definition of the actors, direct their interventions and coordinate their actions (Aoki, 1990; Eymard Duvernay, 1989; Gaffard, 1990).

But recruitment goes beyond the entities physically present during negotiation. To analyze the recruitment process, it is necessary to return to the concept of a similarity network introduced in the discussion of model 1. Take all the Ai and Tk on the list at the beginning of the conception process. A similarity network is associated with each Ai. It determines the population of actors similar to Ai, whether within or outside the conception network. Within the conception network, Ai 'represents' its similarity network. This representative will be all the more accurate where the represented actors have a strong similarity index with Ai: the degree of accuracy depends on the morphology of this network and on the distribution of indices. When the similarity index is high, Ai's choice has a good probability of influencing that of the other actors concerned, whereas when the similarity index is low, the probability of similar choices is low.

Thus, within a conception network when an Ai enters into negotiation with other Aj, he speaks for his similarity network; it should not be forgotten that the socio-technical networks which he defends and promotes are inscribed in the Tk and he does so for all the actors he represents.

The dynamic

The conception network's dynamic can be analysed in terms of increases and reductions in convergence. The concept of agreement was used above, and it was observed that it could be applied to the identification of actors to be recruited (through conventions and organizational rules) but also to the definition of networks that the actors inscribed in the techniques that they wished to negotiate. Thus four configurations can be identified (Figure 11.14).

Case 1 corresponds to perfect agreement. The consensus applies both to the expected socio-technical networks and the list of participating actors (as well as their rights and duties). The conception network which results is convergent. The networks inscribed in the techniques undergoing conception, i.e. the techniques themselves, are identical. Descriptions (no longer) give rise to

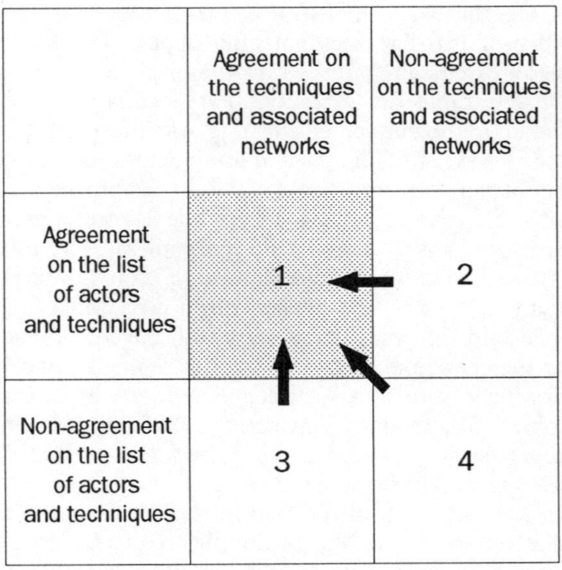

Figure 11.14 Conception dynamic

debate or controversy, nor does the list or role of the actors involved in the development of the techniques concerned.

Case 4 is that of total disagreement which concerns both the definition of the techniques and the expected socio-technical networks related to them, as well as the list of actors involved and their rights and duties.

Cases 2 and 3 are partial disagreements which give rise to differences of opinion.

From the above, it can be said that the degree of convergence or divergence is the (always provisional) result of a process which leads to agreement or disagreement on the description of the anticipated network and the actors involved (or involvable). If the starting point is complete disagreement, agreement can only be reached at the end of a negotiation process during which both the definition of techniques and the list of actors eligible to negotiate have been adjusted and modified. To pass from divergence to convergence mutual adaptation occurs. The dynamics of these adaptations depend, to a large extent, on the forms of organization and the coordination rules which prevail in the conception networks, i.e. who negotiates with who, in what order and according to what sequences. The distribution of competence is also crucial: who are the actors and what are their competences, can they be substituted, and to what degree, by other actors through progressively delving more deeply into their associated similarity network? The organization of these interactions is obviously essential, because the contents of the agreement in particular, i.e. the characteristics of the techniques, depend on their dynamic and as does the determination of the adoption network, as will be shown. It is enough to change the list of actors authorized to negotiate, the order of their intervention, the morphology of the interactions and also the means by which 'represented' actors may be involved (by exploration of the similarity networks) for other techniques to be developed[17] (for an in-depth analysis of the dynamic of research networks linked to conception networks, see Laredo, 1991; Vinck, 1991).

Competition–adoption networks

The conception process ends when network convergence is achieved. The adoption phase then begins which goes hand in hand with manufacture and distribution, etc. When convergence is obtained within the conception network, agreement is reached on Tk technique. But, as has been stressed, the agreement is not on the technique in the strict sense. It also applies to the networks inscribed in the technique which will often have been the subject of long negotiations in the conception phase. The identity of the adopters (or more generally, users), the nature of the manufacturing process, the role of the marketing team, and the task of the suppliers, etc., has also been discussed (sometimes heatedly). The reaching of agreement on the definition of adopters (who they are and their requirements) presupposes the involvement of 'spokespersons' for the adopters in the conception process (using the full range of means available to make known the potential users' requirements: market studies, panels, experiments, etc.) and, in certain cases these spokespersons will themselves be the first adopters. What is true for adopters also

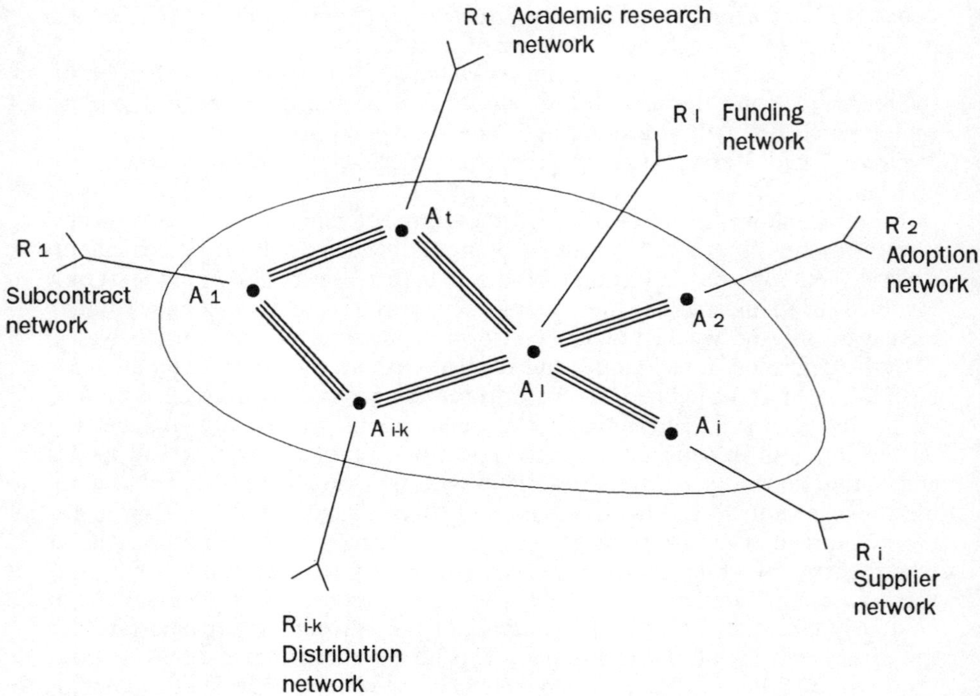

Figure 11.15

holds for all other actors in the conception network, some of whom will be progressively mobilized as production increases. Figure 11.15 shows the conception network and adoption networks together with the related subcontract, manufacturing, research and funding networks.

Tk is fixed and the first adopters can be identified at the moment the adoption process begins.[18] By looking at their similarity networks the potential adopter population and the morphology of the interdependences which link them can be determined. This leads back to the preceding model which describes the adoption of stable techniques. Production, subcontract and funding networks are also among the networks activated in the second phase. Once the adoption process begins, suppliers supply, producers produce and bankers finance. In this model, supply is as important as demand and Tk's adoption network is matched by the financial and production networks of Tk. The final network operates in several directions simultaneously and, in order to win on the adoption front, must mobilize on the manufacturing and subcontract front.

Moreover, at this stage of the analysis it is possible to introduce situations of technological competition. In order to do this, it is enough to imagine other conception networks exist developing in a synchronised fashion the Tj which are linked to adoption networks which overlap (partially or more rarely totally) Tk's adoption network. Adoption and competition of Tk and Tj are described

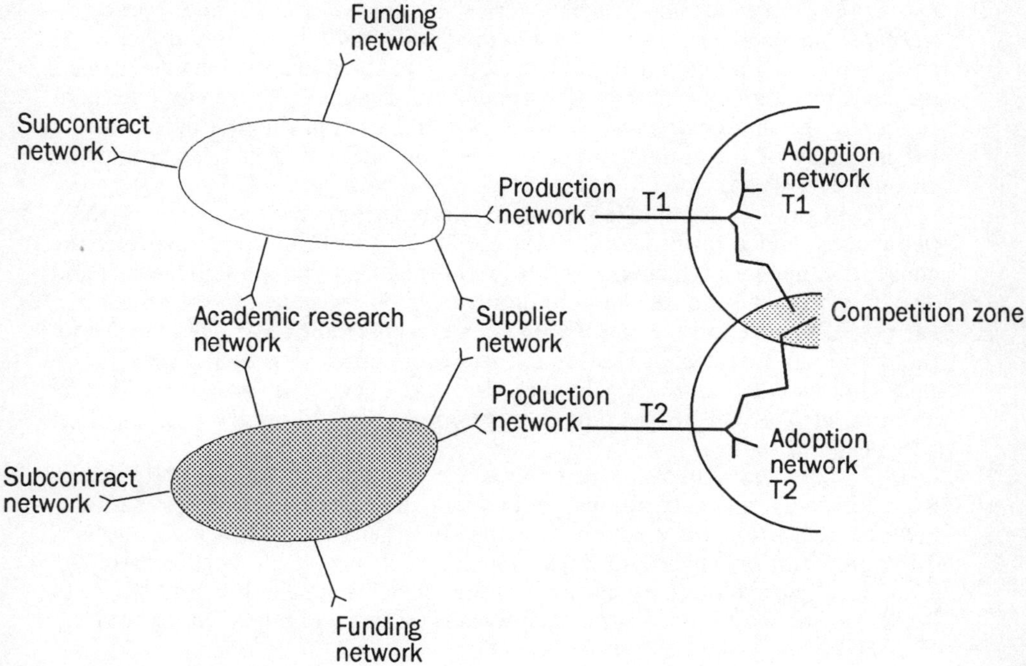

Figure 11.16

as in the preceding models in terms of the techniques (which are stable), the first adopters and their similarity networks (Figure 11.16).

The dynamic of the conception–adoption–competition network

Having looked separately at the dynamic of conception networks and those of adoption, it is necessary to move onto a description which integrates these different stages and analyses their succession.

The succession of the conception and adoption–competition phase

The dynamic of the conception/adoption of a technique is a two-stage process with alternating conception and adaptation phases in the strict sense of these terms. First, the techniques take shape and the actors who will participate in their production and diffusion are defined. In the second phase, which is in a manner of speaking the result of the work concluded in the conception networks, a dynamic develops which is that of adoption–competition. The puzzle of recruitment thus becomes clear. It is during conception, i.e. within the network responsible for conception, that adoption comes into play. The population of those who will adopt depends on the actors involved in the conception process and will be determined by the technical choices made.

During the conception phase the first adopters are negotiated at the same time as the techniques destined for them. It is on this basis that the adoption networks emerge that will be mobilized. The trajectory of these negotiations determines the starting point of the adoption trajectory. Technology strategy (the conception of Tj) and commercial strategy (identification of the first users who by the interplay of the similarity networks set off the adoption process and the ensuing events) are closely implicated.

A conception–adoption sequence is referred to as the succession of these two phases, and is termed t. The sequence begins with the convergence of the conception network which as a result of multiple negotiations stabilizes Tj and Ai. It continues with the mobilization of similarity networks of which the various Ai involved in the conception are the representatives, to be followed by the progressive setting up of adoption and production networks. According to this same model, another sequence can then begin. It continues for the duration of (t + 1*) (with a new conception network followed by an adoption network, and so on).

If R(t*) is the network which closes the sequence t*, it can be said that R(t*) contains the conception, production and adoption network simultaneously. R(t*) is a true creation. It brings together entities which existed in a different form, or which had weaker or different ties, at the beginning of the sequence. The evolutionary dynamic in this model is such that it leads R(t*) to R(t + 1*), then onto R(t + 2*), each succession being organized in two phases (Figure 11.17).

Another distinction is useful to describe the trajectory. It is based on the respective duration of the conception and adoption phases. Two extreme configurations can be identified. In the first case, the conception phase lasts a long time and is followed by a short adoption phase: the technique is evolving and probably compatible with small-scale manufacturing. In the other case, the adoption phase lasts a long time in relation to the (re)conception phase. Techniques are stable and standard. This is compatible with mass production.

Trajectories

Once the succession of the conception and adoption phases which make up a sequence has been understood, the issue of the succession of the sequences needs to be addressed. How does R(t*) move onto R(t + 1*)?

A conception phase always inherits networks formed in former sequences. In the sequence t + 1*, the starting point is R(t*). This means that the conception feeds off pre-formed networks in which techniques and diverse actors (producers, adopters, financiers, marketing personnel and researches, etc.) are tied together and links are stable. In the sequence t*, the number

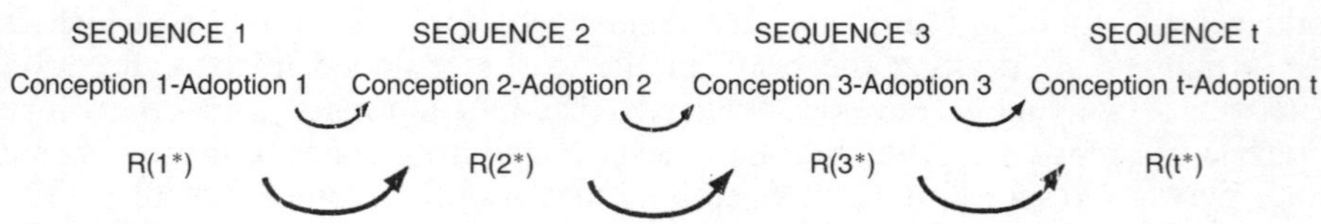

Figure 11.17

of actors involved grows from conception to adoption. Mobilization occurs following the logic of similarity indices. But each mobilization of a new actor (adopter, new banker, subcontractor) opens onto new networks. If the process stops, it is only because a certain stage has been reached; the probability of entry into what will be R(t*) falls below the threshold and is too weak to be significant (or to take an economic metaphor: for the potential adopter the entry costs are higher than the discounted benefits). The wider the network extends the more it multiplies the possibilities of connection with other similarity networks. True boundary objects (Hennion, 1991; Star and Griesemer, 1990) appear which while, of course, belonging to R(t*) also appear in other similarity networks, distinct from those mobilized by R(t*). A conception network profoundly different from that which launched the t* sequence can, thus, be established. The list of Ai who compose one R(t*) (which is constructed differently from the conception network existing at the beginning of the sequence as it has been enlarged by all the actors and all the techniques recruited during the adoption and production phases) forms a reserve of potential participants for the conception network of the following sequence (R(t + 1*). In the new sequence the representatives of the next adopters, who participate in the conception network, will be, for example, selected from the last adopters of the preceding sequence; the researchers enrolled at the end of the sequence by the specialist in monoclonal antibodies, and of whom the similarity networks were different from those of the spokesperson at the beginning of the conception phase, will be chosen as actors in the new conception network. This gives a picture both of a continuity which (R(t + 1*) inherits from R(t*)) and of a transformation principle which rests on the logic of the similarity networks and the permanent lags that they introduce (see Figure 11.18).

The second essential element in the move from R(t*) to R(t + 1*) is linked to the conventions which define access conditions to become an actor. Between t* and t + 1*, it is possible to imagine that the rules can change which define the population of entities who *have the right* to be actors. If R(t*) uses embryos as the material for the development of a medical technique, Tk, and if after Tk's adoption phase, the law forbids the use of these materials (which have been given the status of human dignity), the move onto R(t + 1*) will be profoundly altered. It is possible, on the other hand, that medical experiments

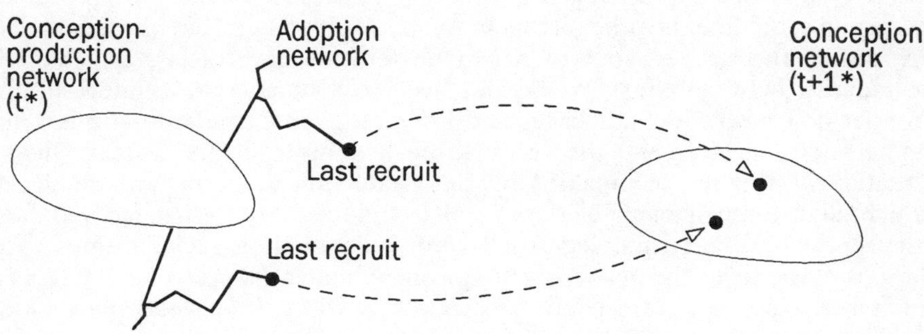

Figure 11.18 Example of the maximum distance between R(t*) and R(t + 1*)

on human beings, forbidden in the past, could be authorized at a later date. There is the rather bizarre case where chimpanzees were used to test certain pharmaceutical products. It was judged that they were not suitably representative of humans and were replaced, in derogation of the law, by groups of voluntary patients supervised by a panel of experts from various scientific organizations. The chimpanzee, once a legitimate subject for manipulation, was replaced by a group of patients and the researcher–actor by a group of dissenting researchers and experts (Bibard, 1991). In any event, the list of actors available to participate in the following conception phase will be modified and it will be necessary to take account of this in explaining R(t + 1*). Variations in conventions are becoming more common with growing concerns about the environment: hence CFCs were suddenly withdrawn from the list of substances which could used in the conception of an aerosol.

If the techno-economic trajectory is the series of transformations, R(t*), R(t + 1*), . . . , it is easy to understand under what conditions the trajectory becomes irreversible and produces, or does not produce, variety.

Irreversibility

In discussing irreversibility, it is enough to look at two extreme situations.

The first is the case where R(t + 1*) reinjects into the conception process all the techniques and all the actors present at the completion of the conception phase t* (the starting point for the conception corresponding to the finishing point of the preceding phase) while adding new actors recruited during the t* adoption, which opens up possible new adoption networks. Thus, learning is accumulated and networks are extended without detriment. Such situations are more common if recruitment has taken place in strongly linked similarity networks. In this case, the latest recruits are quasi-substitutable for the earlier ones. One of the important findings of model 1 (the adoption progresses, the more links multiply and the more the similarity indices reinforce one another) shows that this type of development becomes more and more probable as the sequences succeed one another. In addition, if account is taken of the *necessary* convergence of the conception networks, it results that as the number of conception–production–adoption sequences increases, the networks R(t*), R(t + 1*) become more and more connected and stable. Internal links reinforce one another to the detriment of external links. After numerous sequences, the probability of opening up new networks is low, because each actor has become more and more like the other actors in the network and less like the actors who are not in the network but to whom he remains nevertheless linked. The boundaries between the network and his 'environment' are reinforced, the frontier objects are less numerous and less ambiguous, similarity and internal homogeneity increase and the network breaks away from its context. Such a situation is usually accompanied by the production of norms and standards which align the behaviour of actors and techniques. Strong similarity indices presuppose, in the terminology used here, substitutability relationships. The various elements of the network are equivalent and each speaks for the others. Such a development corresponds to a trajectory which *is becoming irreversible*.

The other situation is that which corresponds to a transition of t* to (t + 1*) in which actors and techniques recruited for the new phase of

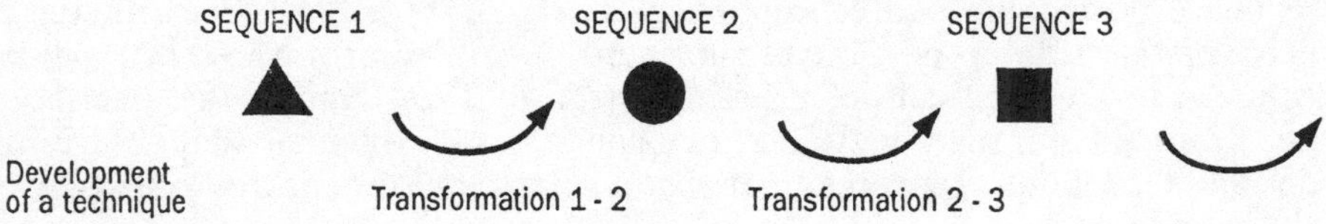

Figure 11.19

conception are different from those of the conception phase of the previous sequence and belong to the population of t*'s last recruits. As a consequence they may not be too similar to the actors who were involved in the previous phase. $R(t + 1^*)$ will have a composition, both of actors and techniques, which places it at some distance from $R(t^*)$. The discrepancies and differences will increase as the sequences follow on from one another. The trajectory is susceptible to reorientation at any time.

In the case of irreversible trajectories the forms of techniques developed during the successive sequences are relatively stable, whereas in the opposite case they can vary significantly from one sequence to the next (Figure 11.19).

The fact that a network becomes caught up in one trajectory rather than another depends on numerous factors. One of them is the morphology of the networks of the recruited adopters.[19] If they are weakly connected, long-standing and with regularly falling similarity indices, the second trajectory has a strong chance of being followed. If, on the contrary, they are strongly connected and the similarity indices show marked discontinuities (frontier effects), it is the first type of trajectory which will be selected from the outset. However this selection is neither automatic nor definitive. Another determinant of the trajectory stems from the nature and form of the coordination mechanisms which play an important role in the recruitment of the conception network's elements at $t + 1^*$. If the actors from the previous conception network are linked by fixed contracts then their involvement is inevitable, and if the equipment, instrumentation or documentary material have not yet been depreciated or replaced, they form fixed resources which must necessarily be exploited. The trajectory will rapidly become irreversible. If, on the other hand, the actors in the conception network at t^* are coordinated through the (quasi-)market, disengagement and re-engagement are easier. Simulations can establish the probability of trends on the basis of knowledge of former transformations.

Variety

This model allows the issue of the generation and maintenance of technical variety to be addressed in a rigorous manner by giving a dynamic interpretation of the technological dynamic.

To take several technological trajectories, the model allows technological competition to succeed non-competitive situations (and vice versa) because through the reconception mechanism, the techniques Tj and Tk are substitutable at t^* but may very well not be at $t + 1^*$.

In the situation where trajectories are not (yet) irreversible, with the sequence t* and the two conception networks of Tj and Tk, each of these networks is made up of a list of actors Aij(t*) and Aik(t*) whose interaction will lead to a consensual definition of Tj(t*), of Tk(t*) and to the identification, through similarity networks, of populations of potential adopters. Why do these adoption networks partially overlap and generate a competitive situation?

For such an overlap to occur it is necessary that both conception networks possess actors, spokespersons of the adopters, who at the end of the negotiation and conception process will be highly similar (because the two similarity networks have numerous elements in common). Such a similarity may be the consequence of a multiplicity of factors:

- Strong connectivity of adoption networks which gives rise to a high probability of strong similarity among their spokespersons within the various conception networks.
- Strong links between the conception networks (recourse to identical scientific specializations, mobilization of the same research teams, subcontracting to the same contract research organizations, using the same consultancy partnership, dependence on the same information sources and studies, participation in the same public programmes, joint distribution networks, etc.).

Situations of technological competition correspond to configurations which may be very diversified. For competition to exist, it is sufficient that adoption networks overlap. But if other mobilized actors and/or techniques (researchers, designers, production engineers, experimental techniques, means of production, etc.) in the two conception networks are profoundly different from one another, then technological competition will not only be based on the intrinsically dissimilar techniques but also on the systems of production, subcontracting university cooperation and funding networks which have nothing in common. Rj(t*) and Rk(t*) do not share anything other than the same commercial target. If, on the other hand, the similarities go so far as to encompass the research actors, the financial circuits, the subcontractors, etc., the configuration is one of firms or groups who are not only targeting the same users but in addition are based on the same techniques. Between a situation of perfect similarity and one where the only things in common are the users, all the intermediary configurations of technological competition are possible. As has been shown, to analyze competition it is necessary to change the perspective to focus on conception networks and the interlinkages which make them more or less dependent.

To move from a competitive to a non-competitive situation, it is enough that from one sequence to the other (and more precisely during the t + 1* conception phase), the networks of adopters selected by the conception networks have been (re)defined in a sufficiently radical manner so as to find themselves without anything in common. Such divergence develops generally over several adoption phases. Generally speaking, the development of competition to a non-competitive situation will be even more probable if the trajectories associated with Tj and Tk are themselves intrinsically divergent, notably in the early sequences. In reality, in line with the rule formulated above, after many adoption sequences, it is more difficult to escape a situation of competition or non-competition. Each technique is taken control of by the (re)conception networks which have a tendency to confirm former choices. On the other hand, in the early sequences choice is possible, e.g. the make up and organization of conception networks can lead to a rapid divergence between initially substitutable techniques (i.e. in competition with the same networks of adopters), or on the contrary to a convergence between techniques which were initially non-substitutable, or again, and probably more frequently to a hybrid situation. The adoption rationale, depending as it does on conception, has in its turn an impact on conception.

Conclusion

On the basis of the above analysis, six conclusions can be drawn:

1. The integration of the conception and adoption phases into a unified analytical framework is not only possible but necessary (Mustar, 1991). It has been shown how the adoption dynamic is generally dependent on the conception dynamic and that, reciprocally, adoption networks permanently redefine the foundations on which the future conception networks will be built. Variety and irreversibility cannot be explained without this integration.
2. The network concept enables the use of the same general categories to analyze the conception and the adoption phases. In order to achieve this, some distance was taken from the usual definition of this concept. As used here, the term network is not considered as an intermediate coordinating mechanism between an organization and the market (Thorelli, 1986; Powell, 1990; Imai and Baba, 1991; DeBresson, 1991; Freeman, 1991). It is taken in a more narrow sense to mean the group of unspecified relationships among entities of which the nature itself is undetermined. The picture of a network is thus reduced to a simple graph. The simplicity of this analytical tool is its principal advantage. It offers a minimal framework to describe interactions in all their diversity and richness. This is why it can be applied equally well to conception and adoption phases.
3. The analysis of conception networks presupposes that the conventions which design the boundary between actors and techniques are specified and that rules and procedures which structure interactions and negotiations are described. The conception phase is completed at the precise moment when network convergence is obtained. In conception networks, the actors are invested with extended competences as they are capable of conceiving, anticipating and reaching compromises. From this point of view they contrast strongly with adoption networks where the only competence which actors have is the choice between adoption and non-adoption. However, by introducing the similarity concept, measured by the degree of resemblance between association profiles, an analysis of the decision-making process is side-stepped. It is enough here to compare the series of former choices to infer the probability of the compatibility of future choices. Such a perspective fits with whatever theory of decision-making one cares to use, including the most sophisticated (an uncertain universe, for example), as it is confined to measuring similarities among actors.
4. The adoption model highlights the essential role played by diffusion of stabilized techniques in the dynamics of social relationships. It has been shown that similarity networks are transformed as the adoption process progresses. Three variables are vital to explain this dynamic: the morphology of the similarity network which organizes the population of potential adopters; the position of the first adopters in

the networks; the form of the probability function linking the similarity indices and adoption decisions. According to the configuration of these variables, trends may be traced which reinforce existing structures, or on the contrary weaken the existing links. Competition among several techniques is only one individual variant of this model.

The conception–adoption model follows actors and techniques in their joint development. Throughout the process of conception, the identity of actors varies, the characteristics of the techniques transform themselves. This momentum contributes to the definition of the various networks which take up the baton once conception is completed. The population of potential adopters is thus defined once the conception process is finished, and at the same time as the first users are selected. Thus the adoption process begins, which is analyzed using model 1 (once the techniques, similarity networks and position of the early adopters are known).

5. The conception–adoption–competition model enables the production or disappearance of technical variety, as well as the creation of irreversibilities, to be taken into account. Some of the factors which are responsible for these trends were identified: the links between conception networks, which have a general impact on local decisions – but also the form of adoption network – which may sometimes update the list of actors involved (involvable) in the conception process, constitute some crucial variables, along with conventions establishing the foundations between techniques and actors. The algorithms used for simulation should be designed to highlight the impact of these different variables in the overall dynamic.
6. The models advocated enable the emergence of collective structures to be mapped resulting from a large number of decisions taken by agents within a network of fluctuating relationships. These emerging structures are responsible for the extension or disappearance of variety as well as for establishing irreversibilities. Linked to the dynamic of conception and diffusion of techniques, they encompass not only markets but also organizations. Given the complexity of these dynamics, simulation is the only possible approach for identifying the different families of trajectories as well as the variables which account for the emergence of a particular structure. Computers are coming to the rescue of Leibniz's god and making up for market deficiencies. The models presented have been designed to make such simulation possible.

Notes

1. 'In the absence of any other determining factor, what occurs is the maximum possible with regard to the capacity of time and space'.
2. This chapter is a resumé of a longer text which, in particular, takes account of complementarities among techniques and suggests simulation algorithms.
3. This population is obviously subject to change as techniques evolve, in particular as the process of diffusion progressively advances. There are multiple reasons for these linked variations and they stem as much from transformations/adaptations of techniques as from the initiatives of suppliers or from changes in the user population.
4. The well-known criticisms of the different models will not be gone over here. Even if it is accepted that techniques and the supplier's strategy may be modified during the diffusion process, the issue remains the same. While the use of these refinements allows for the potential adopter population varying according to the state of the technique or the suppliers, this does not mean that at any given moment it will not be possible to identify the population or, more fundamentally, its very existence.

5. A mainstream economist would no doubt say that this results from the constraint imposed by the necessarily limited universe of available merchandise which compels consumers to choose from the same goods. If, in theory, it is possible to have a world where each consumer may choose different goods, in practice it is impossible because resource scarcity means that the inventory of goods is necessarily limited. In this context, it is the scarcity constraint that leads to the similarity of certain utility functions and not some sort of interdependence.
6. This interdependence might go as far as the coordination of expectations.
7. Any algorithm can be used, e.g. STRUCTURE, CONCOR, etc.
8. For an Ai and the associated Sn(Ai) two extreme configurations are possible. The first is a star-shaped graph for n as large as required. If Sn(Aj) has the same morphology it could be that Sn(Ai) and Sn(Aj) have numerous elements in common even though they have very different association profiles. In this case each actor simultaneously occupies several sharply demarcated worlds. In their shared world(s), Ai and Aj are similar but in all other worlds they are totally unalike. Ai and Aj play the 'weak ties' role, the importance of which has been shown by Granovetter. In the other extreme case, Sn(Ai) and Sn(Aj) are closely connected; Ai and Aj have coherent identities and are very similar.
9. Until this point, the elements compared have been human. Obviously, this is an unnecessary restriction. Ai and Aj could be technical objects, and the associations which they form could be human beings (who conceive, manufacture, distribute, adopt and contest them) or other technical objects with which they interconnect or plug into. The concept of similarity thus corresponds to that of substitutability.
10. It is clear that any increasing function for which the values are between 0 and 1 will do the trick.
11. In other words, Bi + m has a similarity index which is not 0 with at least one element of R(Ai).
12. This applies on condition that n > or = 2.
13. A complete analysis of the adoption dynamic in competitive conditions supposes, (a) that three properties are taken into consideration: substitutability, compatibility and complementarity, (b) that it is accepted that these properties vary during the adoption process (with the exception of intra-generational substitutability). For reasons of space, I cannot develop this important point here. Techniques are considered to be substitutable, incompatible and non-complementary. For a complete analysis see Callon (1992).
14. n is selected in function of the morphology of the networks.
15. The model can be elaborated by making technologies evolve, i.e. by varying their degree of compatibility and complementarity. Simulations which take into account evolutions in the competition/adoption process would be more complicated, but would not introduce really new variables. Besides, the question of the actors' expectations may be considered. In weighing up T1 and T2, it is likely that Bi + m would take account of future adopters (e.g. if he thought that other potential adopters would opt for T2, it would choose this technique even if his similarity network selected T1). Here the hypothesis is made that the degree of similarity of expectations is in line with the similarity network, which automatically resolves this issue in this case.
16. This presentation is somewhat stylized. For brevity, not all the hypotheses of translation sociology, on which the following is based, have been discussed here.
17. The dynamic of the conception network corresponds to what economists call learning.
18. In this model the identification of the first adopters is not an exogenous 'small event' (Arthur, 1989) but the endogenous result of the conception process (see Mangematin and Callon, 1991).
19. The morphology of the subcontracting, funding, research, etc. networks should

also be taken into account. For reasons of brevity, only adoption networks are discussed here.

References

Akrich, M. (1989a), 'La construction d'un système socio-technique. Esquisse pour une anthropologie des techniques', *Anthropologie et Société*, 12.

Akrich, M. (1989b), 'De la position relative des localités; systèmes électriques et réseaux socio-politiques', in *Innovation et ressources locales, Cahier du CEE*, vol. 32.

Akrich, M. (1991), 'The description of technical objects', in Bijker and Law (ed.).

Akrich, M. (1992), 'From sociology to semiotics', in Frost and Pfaffenberger (eds).

Aoki, M. (1990), 'Toward an economic model of the Japanese firm', *Journal of Economic Literature*, March.

Arthur, B. (1989), 'Competing technologies, increasing returns and lock-in by historical events', *The Economics Journal*, vol. 99, March.

Bibard, L. (1991), 'La place et le rôle des sciences dans les innovations techniques: quelques cas en biotechnologie', Thèse de socio-économie, EHESS.

Bijker, W.E., Hughes, T.P. and Pinch, T. (eds) (1987), *Constructing Networks and Systems, Case Studies and Concepts in the Sociology and History of Science*, MIT Press.

Bijker W.E. and Law, J. (eds) (1992), *Shaping Technology, Building Society: Studies in Sociotechnical Change*, MIT Press.

Bruno S., Cohendet, P., Desmartin, F., Llerena, D., Llerena, P. and Sorge, A. (1991), 'Modes of usage and diffusion of new technologies and new knowledge', *Fast*, June.

Burt, R. (1987), 'Social contagion and innovation: cohesion versus structural equivalence', *American Journal of Sociology*, vol. 92, May.

Callon, M. (1986a), 'Some elements of a sociology of translation: domestication of the scallops and the fishermen of St Brieuc Bay', in Law (ed.).

Callon, M. (1987), 'Society in the making', in Bijker, Hughes and Pinch (eds).

Callon, M. (ed.) (1989), *La science et ses réseaux*, La Découverte.

Callon, M. (1991), 'Réseaux technico-économiques et irréversibilité', in Boyer, Chavance et Godard (eds), *Les figures de l'irréversibilité en économie*, Editions de l'Ecole des Hautes Etudes en Sciences Sociales.

Callon, M. (1992), 'Modèles de simulations pour la dynamique des réseaux de conception et d'adoption', CSI.

Callon, M., Law, J. and Rip, A. (eds) (1986), *Mapping the Dynamics of Science and Technology*, Macmillan.

Courtial, J.P. and Cahlik, T. (1991), 'Simulation des réseaux de conception et d'adoption des techniques', CSI.

David, P. (1975), *Technical choice, innovation and economic growth*, Cambridge University Press.

David, P. (1986), 'Understanding the economics of QWERTY: the necessity of history', in Parker (eds), *Economic History and the Modern Economist*, Basil Blackwell.

David, P. (1987), 'New standards for the economics of standardization', in P. Dasgupta and P. Stoneman (eds), *Economic Theory and Technology Policy*, Cambridge, Cambridge University Press.

David, P. and Foray, D. (forthcoming), 'Self organized criticality and the economics of document standards diffusion: the case of EDI'.

DeBresson, C. and Amesse, F. (1991), 'Networks of innovators: a review and introduction to the issue', *Research Policy*, 20.

Degenne, A. (1991), 'Présentation du dossier', *Sociétés Contemporaines*, no. 5, March.

Doreian, P. (1988), 'Equivalence in a social network', *Journal of Mathematical Sociology*, vol. 13.

Dosi, G. (1988), 'Sources, procedures and microeconomic effects of innovation', *Journal of Economic Literature*, 26.

Dosi, G., Freeman, C., Nelson, R., Silverberg, G. and Soete, L. (eds) (1988), *Technical Change and Economic Theory*, Frances Pinter.

Dosi, G. and Metcalfe, J.S. (1991), 'Approches de l'irréversibilité en théorie économique', in Boyer, R., Chavance, B. and Godard O. (eds), *Les figures de l'irréversibilité en économie*, Editions de l'Ecole des Hautes Etudes en Sciences Sociales, Paris.

Elster, J. (1975), *Leibniz et la formation de l'esprit capitaliste*, Aubier Montaigne.

Eymard, Duvernay F. (1989), 'Conventions de qualité et formes de coordination', *Revue Économique*, 40.

Foray, D. (1991), 'Dynamique économique et nouvelle exigence de l'investigation historique', *Revue Economique*.

Foray, D. (1991), 'Rendement croissant, invention et progrès technique: vers un nouveau programme de recherche', mimeo.

Freeman, C. (ed.) (1990), *The Economics of Innovation*, Edward Elgar.

Freeman, C. (1991), 'Networks of innovators: a synthesis of research issues', *Research Policy*, vol. 20, 5, October.

Frost, R. and Pfaffenberge, B. (eds) (in preparation), *Material Discourse, The Meanings of Human Artefacts*.

Gaffard, J.L. (1989), 'Marché et organisation dans les stratégie économiques des firmes industrielles', *Revue d'Economie Industrielle*, 48.

Hennion, A. (1991), 'La médiation musicale', Thèse de Sociologie, EHESS.

Hughes, T. (1983), *Networks of Power, Electrification in Western Society 1880–1930*, Baltimore, Johns Hopkins University Press.

Imai, K. and Baba, Y. (1991), 'Systemic innovation and cross-border networks: transcending markets and hierarchies to create new techno-economic systems', in *Technology and Productivity*, ACDE.

Kidder, T. (1981), *The Soul of a New Machine*, Allen Lane.

Larédo, P., Meyer, J.B. and Vinck, D. (1991), 'Les réseaux de recherche dans les programmes comunautaires: le cas des actions concertées dans le domaine de la santé', CSI-CCE.

Latour, B. (1987), *Science in Action*, Open University Press.

Latour, B. (1988), 'Mixing humans and non humans together: the sociology of a door-closer', *Social Problems*, 35.

Latour, B. (1989), *The Pasteurization of French Society*, Harvard University Press.

Latour, B. (1991), *Aramis or the Love of Technology*, Harvard University Press.

Latour, B. and Woolgar, S. (1979), *Laboratory Life*, Sage.

Law, J. (ed.) (1986a), *Power, Action and Belief: A New Sociology of Knowledge*, Sociological Review Monograph 32, Routledge.

Law, J. (1986b), 'On the methods of long distance control: vessels, navigation and the Portuguese route to India', in Law (1986a).

Law, J. (1987), 'Technology and heterogeneous engineering: the case of the Portuguese expansion', in Bijker, Hughes and Pinch (eds).

Law, J. (ed.) (1991), *A Sociology of Monsters*, Sociological Review Monograph 38, Routledge.

Law, J. (forthcoming), 'The Olympus 320 Engine: a case study in design, autonomy and organisational control', *Technology and Culture*.

Law, J. and Callon, M. (1988), 'Engineering and sociology in a military aircraft program: a network analysis of technical change', *Social Problems*, 35.

MacKenzie, C. (1991), *Inventing Accuracy: A Historical Sociology of Nuclear Missile Guidance*, MIT Press.

Mangematin, V. (submitted), 'Proposition d'une méthodologie pour l'analyse de la technologie'.

Mangematin, V. and Callon, M. (submitted, 1991), 'Technological competition, strategies of the firms and the choice of the first users: the case of road guidance technologies'.

Mustar, P. (submitted, 1991), 'Processes of integrating science with market: the creation of technology-based business'.

Powell, W.W. (1990), 'Neither market nor hierarchy: network forms of organization', *Research in Organization Behaviour*, vol. 12.

Rabeharisoa, V. (1990), 'La construction de réseaux technico-économiques dans le domaine de la régulation thermique', CSI-AFME.

Rosenberg, N. (1982), *Inside the Black Box*, Cambridge University Press.

Saviotti, P. (1991), 'Technological evolution, product characteristics, competition and variety, colloquium management of technology: implications for enterprise management and public policy', Conférence des Grandes Ecoles, Paris, May.

Star, L. (1991), 'Power, technology and the phenomenology of conventions: on being allergic to onions', in Law (ed.).

Star, L. and Griesemer, J. (1989), 'Institutional Ecologies, Translations and Coherence: Amateurs and Professionals in Berkeley's Mauseumn of Vertebrate Zoology', *Social Studies of Science*, 19.

Stiglitz, J. (1987), 'Learning to learn, localized learning and technical progress', in Dasgupta, P. and Stoneman, P. (eds), *Economic Theory and Technology Policy*, Cambridge University Press.

Stoneman, P. (1983), *The Economic Analysis of Technical Change*, Oxford University Press.

Teil, G. (1991), 'Candide, un outil de Sociologie Assistée par Ordinateur pour l' analyse quali-quantitative de gros corpus de textes', Thèse de socio-économie de l'innovation, Ecole des Mines de Paris.

Teil, G. (1990), 'Halfway between connectionism and symbolism: a sociological machine', CSI.

Thorelli, H.B. (1986), 'Networks: between markets and hierarchies', *Strategic Management Journal*, 7.

Vinck, D. (1991), 'La coordination scientifique, Etude de deux formes spécifiques: le laboratoire et le réseau', Thèse de socio-économie de l'innovation, CSI.

Part IV
Institutional and technological change: the importance of diversity

Introduction to Part IV

Arndt Sorge

It is not common to think of technology and social institutions as basically similar phenomena. Observers may well agree that both are important for competitiveness and there is a growing discussion about which institutions and kinds of technology are competitive under which conditions, whether forms of technology or institutions tend to be complementary or mutually substitutive, which is more important under which conditions, and other issues. In spite of the growing awareness of relations between technology and institutions, and between these and competitiveness, still people tend to consider technology and institutions as rather different breeds. Anyone who insists on the similarity runs the risk of being called a technocrat, if you liken institutions to technology, or a naive fool if you argue that technology could be subject to the same arbitrary or politicised design which characterizes institutions.

I would like to run both risks here and argue that

- technology and institutions are very similar in type;
- awareness of this similarity is important for realizing in which way they contribute to competitiveness; and
- this understanding opens the way towards grasping the importance of diversity, both in technological and institutional development.

First, technology and institutions are similar in being *artefacts*. They are constructed by human beings, rather than found in nature. Both technology and institutions are cultural artefacts, and they are designed – consciously or unconsciously – with an instrumental, expressive or ethical purpose in mind, and usually a mixture of these (Sorge, 1985).

The constructedness of technical and institutional artefacts has an important implication. It refers to a characteristic bundling of instrumental, value-rational, traditional and expressive design motives, and to a coupling of economically utilitarian and socially arbitrary motivations. Take a simple example to illustrate a general point: Soviet manufacturers of aero jet engines have not evolved by-pass and fan jet engines, which have become the state of the art in Western civil jet engine technology. This is not because the Soviets are technically inferior, or not able to see the advantages of by-pass engines. The simple fact is that by-pass engines are not used in jet fighters or bombers; civil aviation and technology were much more in the grip of the military establishment in the Soviet Union compared to the West (Aeroflot is a division of the Air Force), civil air traffic is heavily regulated and protected and noise and fuel consumption are less of a problem.

Thus, institutional diversity between the United States and the Soviet Union explains, and is of a basically similar nature as, technical diversity between airliners of the two countries. One should not exaggerate the differences between capitalist dynamics and socialist sloppiness, in order to explain the case in terms of capitalist progress and socialist lagging-behind. In fact, the Soviet Union had the first operational civil jet aircraft (the Tupolev 104) flying safely at a time when the first Western civil jet aircraft (the DeHavilland Comet) had been withdrawn after disastrous crashes due to metal fatigue, and long before the French Caravelle or the American Boeing 707 were in service.

The Tu 104 had been a spin-off product, developed out of a military aircraft. Likewise the competitiveness of Boeing in civil jet aircraft owes a great deal to having won the first order for a military tanker jet, to refuel strategic bombers in mid-air. Boeing took the opportunity to redevelop that plane into the father of its series of civil jets (the Boeing 707). The difference between Boeing and Tupolev is merely that Boeing was from then on forced to compete, whereas Tupolev had a secure market. Thus, the ensuing difference in competitiveness is due to technical developments which became diverse because of institutional diversity.

But this is not merely a game between market orientation and governmental control, capitalist dynamics and stagnation in socialist civilian economies. The first wide-bodied jet with by-pass jet engines owes its existence, not only to foreseeable civil market demand, but equally, military imperatives. The US forces had much larger problems of logistically supplying and reinforcing their army and air force in Europe than the Soviets. Across the Atlantic, road and rail transport are not viable, and transport of troops by ships takes much longer and is vulnerable. This gave a decisive impetus to the development of larger military transport aircraft which required by-pass engines more urgently. As a competitor that had lost out in the battle for a giant military transport aircraft, Boeing converted its military project to become the 747, which was again at the origin of a series of wide-bodied planes with by-pass engines, made by several manufacturers, such as the DC 10, the Lockheed Tristar and the European Airbus. Thus, even in fiercely competitive civil aviation, institutional arrangements, government projects and other influences outside the grasp of the civil market economy, continue to play an important part (Hayward, 1986).

The interplay of economic motives with social, institutional and political motives is not only characteristic of an industry which is particularly concentrated and marked by government – notably military – influences, such as the aerospace industry. The same pattern can be detected in other industries, albeit in a more refined way. In one of the chapters to follow, Watanabe argues that participation in Japanese quality control circles is explained by the economics of internal labour markets as well as by the socio-cultural fabric of Japanese society. Arguments of the type 'it is not economic but socio-cultural and political factors that explain why . . .', or the reverse, are fairly widespread. However, it is not a form of argument which takes us much further. What very often happens is that economic motives or mechanisms take a specific form because of socio-cultural and political factors and, the other way round, that economic factors are part of the socio-cultural heritage and the political game.

This can be expressed by the concept of interpenetration, which comes from the later writings of the sociologist Talcott Parsons: differentiated sub-systems of society are never entirely separate but part of each other, such that the political sub-system has an economic sub-sub-system and vice versa, the technical sub-system has economic and political sub-sub-systems, etc. The logic of this approach is to insist that economic, social, political and technical phenomena take form, through reciprocal conditioning and interaction; the consequence is that a kind of playing-off of factors from these different domains against each other is usually not meaningful. The interesting thing about this reciprocal conditioning and interaction is that it maintains the distinctiveness of both technical and institutional characteristics in an economy and society. It is a diversity-maintaining process that takes effect notwithstanding tendencies of technology and institutional transfer and change. It does therefore not exclude innovation and change but emphasizes that technical and institutional change serve to innovate existing characteristics of an economy and society, rather than doing away with their specificity.

More recently, an evolutionary approach has been developed which suggests that economic opportunities and constraints, streams of technology development and application, and social, political and institutional characteristics of a society and economy tend to interact and evolve jointly. This interaction and reciprocal conditioning brings about relatively stable traditions, crystallized into industrial structures, societal institutions and technical design and development conventions, habits and capabilities. This approach is put forward, for instance, in Dosi *et al.* (1988). It is also inherent to the growing literature on the social construction of technology. More recently, the usefulness of the approach was asserted in a comparison of selected technological trajectories in European countries (Cohendet *et al.*, 1991). This showed how the distinctiveness of existing national patterns is asserted in the diffusion and development of identical basic or generic technologies.

Another related approach with a similar line of argument is the societal effect approach, which had emerged from cross-national comparisons of organization, manpower, education and training, professional careers, remuneration and industrial relations (Maurice *et al.*, 1982). This approach can be used to show that a basic technology, such as computer-numerically controlled machine-tools, is developed in different ways, follows different diffusion patterns, fits into different national systems of organizing work and socializing workers, is designed with different types of users in mind, and has very different impacts on employment and industrial performance in specific sectors, even when two roughly similar countries such as France and Germany are compared (Maurice and Sorge, 1989). The contrast between these countries does carry connotations different from those between Soviet and US aircraft industries, mentioned earlier; it is not lumbered by the connotations of capitalist advance and socialist stagnation.

Such approaches are borne out by the contributions which follow in this part of the book. What they have in common are the following basic arguments:

(1) Firms do not respond with infinite elasticity, to perceivable market and technological environments. Actors are typically inelastic, i.e. they respond in the way in which they have been conditioned by past investment into capital and manpower, and by the nature of their specific social, political and institu-

tional environments. To that extent, their behaviour is dependent on specific contexts and hence diverse. This argument implies that institutional and technical diversity perpetuates itself in technological and social history, rather than being assimilated by universal pressures, and elastic and identical response.

(2) Technical and institutional design do not happen through adaptation of actors to, supposedly objectively given, tendencies in the evolution of economic, scientific and technological factors. Actors are constructed and organized in specific ways, which makes them assume particular tendencies, and enables or disables them to pursue particular paths of institutional and technical development. Lundvall's chapter describes this variety of institutions as 'national systems of innovation'.

The variety which is constituted by alternative institutional patterns may be summarized to include in particular the following dimensions. These are more concerned with institutional aspects than with types of technology. But it can be added that they have implications for types of technology development and usage. Here, then, are the institutional dimensions of variety between countries, as they come across from the chapters that follow:

Relations between *competitors* in roughly the same industry may differ between adversarial competition without sharing of certain functions or co-operation on the one hand and cooperation, notably within industrial districts, such that certain marketing, development, production, servicing or vocational training efforts are pooled or shared. The social construction of such relationships is crucial to the competitive advantages of types of enterprise and industrial concentration. Externalization and pooling of functions increases the competitiveness of smaller firms, whereas non-pooling and non-sharing increases the advantages of large competitors. Furthermore, institutionally rich industrial districts may permit a greater amount of product variety or product differentiation, through greater entrepreneurial decentralization of decision-making, than would arise in concentrated enterprises. Notably, this point is made in the chapters by Piore, Hollingsworth and Sabel.

Relations between *personnel categories* and *jobs in the enterprise* may imply greater specialization of functions or more overlapping work roles. Specialization of functions is associated with large enterprises, high technology, large batch or mass production and cost minimization strategies based on increasing returns to scale. On the other hand, overlapping and enriched work roles are characteristically matched by enterprise strategies based on achieving product quality or more customized production and product innovation. This contrast is fundamental to the chapters by Watanabe and Marsden. Japanese enterprises are competitive by focusing on product quality, which they can do because quality circles are more systematically used, as a form of overlapping work roles and enriched production tasks. British enterprises, by contrast, similar to the American firms mentioned by Piore, feature more specialized responsibilities, impoverished production tasks, and greater prevalence of cost-cutting strategies.

Relations between *enterprises* and *their institutional environment* vary between institutional richness, with many attachments to manufacturers' associations, governmental agencies and education and training institutions on the one hand, and institutional isolation on the other. The more institutionally attached and embedded firms are, the more they are able to pursue strategies emphasizing product innovation, product quality and customization of products. This contrast is sharp with regard to the relative advantages of smaller firms, whereas larger firms are presumably better able to create the institutional embeddedness required for innovation and product quality or to institute appropriate patterns internal to the firm itself. This point is central to almost all the following chapters.

The relations between the actors mentioned above are not only important for the competitiveness of individual firms, types of firms and industrial structures, considered at micro or meso levels. In addition, they are tied into characteristics of economic development at a highly aggregated macro level. With a long-running shift from mass markets to more fragmented and individualized markets in Western economies, prospects for macroeconomic growth depend on the extent to which a society already has, or turns to, institutions that facilitate an emphasis on product innovation and quality, viable small and medium-sized firms, and customized production.

These institutional dimensions are now shown to be related to the technological variety between countries. From the chapters of both Watanabe and Hollingsworth, it emerges that the Japanese are better at continuous innovation and improvement after the first pioneering stages in a longer technological trajectory have been completed. This is facilitated by within-firm labour mobility which cuts across, and links diverse functions such as research, development, production engineering and production. Consequently, piecemeal improvement is very important for the competitiveness of Japanese companies, and too often neglected by observers who underline the sheer presence of large numbers of robots as a productivity-enhancing device. The United States by contrast, appears better geared to perform radical basic innovation, because of its more entrepreneurial culture, industry–research linkages in that domain, and elastic professional and industrial labour markets.

In a similar way, a bridge can be built from the concepts of institutionally rich attachments and enriched or overlapping work roles, over to a form of technology which allows flexibility of utilization. By contrast, there is also a relationship between specialization of functions in the enterprise and single-purpose automation. Every time, it would seem that a country tends to be competitive in a variant of technology or stage of technical development, which roughly matches its institutional characteristics. Thus, although technology diffusion has strongly universalistic and convergence-enhancing elements, it is also subject to a remarkable diversity-enhancing influence. These countervailing influences appear to play one against the other. Moreover, they are also linked with each other, such that the tendency in one direction depends on a countervailing tendency in the opposite direction. The consequence is that even internationalization of technology makes countries develop relative competitive advantages in specific stages and variants of technology development and diffusion.

This leads to the conclusion that diversity in institutional patterns and technology development is not an idiosyncratic and largely irrational aberration from the supposedly one best way. Evolution towards diversity is, instead, an important phenomenon in the development of regional or national competitive advantages. By the same token, it is an important mechanism in the international division of labour. So although the following chapters document highly interesting possibilities of one region or country learning from another, this learning should not be interpreted as a form of assimilation. Other nations do adopt Japanese-style quality circles, admire North Italian industrial districts, dream about Silicon valleys, or try to emulate German-style vocational training. But such instances of technology or institutional transfer are, first, specific to particular phases of economic development, and second, they tend to lead to modified forms applied in different situations with different outcomes. In institutional and technical innovation, it turns out that once the trend-setting pioneers have made their mark, the followers have to live by continuous improvement, rather than straightforward copying which does not provide secure advantages.

In that way, the implication is that rather than promoting technology and institutional transfer *tout court*, we should be discussing ingenious technology selection and improvement, and see this related to the evolution of the institutional portfolio of a society. This has important consequences for policies of private and public actors, which cannot be discussed here.

The overall argument in this part of the book naturally still has some open ends, loopholes and contradictions. But its central message is fairly robust and deserves attention:

- Diversity is an asset, whereas assimilation and copying is a liability.
- Competitive advantage accrues through reciprocal matching of institutions and technology.
- Internationalization means division of labour and, hence, diverse evolutions of institutional and technical portfolios.

References

Cohendet, P., Llerena, P. and Sorge A. (1991), *Diffusion and Mode of Usage of Science and Technology. Report of a Study for the FAST Programme*, Strasbourg/Maastricht, Research Report.

Dosi, G., Freeman, C., Nelson, R., Silverberg, G. and Soete, L. (eds) (1988), *Technical Change and Economic Theory*, London/New York: Pinter.

Hayward, K. (1986), *International Collaboration in Civil Aerospace*, London: Frances Pinter.

Maurice, M., Sellier, F. and Silvestre, J-J. (1982), *Politique d'éducation et organisation industrielle en France et en Allemagne. Essai d'analyse sociétale*, Paris: Presses Universitaires de France.

Maurice, M. and Sorge, A. (1989), *Dynamique industrielle et capacité d'innovation de l'industrie de la machine-outil en France et en RFA*. Aix-en-Provence: Laboratoire d'économie et de sociologie du travail, document 89-1.

Sorge, A. (1985), *Informationstechnik und Arbeit im sozialen Prozess. Arbeitsorganisation, Qualifikation und Produktivkraftentwicklung*. Frankfurt/M.: Campus.

12 User–producer relationships, national systems of innovation and internationalization

Bengt-Åke Lundvall

Introduction

Carlsson's chapter in Part I has already pointed to the importance of 'national systems of innovation'. However, in order to see why this system is a useful analytical concept a radical revision of the basic neo-classical assumptions is necessary. In a world where all agents are perfectly rational (maximizers of utility and profit, with unlimited access to information, and an unlimited capacity to gather and process information) and all transactions take place in pure markets, with anonymous relationships between buyers and sellers, national borders would play a limited economic role. In such a world, institutional and cultural differences between nations would not, by assumption, interfere with economic processes and structural differences would reflect nothing but historical investment decisions.

Only when the assumptions of homogeneously, perfectly, rational agents and pure markets have been revised does the relevance of national systems of innovation become obvious. What makes national systems of innovation important is that the organized markets of the real world may be organized differently in different national systems and that the behaviour of agents, rooted in different systems, may be governed by different rules and norms.

Introducing innovation and interactive learning

Conversely, the need for a radical revision of the assumptions on perfect rationality and pure markets becomes clear when the focus is shifted, away from allocation and commodity exchange towards innovation. In a stationary world, where the central activity of agents is the exchange of use values with constant characteristics a theoretical construct, assuming perfect rationality, might be regarded as a useful abstraction. In such a stationary context, where the most important activity is economic exchange, it would also be legitimate to assume all transactions to take place in pure markets with anonymous relationships between buyers and sellers.

* A slightly different version of this chapter will be published in, *National Systems of Innovation* (Lundvall, B-Å, ed., Pinter Publishers, forthcoming in 1992).

An 'innovation as an interactive process-perspective' brings two crucial new elements into the analysis. The first is, of course, uncertainty, reflecting change and growing complexity. Innovation is, by definition, the creation of qualitatively different, new things and new knowledge. Therefore, agents involved in the creation and adoption of innovations cannot reasonably be assumed to know the probability of all possible outcomes of this process. And, when uncertainty is introduced it becomes obvious that agents must be characterized by bounded rationality rather than by perfect rationality.

Differentiated rationality in the innovative economy

But if rationality is bounded we cannot insist upon homogeneity in the rules of behaviour, followed by agents. And, if we want to specify the behaviour of an agent it is not sufficient to characterise him/her/it as 'rational' any more. Behaviour must be further specified and this specification implies that agents might be different in several respects.

Their attention might be directed differently. For example, some agents might focus upon financial variables while others focus upon the use-value dimension of the economic process. The time horizon of agents might differ; some agents might focus upon the near while others focus upon the far away future. Motivation and effort might, even when material incentives are identical, be more or less strong when entering different kinds of economic activities. The competence of agents might be different, reflecting specialized talent and cumulated learning and experience, etc. Finally, agents might be more or less honest/opportunist in dealing with other agents. In an uncertain world it might be difficult to say what is the most (or the least) rational behaviour in these different respects. Typically it will depend upon the context in which agents act and interact.

Interactive learning and the limits of instrumental rationality

The second new element, introduced by the innovation perspective, is a change of focus from a process of calculation and decision-making towards a process of learning and creation. If technologies were stationary it would be reasonable to assume agents to concentrate on making successful transactions and to regard them as calculating 'Cartesians'. Further, the market institution could be analyzed and assessed exclusively from the viewpoint of transaction.

As all the chapters in Part I have amply demonstrated, introducing interactive learning will affect indirectly the assumptions on the rationality of agents. If what economic agents do in the economic sphere is mainly, to react to price signals and assumptions, presenting them as calculating and maximizing individuals might catch the most important aspects of their behaviour (this is the argument of Hicks – see Klamer, 1989). But if they are permanently involved in processes of interactive learning sometimes demanding cooperation and sometimes involving the collective creation of complex new knowledge, other aspects of behaviour become as important.

Actually interactive learning is undermined seriously when parties act exclusively from the viewpoint of calculation and maximizing. (Think of the efficiency of an inter-firm cooperative R&D lab where all scientists use most of their limited intellectual and spiritual energy on maximizing the return for their mother company.) Both individual actors and organizations less rational (instrumental) in this sense, acting according to sets of social norms including for example mutual respect and mutual trust, might be more successful than the purely calculating ones.

To say that it is 'rational' (for the individual or for collectives of individuals) not to be 'rational' does not solve the complexity of the matter, however. The reason why norms survive as social norms is that they are selected, not exclusively because of their usefulness but that they survive, mainly for extra-economic and 'irrational' reasons (Elster, 1989). As pointed out by Arrow (1971), it is impossible to buy trust, and if you could buy it, it would be of little value. The point to be made here is that the central importance of processes of interactive learning in the economy may support the viability of social norms and vice versa.

Not only does innovation, then, imply uncertainty, and thus bounded and differentiated rationality. Actually, it points towards a break with a purely instrumental and strategic rationality. 'Discursive rationality' (Habermas, 1984) might be more characteristic for processes of interactive learning than instrumental rationality. This would imply a strong impact of culture on the economic agents and for the possibility that different national systems foster agents, with different characteristics. In the last part of the chapter, this point will be illustrated through an analysis of two interacting national systems, one characterized by honest and trusting agents and the other by opportunism and distrust.

Innovation, culture and international differences

Of course, no economist would deny the existence of institutional and cultural differences between nations. What is controversial, however, is if such differences matter when it comes to the analysis of national economic performance, and if the costs in terms of complexity and loss of generality for bringing them into the analysis should not be regarded as prohibitive. This chapter argues that models ignoring such differences will be misleading when it comes to explaining important current phenomena in the international economy.

It will be shown that product innovations tend to take place in organized markets and in an interaction between users and producers. Important in this interaction is the flow of qualitative information. Unlike price signals and other one-dimensional concepts such information cannot be transformed into 'bits of information'. Therefore communication and the 'distance' between users and producers will affect the processes of interactive learning involved.

More important than geographical-physical distance might be cultural distance. For reasons already alluded to, communication and cooperation between parties belonging to different social and cultural systems will always be difficult. This will especially be the case when it comes to such complex and uncertain processes as innovation. This is one fundamental reason why it is meaningful to define, and analyse, *national* systems of innovation.

An outline of the chapter

The chapter is divided into five main parts. In the first part, it is demonstrated why product innovation implies organized markets. In the second part, some of the elements of organization constituting the organized market are specified. In the third part, different dimensions of space and distance are introduced, and in the fourth part the role of distance is discussed in relation to, respectively, stable technology, incremental and radical innovation, and technological revolution. In this context it is concluded that only standardized and stable technologies can be regarded as being close to the internationally footloose technology, assumed in neo-classical models of international trade. In the fifth and last part of the chapter untraditional assumptions regarding the rationality of agents are introduced into our analysis of international user–producer interaction. What happens if national systems differ, in terms of how the rationality of agents is specified? In order to illuminate this question two different specifications are introduced: one involving honest and trusting rational behaviour (characteristic of the H-economy) and another involving opportunistic and suspicious rational behaviour (characteristic of the O-economy).

In this context it is concluded, first, that the innovative capability would be higher in the H-economy and that the degree of vertical integration would be lower. Second, it is argued briefly that international differences in rationality, and their impact upon international user–producer interaction, may actually explain and illuminate a series of phenomena in the international economy which is not easily grasped by standard economics, including the role of multinationals and the process of establishing the European Single Market.

The general conclusion from this analysis of innovation as an interactive process involving users and producers with differently specified rationalities in learning, is that the internationalization of the economy and the weakening of the national systems of innovation is a contradictory, rather than a uni-linear and harmonious, process and that the specific institutional set-up of national systems cannot easily be substituted for by transnational institutional set-ups.

Product innovations take place in organized markets

Product innovations are not easily integrated in formal economic models and this is one reason why they have been largely neglected in economic theory. Here, however, product innovations will be put at the very centre of the analysis. One of the problems to be discussed, is how such a change in perspective affects our understanding of markets versus organizations.

It should be observed that the analysis refers primarily to interactive processes where both parties are professional units (private firms or public organizations). One reason for this restriction is that 'needs' is a fuzzy concept when private consumers are involved. Another is that private consumers are seldom actively involved in the interactive process of innovation. It should also be noted that the perspective is rather abstract and that many complications have been neglected. For example, it is obvious that importing units often are trading companies, mediating between the producer from abroad and the domestic users. Here it has been assumed that the buyer and the user are one and the same.

Two fundamental starting points

Behind the following lie two observations. The first is that innovation is an ubiquitous process going on everywhere, and all the time. The second is that modern society is characterized by a highly developed, vertical, division of labour. Most activities aiming at innovation take place in the formal sector of the economy where producers produce, not for themselves, but for others.

As we shall see, by combining these two elementary observations we can already raise some interesting questions regarding the institutional set-up of the economy. Will an institutional set-up, ideal in terms of its impact upon the allocation of given resources, be ideal also in bringing to life new ideas and materializing them into new products adopted to the needs of potential users? How should markets and organizations be combined in order to fulfil one or both of these sets of functions? Are there any mechanisms which automatically select the optimal combination of markets and organizations? What need and room is there for 'institutional design'?

Innovation as a collision between needs and opportunities

The scientist, developing instruments for his own use in order to be able to pursue a certain experiment in his laboratory, knows better than anyone else which needs the instruments should fulfil. He/she will also have a better insight into the new use-value characteristics of the resulting innovation than anyone else.

The R&D laboratory engaged in the development of new process equipment and the production department in the same firm are already in a more difficult situation. A reciprocal flow of qualitative information regarding technical opportunities and user needs must connect the two departments. To find organizational forms well suited to solving this kind of mutual information and learning problem is a far from trivial task for management. Several recent contributions comparing the intra-firm organization in the United States and Japan actually point to the importance of this problem of institutional design (see for example Aoki, 1989 and Lester, 1989).

But the problem of organising coordination and exchange of qualitative information will be even more difficult when the innovating producer and the potential user belong to two different organizations, separated by a market. Therefore, it seems paradoxical that such a big proportion of innovative activities, measured both as input (the proportion of R&D oriented towards product innovations, registered in the OECD statistics) and as output (proportion of important innovations used by firms and sectors, separate from the innovating units – see Pavitt, 1984), aim at product innovations. How can the producer know the needs of potential users when markets separate users from producers? And, how can the potential user get information about the specific characteristics of a new product?

In order to answer these questions it is useful to proceed step by step, starting with the pure market as it is presented in textbooks in standard economics.

The pure market and the transaction cost approach

It is obvious that product innovations would be rare and accidental if markets were characterized by anonymous relationships between producers and users. Producers would have difficulties in observing new user needs and users would lack qualitative information on the characteristics of the new products.

Although the pure market figures as the ideal norm in the neo-classical analysis of allocation it represents an institutional set-up which is hostile to innovation. The neo-classical theoretical scheme is, perhaps, best understood as reflecting an economy where innovation is a marginal and accidental phenomenon, only affecting the production process. In such an economy the pure market might survive. And, conversely, the pure markets would effectively block product innovations in such an economy.

Another way to analyse the institutional implications of product innovations is in terms of transaction costs.

In a situation where new products were continuously introduced in markets characterized by small numbers of sellers and buyers, uncertainty would be quite fundamental. The uncertainty would be built into the commodity itself and information impactedness/asymmetry, reflecting that the producer knows more about the characteristics of the new product than the user, while lacking knowledge about user needs, would be a typical phenomenon.

As a consequence, according to the original Williamson approach to transaction costs (Williamson, 1975), we should expect vertical integration to take place whenever product innovations were frequent and important. Product innovations would disappear, as they were gradually transformed into process innovations through vertical integration. This analysis leads us, as did the pure market assumption, to believe that product innovations are unimportant and infrequent.

Fundamental for the conclusion is the Williamsonian assumption of 'opportunism' (self-seeking behaviour with guile) as a generalized or frequent type of behaviour. It is opportunism which gives rise to prohibitive transaction costs in situations characterized by uncertainty. The last section of the chapter goes into a critical discussion of the general validity of this assumption.

The organized market as the institutional response

Paradoxically, in the real world product innovations are not rare and efforts to develop such innovations are common. How come? The simple answer is that most markets are not anonymous and pure. Most markets involve an element of mutual exchange of qualitative information, and sometimes direct cooperation between users and producers in the process of innovation. The relative importance of product innovations imply that most markets are organized markets. This implies that the modern economies are 'mixed' in a fundamental sense; not only is the private sector coexisting with a big public sector; the relative success of the market economies in terms of technical progress has its roots not in the purity of the markets but rather in their impurity.

Why not just process innovations?

But how can innovative activities aiming at product innovations survive? Which are the relative advantages of organized markets as compared to the pure market and the pure hierarchy? Here, it might be useful to repeat an argument put forward by Arrow (1974). According to Arrow, the formation of an organization might be regarded as a process involving both growing efficiency and growing inflexibility. The negative aspect of bringing more and more activities into organizations is that they get locked into a network of communication codes and channels – very difficult to adjust when faced with radical change in the environment.

Inspired by this argument, I shall argue that the organized market might be regarded as a compromise, taking into account both the advantages of acting collectively – in our context primarily, to exploit the fruits of learning by interacting – and the costs of rigidity. The organized market represents a degree of rigidity which is necessary to produce innovations, but a rigidity of lesser degree than the one represented by the pure organization.

Second, we must take into account how vertical integration affects the participation of the integrated units in the processes of interactive learning taking place outside the pair of integrated units. Integrating a producer will give a user more easy and direct access to technological know-how. And, at the same time, the integrated producer will get more easy and direct access to knowledge about the changing needs of this particular user.

But the price paid for these advantages, supporting product innovation, might be high in the long run. Users and producers not integrated will be reluctant to give away sensitive information to the pair of integrated units. Non-integrated users will be reluctant to give away information about their strategic bottleneck problems to a producer integrated with a competitor. And non-integrated producers will be unwilling to give away their most advanced technical know-how to a user integrated with a competing producer.

As a result, the input to the process of innovation from interactive learning becomes specialized and limited in scope. Thus, there is a trade-off between the short-term advantages of a closer interaction between the integrated units and the long-term cost of their isolation from the broader process of interactive learning involving several formally independent users and producers. The informal organization, characterizing the organized market, is less exclusive in this respect and keeps open a more encompassing network of information, cooperation and interactive learning. This is one reason why firms may prefer to go on operating with relatively high 'transactions cost' in the organized market rather than joining a pure hierarchy, locking them into a much more narrowly defined space for interactive learning.

This somewhat functionalist explanation of why organized markets tend to survive and make product innovations possible should be complemented in two respects. First, we must discuss the compatibility of the organized market with the behaviour of individual agents. Second, we must ask if the functional perspective implies that the actually existing institutional set-up is optimal, satisfactory or unsatisfactory.

Opportunism and transaction costs

If most agents acted opportunistically we should expect the organized markets to become transformed into hierarchies. The efforts put into product innovation and their frequency in the real world may be regarded as evidence that agents are able to restrain their opportunistic tendencies – at least when involved in processes of interactive learning. Actually, there might be several good reasons for such a restraint. Some of these reasons might be found in fundamental behavioural propensities, at least as strongly rooted in our culture as the tendency towards unrestricted egoistic instrumentalism (Giddens, 1984):

- People prefer to live and act in environments of security and trust. Their own contribution to the creation of such environments might be to act honestly and to react negatively to opportunist agents.
- People hate to lose face. In most societies children learn that being caught in a lie, or in a conscious misrepresentation of reality, is a way to lose face.
- The context of a specific kind of interaction may be more or less inviting to opportunist behaviour. In markets for old horses and used cars, opportunism might even be a generally expected and accepted kind of behaviour. Processes of interactive learning and innovation may be less open to this kind of behaviour. Specifically, professionals representing engineering and science might be less opportunist-prone than economists. (Company lawyers might actually be extremists in supporting opportunism because their existence is based upon it.)
- This does not rule out that cultural differences may be reflected in different degrees of opportunism and, in the last section of the chapter, the implications of international differences, in this respect, will be discussed.

Are established user–producer relationships always satisfactory?

Another question regards the functionality of the existing institutional set-up in terms of markets, organizations and organized markets. Above, some reasons why agents might prefer organized markets to vertical integration and to pure markets were mentioned, and we have actually argued that user–producer relationships are necessary prerequisites for product innovations. These arguments do not, however, imply any mechanism of selection, ascertaining that the actual network of user–producer relationships is always optimal or even satisfactory. In another context I have pointed to two different characteristics systematically fostering unsatisfactory innovations (Lundvall, 1989).

One characteristic is asymmetrical power relationships between users and producers, resulting in biased technical change. When producers dominate, both in terms of financial resources and in terms of technical competence, there is an inherent tendency to develop costly innovations, which is not well suited to the needs of users. This seems, for example, to be the case when a few big producers produce systemic capital equipment for a large number of small users (i.e. dairy factories, cement factories, nuclear plants). But the dominance of a few big users might also give rise to unsatisfactory innovations. The US automobile industry, and its interaction with domestic producers of machine tools, seems to illustrate this kind of mechanism.

The second characteristic is the tenacious character of user–producer relationships. This problem will be most severe in periods of technological revolution, bringing fundamental changes in technological opportunities and in user needs. In such a period, the prevailing pattern of user–producer relationships will adjust, only slowly, to the emerging new needs and opportunities.

Specifying user–producer interaction

In this part of the chapter we will specify briefly some of the elements of organization constituting the organized market.

Exchange of qualitative information

The most basic function of the user–producer relationships in relation to product innovations is to communicate information about, respectively, technological opportunities and user needs. Here, some of the central concepts in Arrow (1974) are useful.

The development of codes of information and of channels of information in organizations might be regarded on line with investments in tangible capital. The more codes and channels are used, the more efficient they become. Similar mechanisms will be at work in the organized market. The user and the producer will gradually develop channels of information and a common code, making the exchange of information more efficient. To leave a well-established user–producer relationship becomes increasingly costly and involves a loss of information capital.

Cooperation

The relationships between users and producers will sometimes involve direct cooperation. For example, a user might invite a producer to take part in solving a specific problem within the organisation – a problem which cannot be solved without innovation. Cooperation might take place both while developing the solution, and while introducing it in the user organization.

Hierarchy and mutual trust

User–producer relationships will often involve elements of hierarchy, and the direction of the innovations will reflect who is the dominating party. An uneven distribution of power may reflect differences in the command of financial resources, as well as differences in technical competence.

But in most user–producer relationships we find other social elements combined with hierarchy and dominance. Without a certain degree of loyalty, mutual respect and trust, transaction costs would, as already pointed out, become prohibitively high and vertical integration would take place.

On the durability, selectivity and intensity of user–producer relationships

It follows from what has been said that user–producer relationships tend to be durable and selective. As pointed out, it takes time to develop efficient codes and channels of information. This might be even more so when it comes to establishing relationships of dominance, trust and common codes of conduct.

The main reason why markets tend to become organized pointed to here is innovative activities. It follows that the intensity and character of user–producer relationships might be very different in different parts of the economy. In parts where the product technology remains almost constant the relationships might become close to the ideal of the pure market. In other areas a high degree of complexity and radical change in technologies might result in a full formal vertical integration. But most real markets will fall in between these two extremes.

User–producer interaction in space

We can now discuss the spatial aspects of user–producer interaction. Before entering this analysis, the concepts of 'space' and 'distance' must be discussed.

In the present context, it is useful to operate with four different dimensions of space: economic, organizational, geographical and cultural space will be introduced in our analysis of user–producer interaction. There are important interconnections between these four dimensions. A long distance between users and producers, in one of these dimensions, might, for example, be overcome by a close distance in another dimension.

Economic space

Economic space relates to how different economic activities are located in a system of production where the vertical division of labour is highly developed. In an extended input–output table (a table where deliveries of capital goods are included and where organizations, producing science and technology, are represented as sectors), the economic distance between activities will be indicated by the prevailing set of input–output coefficients.

Organizational space

Organizational space refers to horizontal and vertical integration. In the traditional analysis, the 'organizational distance' between economic activities is presented as discontinuous. Either there is full integration, and the distance is zero, or there is no integration and the distance is infinite. Introducing organized markets, and the possibility of more or less distant relationships between the branches in conglomerate corporations, allows for a more continuous concept of organizational distance.

Geographical space

Geographical space can be measured, unequivocally, in terms of distance, when activities can be assigned to distinct locations. This dimension of distance might be more or less relevant, depending upon which types of interactions are studied and depending upon the prevailing systems of transportation and communication. Measured in terms of costs and/or time, an economic–geographical space might be constructed, related specifically to the kind of interaction in focus.

Cultural space

Cultural space is a loose and much more complex concept but it relates to important characteristics in the real world. Especially when studying learning processes and communication processes, involving complex and ever-changing messages, this dimension is important. The general institutional framework – including norms and codes – represents a context for communication, and individuals and organizations will decode information in accordance with this context. When cultural differences are present, certain types of messages will be difficult to transmit. Complex and ever-changing messages combining explicit information with tacit assumptions regarding mutual obligations will often be required in interactions involving innovative activities. Here, cultural differences between user and producer may block the interaction.

This is one reason why nations still play an important role as economic entities having a relative autonomy. The cultural distance between domestic activities is short. Certainly, a cultural convergence towards an American lifestyle in the post-war period has weakened the impact of cultural distance in this sense, but the fact that national languages and national differences in the institutional set-up persist, implies that this factor cannot be ignored.

What makes this dimension complex is, of course, that qualitative differences are not easily transformed into one quantitative variable. What matters in relation to interactive learning is the difficulty involved in understanding and decoding complex information. For example, one complication is that it is not always the case that interaction between culturally very different systems is more difficult than an interaction involving systems which are apparently more akin to each other. If the differences are drastic the parties involved will recognize them and take them into account. More subtle differences might be less visible and create more serious problems for the agents involved. Cultural distance must take into account the kind of interaction involved, and given the kind of interaction it might be defined as the costs of learning to understand (and use) the codes of the other party.

Interconnections between the dimensions of space

The different dimensions of space and distance are interconnected in a complex way. Economic space does not play any independent role in our analysis. Actually, the unit of analysis discussed so far, the user–producer dyad, is defined as the combination of units, as close to each other as possible, measured in this dimension.

Organizational space has already been discussed in relation to vertical integration and organized markets. Here, it will be confronted with geographical and cultural space. It is obvious that a short organizational distance may substitute for geographical and cultural proximity. The multinational corporations tend to develop their own internal channels and codes of information and they may also develop a specific, transnational, corporate culture. Why this type of substitution will take place in the form of vertical integration – eliminating markets for product innovations – will be discussed at the end of this chapter.

User–producer interaction and distance

In this section, it will be discussed how the character of technical change affects the relative importance of spatial, cultural and organizational distance between users and producers involved in interactive learning and innovation. In this context, a distinction will be made between four different forms of technical change: stationary technology, incremental innovation, radical innovation, and technological revolution. The distinctions reflect the taxonomy developed in Freeman and Perez (1988).

Stationary technology

First, let us assume that technological opportunities and the needs of users are fairly constant. The engineering activities going on in firms and institutions will develop norms, standards, and terminologies giving a full description of the technology. A high degree of standardization means that communication between user and producer can be performed over long distances and it might even be transformed into 'bits', cheaply transferred between computers by telecommunication. In this case distance – geographical as well as cultural – will play a very limited role. This is a case where industries become virtually 'footloose'. Activities can be located anywhere and one should expect 'comparative advantage' in terms of the relative scarcity of factors of production (defined in a broad sense) to play an important role in determining the spatial division of labour.

Perhaps we might argue that some pre-industrial societies were technologically stationary for periods, but when analyzing modern societies this case is only a theoretical construct and may be regarded primarily as a reference for the other and more realistic cases.

Incremental innovation

Second, let us assume user–producer interaction where incremental innovation is a permanent activity. The codes and channels of information developed between users and producers must be flexible in order to take into account the change in technological opportunities and the change in user needs. Stable user–producer relationships will develop in order to overcome uncertainties, at both sides of the market. The drive towards standardization will be countered by recurrent changes in the technology. Messages will be complex and changing and the information cannot readily be translated into 'bits', transferable by telecommunication. In this case distance will play an important role. Being close to advanced users will form a comparative advantage for the producers, and vice versa.

In small open economies we find many examples of patterns of specialization, reflecting comparative advantages based upon such a close interaction between competent users and producers. Dairy technology in Denmark, wood cutting and metal working technologies in Sweden and technologies related to shipping and fishery in Norway have all developed in a historical process of interaction between industries where users have been export-oriented and, for this reason, quite competent and demanding. This kind of interaction has resulted in lasting, strong positions in the world market for users as well as producers. Such industries, often forming parts of national industrial complexes, are not footloose. On the contrary, they base their comparative advantage upon geographical and cultural proximity (Andersen *et al.*, 1981).

Transnational corporations might locate in such national strongholds in order to reap the fruits of the interaction between national users and producers if the potential market is big enough or growing rapidly. But it is not always a simple operation to enter into such close-knit organized markets where non-economic relationships may have infiltrated the relationships between users and producers. A takeover of a national producer might, for example, trigger negative reactions among users preferring a domestic producer for loyalty reasons.

Radical innovation

The third form of technical change to be considered is one of radical innovation. In this context, the 'interpretative grid' developed in Dosi (1982), and the further discussion in Archibugi (1986), are useful. The transposition of the 'paradigm' concept from the world of science to that of technology illuminates important aspects of the user–producer interaction. One important activity in 'normal science' is the standardization of terminology, increasing the possibility for communication over long distances between scientists. Another is the development of a common and accepted set of criteria for success. New paradigms will be characterized by an unstable and mixed terminology and the existing standards of success will not be easily applied to them. In a period when new paradigms develop, world-wide communication will be difficult as a consequence. Formal procedures of testing will be lacking and 'unscientific' criteria, such as group solidarity and mutual trust, will play an important role in the formation of a new research programme.

These characteristics of competing new paradigms in science will appear in an akin, but somewhat different, form when we observe radical innovations. A radical change in technology implies that the codes developed to communicate a constant, or a gradually changing, technology will become inadequate. Established producers, following a given technological trajectory, will have difficulties in evaluating the potentials of the new paradigm. Users will have difficulties in decoding the communication coming from producers, developing new products built according to the new paradigm.

In this case, geographical and cultural distance might play an even more important role than in the case of incremental innovation. The lack of standard criteria for sorting out what is the best paradigm implies that 'subjective' elements in user–producer relationships – as mutual trust and friendship – will become important. In an experimental phase the lack of communicability might make 'hands-on-experiments' necessary for potential users. In this phase, eye-to-eye contact between users and producers might be necessary in order to communicate new technological opportunities and user needs. The accelerating growth of 'Silicon Valleys' around the world can be interpreted in those terms.

A new cluster of users and producers might be 'footloose' *ex ante* – accidents may determine where the first units are locating – but *ex post* they may become strongly rooted in regional or national networks of user–producer relationships, giving them a comparative advantage in national and international competition. Interesting modelling exercises, illustrating stochastic allocation in space when increasing returns are involved, have been developed by Brian Arthur (1988). The agglomeration effects assumed in those exercises may originate from different sources – here it is argued that one of the most important sources of agglomeration effects explaining Silicon Valley phenomena is the need for proximity between users and producers, – in the phase of radical innovation.

Which role will multinational corporations play during such a phase? It depends upon how strongly such corporations will be linked to the old natural trajectories. Provisionally, one might argue that the vertically integrated multinational corporation will be quite resistant to new technologies and ideas emanating from new paradigms. This might, to a certain degree, be compensated for by a long-term perspective, and strategical planning, taking radical technical change into account. In the field of microelectronics, the biggest firms operating in the computer market were rather slow in recognizing the potential of the new paradigm, in terms of the PC technology. The current activity of multinationals in the field of biotechnology (Chesnais, 1988) might reflect a learning effect from that experience.

In a later phase, when the new paradigm has proven its worth, it becomes, of course, an attractive field for takeovers and investment. The flexibility in space, originating in the peculiar form of organization characterizing multinational corporations, facilitates simultaneous location in several high-tech strongholds and the sourcing of innovations from such strongholds. If the multinationals become more dominating in this phase it is not because they are the units best suited to introduce radical innovations, however, but rather because of their financial and marketing power. At least in the short and medium term, it can be argued that radical innovations will tend to weaken their relative position.

In a long-term perspective, some paradigms will establish themselves as 'normal technology' and establish new, rather stable, trajectories. Terminology, and criteria of success, will become standardized and incremental innovations will take place along a new trajectory. In a limiting case the technology becomes stationary. The process of standardization will weaken the need for proximity between users and producers but proximity will still have a role to play as long as product innovation is going on.

The limiting case of stationary technology is not very interesting, in an industrial society. Within manufacturing, the clothing industry might come closest to this theoretical construct. If we scrutinize the development in this sector we shall find that, even here, technological opportunities as well as user needs have been changing considerably during the post-war period. On this basis, it can be concluded that proximity will play a role in almost all user–producer relationships in an industrial society and, thereby, counteract tendencies towards 'footloose' industry.

Technological revolution

Finally, the implications of a cluster of basic, radical innovations should be discussed. A basic innovation is defined as one with a potential use in most parts of the production system. Microelectronics and, potentially, biotechnology, reflect basic, radical innovations in this sense. Microelectronics can be used in almost all production and administrative processes in a modern society, and it is an important component in many new consumer goods.

In our framework, the most important consequence of such a cluster is that it tends to break up the traditional network of user–producer linkages and to constitute the basis for new user–producer linkages. This will affect vertically integrated corporations as well as organized markets. In both cases, inertia, and the 'costs' involved in breaking up existing codes and channels of information, will tend to cement the existing structure of user–producer relationships. The force of resistance might be strongest where the interaction has been most effective, in establishing strong poles of competitiveness, both at the user side and at the producer side.

Such a 'technological revolution' may have a drastic impact on the spatial division of labour. Regions and countries which are 'latecomers' can get a competitive edge before regions, characterized by historically rooted networks of users and producers (Perez and Soete, 1988). The radical change in the global localization of manufacturing, shifting production from the 'Atlantic region' – Western Europe and the eastern part of the United States – towards the Pacific region, reflects such a mechanism. Twenty years ago, 60% of the value of production was located in the Atlantic region and 40% in the Pacific region. Today, the relative shares have shifted so that the Pacific region produces 60% and the Atlantic 40%.

Such processes of reallocation of labour and production, in space, may persist both at the global and the local level. But it does not imply that industries have become 'footloose'. Rather, it reflects that the roots of old industries, in historically established networks, have become too strong and that new industries are now in a process of rooting themselves in new networks.

Empirical illustrations

In some of the literature on industrial networks, a distinction is made between domestic and international user–producer relationships. The empirical results suggest that national borders matter in this context.

1. The study by Håkansson (1989) shows that 80% of all cooperation activities involving user–producer interaction are domestic (*op.cit.*: 110). Local and district relationships are not very frequent, however (less than 20%). This supports the hypothesis that systems of innovation tend to be national rather than regional or transnational.

2. In a comprehensive study of user–producer relationships, focusing upon British, German and Swedish firms, the characteristics of domestic relationships are compared with international relationships (Hallén *et al.*, 1987). The authors test a series of hypotheses, and one of the main conclusions is worth quoting at length:

> The results thus reveal a basic difference in the working of business relationships in domestic versus export markets. The strength of export relationships develops as far as the product exchange requires, i.e. initial adaptations are made and information exchanged to cope with these. However, in domestic relationships the adaptations in addition to this constitute a viable framework for expanding interaction processes, which both permits and requires more information exchange and new adaptations. Thus a cyclical strengthening of the relationship is obtained. (*op.cit.*: 36)

It is important to take into account that the firms in this sample are internationally oriented, manufacturing firms. Thus, the conclusion should be expected to be valid, *a fortiori*, for less internationalized units. Again, the result supports the relevance of national systems of innovation.

Introducing differently specified rationality in international user–producer interaction

It has been argued that 'cultural distance' might have an important impact upon the patterns of national and international user–producer relationships and upon innovation. Culture is a broad and vague concept and it might be useful to go into some detail, with an illustration of why culture invades the core of the neo-classical economic system – the market and the rationality of agents.

In this last section, the focus will be upon what happens when one of the core elements in the economic system – the rationality of agents – is invaded by culture. How will international differences in the specification of the 'rationality' of agents affect national performance? What are the implications of internationalization for the process of innovation when rationalities are specified differently between national systems?

Homogeneously specified rationality

In standard economics, all agents are assumed to be equipped by a rationality identically specified. Even when the assumption of hyperrationality is challenged, all agents are assumed to follow one and the same set of behavioural rules. In this respect, the owner of the hot-dog stand on the corner is assumed to be identical with the management of IBM. In both cases it is assumed that the motives, the ambitions and the competence are the same. One, and only one specification of rationality across agents is an assumption economists are very reluctant to sacrifice when introducing more realistic assumptions in their models. Without it, mathematical models would become less neat, and aggregation more complex.

One of the few economists who has attacked, explicitly, the principle of one specification of rationality and also pointed to concepts making it possible to operate with differently specified rationality is Kornai (1971: 174–5 *et passim*). Another contribution has been made by Peter Allen (1988) who in his very interesting model of fishery in Canada differentiates between fishermen who are respectively 'Stochasts' and 'Cartesians'. This model demonstrates the fruitfulness of differentiating when specifying the rationality of agents. The resulting system dynamics change radically depending upon the assumptions made in this respect.

Honest agents vs. opportunist agents

One of the interesting contributions of the transaction-cost approach, as developed by Oliver E. Williamson, is the introduction of 'opportunistic behaviour'. Arrow and others have pointed to the 'free rider' problem and to 'moral hazard', illustrated by insurance contracts, showing how rational behaviour might imply strategical behaviour and even when instrumental cheating the other agents. Williamson went a step further in this direction in introducing opportunism – acting for one's own interests with guile – as a kind of behaviour of rational economic men when uncertainty is ruling.

In doing so, he actually pointed to a need for a further specification of the assumptions of rational behaviour in this specific 'moral' dimension. We might specify the behaviour of 'rational economic men' (REM) either as building upon honesty and trust or as building upon opportunism and mistrust.

In order to illustrate the importance of international differences in rationality, the distinction between opportunism and honesty is going to be used as the illustrative example. Closer studies of international differences might point to important differences in the specification of rationality along other lines – for example regarding the time horizon of agents, the relative orientation towards and focus upon, respectively, finance variables in and use-value aspects of production, etc.

Let us assume two coexisting systems; one populated by honest agents (Hon-REM system) and the other by opportunist agents (Opp-REM system). If an agent tries to behave opportunistically in the Hon-REM system the sanctions, in terms of exclusion from further interaction, might be very hard. If, on the other hand, an agent in the Opp-REM system tries to behave honestly, and expects others to be honest to him, he would come out as the loser. The others would not believe in his good intentions and his own expectations would not be fulfilled. In both systems the single agent would be best off following the dominating rule of behaviour and, in this sense, both systems are REM-societies – they are both compatible with consistent goal-seeking behaviour.

What determines if a system becomes Hon-REM or Opp-REM? Theoretically, the question might be analyzed through game theory, comparing the viability of cooperative vs. conflictual rules of behaviour (Axelrod, 1984). In the real world the dominance of, respectively, Honesty and Opportunism at the national level will reflect historical and, perhaps, even pre-capitalist developments. As demonstrated by Arthur (1988) small historical events may impose a dominance of one single technology when increasing returns are involved. Similar mechanisms might have been at work in selecting rules of behaviour at the national level.

Coexisting sub-systems with differently specified rationalities within a nation

What characterizes the two different systems? Here, the focus will be upon how transaction costs and interactive learning involving users and producers are affected. In Table 12.1 the two subsystems are contrasted in these respects.

The fact that opportunism results in high transaction costs is a well-known result of the work by Williamson. Agents, acting with guile, will create an atmosphere of mutual distrust between users and producers and as a result it will be very difficult and expensive to formulate and monitor contracts acceptable to both parties. In the system where agents are honest and trusting the transaction costs will be greatly reduced. The high transaction cost will make vertical integration a much more common phenomenon in the Opp-REM system and, *ceteris paribus*, at any point of time we would expect the degree of vertical integration to be higher in the Opp-REM system; a smaller proportion of the flow of products would involve market transactions.

Table 12.1 Sub-systems characterized by different rationalities

	Opp-REM	Hon-REM
Transaction costs	High	Low
Learning capability	Low	High

The impact upon interactive learning might be even more dramatic. It is extremely difficult to engage successfully in interactive learning if both parties act with guile. Generally, processes of learning will be stimulated by an environment characterized by honesty and mutual trust. (The strong sanctions against pupils cheating in school and the strong expectations of teachers truthfulness reflect this fact.)

Again, the lack of trust might be overcome by vertical integration. But the integration will not guarantee a high learning capability. As far as intra-firm relationships involve opportunist agents, the negative impact upon learning will remain. It may be possible to command opportunist agents to pursue repetitive operations and to use economic threats to force them to perform simple tasks. It is much more difficult to force them into processes of complex and interactive learning, however. Even very ambitious monitoring schemes might be of little use in this context. So, even if the transaction costs can be reduced through vertical interaction the effect of integration upon learning is much more dubious.

At this point, it becomes clear that the transaction-cost analysis sometimes involves a contradiction. The approach is often used to explain the actual institutional set-up in terms of markets versus vertical integration. It is assumed that the actual set-up reflects a selection process, tending to select the most efficient institutionally, i.e. the set up involving the lowest production and transaction costs. But, when it comes to behaviour only one specification of rationality (opportunism) is stated. And, this specification is obviously very 'inefficient' both in terms of transaction costs and in terms of learning capability. One reason for this contradiction is, of course, that the transaction-cost approach keeps the focus upon transactions of given bundles of goods, joining the neo-classical analysis, otherwise criticized, in a neglect of product innovation.

One alternative interpretation of the Williamsonian transaction-cost approach, is to regard it as a theory strongly rooted, in US culture, where the *quid pro quo* logic may have a stronger grip on all kinds of social relationships than anywhere else. If this is true, the Williamson analysis takes for granted the most problematic institutional aspect of the US system, while trying to demonstrate that the system selects a reasonably efficient institutional set-up on this basis.

Interaction between users and producers belonging to different sub-systems

Within a national system users and producers, belonging to different sub-systems, might get involved in an interaction. The opportunist producer will be looking for honest and trusting customers whose honesty and trust he may exploit. But within national systems, the professional user, acting as an honest and trusting customer, will develop methods and techniques to identify and avoid domestic opportunist producers. Information-sharing between firms belonging to the same industry regarding the trustworthiness of different producers will often be efficient, especially in a small country and in areas where few potential producers are involved.

The greatest difficulties with identifying opportunist producers might be in new, growing industries, and in industries where quality control of the product is difficult, even *ex-post*. The increasing number of firms offering rather diffuse products and services under the labels of information services and consultancy might leave room for opportunist behaviour, difficult to identify and avoid even for professional users.

But, normally, domestic interaction between agents from the two sub-systems would be limited. The honest professional user will avoid the opportunist for two different reasons. One is, of course, the risk of being cheated and the other might be that any kind of collaboration with opportunist agents might be regarded as negative, in the sub-system of honest agents.

International user–producer relationships involving differentiated rationality

Let us now assume that two different national systems differ in terms of the rationality governing the behaviour of agents in the two systems. How will inter-firm relationships develop between an economy dominated by opportunism (the O-economy) and another dominated by honest and trusting agents (the H-economy)? The agents of the H-economy would become seriously disappointed with the behaviour of the agents of the O-economy. The agents of the O-economy would not be able to understand the behaviour of the agents from the H-economy. If both sets of agents stick to their original pattern of behaviour, transaction costs would become extremely high and interactive learning involving partners from the two systems would remain very limited.

The other possibility involves 'institutional learning' and adaptation. The H-firms may differentiate their behaviour and begin to act as opportunists in their interaction with the foreigners from the O-system while sticking to honest behaviour domestically. This would not reduce transaction costs nor would it increase the potential for interactive learning at the international level. And gradually selective opportunism may become generalized and include domestic transactions. If such a version of 'Gresham's law' were at work, any small germ of opportunism might spread and contaminate the global economy as a whole.

Alternatively, the firms from the O-system might learn to behave honestly when interacting with H-firms, while sticking to opportunistic behaviour at home. In this case we might get lower transaction costs and a higher learning capability in the international user–producer relationships than in those internal to the O-system. Axelrod's (1984) experiments confronting different rules of behaviour actually point in this optimistic direction, showing how a simple and potentially cooperative 'tit-for-tat' rule tends to win the game.

All the most probable outcomes would involve high transaction costs and low interactive learning capabilities when users and producers are rooted in different national systems characterized by differently specified rationalities. To overcome this problem vertical integration would be a frequent outcome. Instead of international user–producer interaction we tend to get multinational corporations. Instead of sales between users and producers of different nationality we tend to get commodity flows across national borders but inside the same corporation.

Implications at the international level

High transaction costs, and a low learning capability at the national level, are serious problems which might reflect the presence of opportunistic behaviour inside the national economy. But we can now see why they become much more severe when agents from different nations are involved. In order to identify and cope with opportunistic agents it is extremely important to know, and understand, the culture, language and the codes of conduct of the other party. In cases of disagreement it is also important to understand and have equal access to institutions solving legal problems. These observations might inspire a reassessment of different phenomena in the world economy:

- The dominating role of multinational corporations.
- The problems for small firms in becoming international.
- The limits of internationalization.
- The importance of product standards and standardization.
- The impact of the European Single Market upon innovation and growth

One important reason why multinational firms tend to organize such a big proportion of the international commodity flow might be that user–producer relationships, involving foreign users, are expensive in terms of transaction costs and that opportunism and international differences in rationality will tend to block interactive learning. If this is so, the multinational firms may be regarded as 'second-best' responses to this problem.

Among the reasons why export-oriented, single-nation firms will have problems when going into the international trade of commodities which are characterized by technical change might be high transaction costs and a low capability of interactive learning. Many of these firms have developed their specific competence and products in an interaction with competent and demanding domestic users. As they become oriented towards export niches, and weaken their ties to the domestic economy, they might lose their innovative capability and they will also be burdened by high transaction costs. In national systems (Denmark, Ireland, Portugal, Australia) which have few multinational champions the competitiveness of the whole economy may suffer for the reasons given above.

We can also see why internationalization and the dissolution of national systems of innovation must be regarded as a contradictory, and complex, process. The multinational firms will often represent institutional frameworks which are less flexible than the organized markets. A process of internationalization based upon multinational corporations might, actually, weaken the innovative potential not only of single national systems but also of the global economy as a whole.

The analysis in terms of interactive learning and transaction costs points towards the importance of standardization. Especially when international institutional differences are involved, standardization becomes crucial to the pattern of international user–producer relationships. Standardization between countries, in terms of business procedures, technology and product quality, reduces the uncertainty of a foreign user and limits the room for opportunism on the producer side. Standardization reduces transaction costs and in some cases it might stimulate international interactive learning.

The European Single Market is interesting in the light of the foregoing discussion. Obviously, this project aims at a standardization in terms of products and technology. If the project is a success in this dimension we should expect uncertainty and transaction costs to be reduced. How it will affect the process of innovation is a more open question, however. The actual design of standards might be crucial in this context. A more fundamental question concerns the long-term trends in the rationality of economic agents at the European scene. Is the single market project designed to influence basic cultural values reflected in the rationality of economic agents?

Empirical illustrations

The cultural and institutional differences analyzed in the literature on innovation are foremost when referring to comparisons between Japan and United States. This literature supports the importance of institutional and cultural factors relating to the organized market and to differentiated rationality:

1. There are important differences in inter-firm relationships between the United States and Japan. Japan is characterized by closer and more long-term user–producer relationships (Freeman, 1987: 49 ff. and Dertouzos *et al.*, 1989: 99 ff.).
2. The Japanese inter-firm relationships are more characterized by loyalty and trust and, at the same time, the Japanese economy is characterized by a much lower degree of vertical integration than the US economy (Dore, 1986).

These observations, repeated in many different comparative studies, do not contradict that rationality is differentiated between the two countries. Whether or not such a difference should be summarized by the juxtaposition between opportunism in the United States and honesty in Japan remains an open question. A recent study pointed to Japanese 'cultural and institutional constraints on opportunistic behaviour' in connection with an international comparison of interaction between users and producers of nuclear plants. The practising of honesty and trust might, however, be quite selective and refer primarily to members of the same industrial group.

Generally, there is a strong need for international comparative studies involving countries closer to each other in terms of geography and culture than the United States and Japan, and aiming at understanding international differences in the organization of markets and in the specification of the rationality of agents.

References

Allen, P. (1988), 'Evolution, innovation and economics', in Dosi, G., Freeman, C., Nelson, R., Silverberg, G. and Soete, L. (eds), *Technology and Economic Theory*, London, Pinter Publishers.

Andersen, E.S., Dalum, B. and Villumsen, G. (1981), 'The importance of the home market for the technological development and the international specialisation of manufacturing industry', in *Technical Innovation and National Economic Performance*, IKE Seminar, Aalborg, Aalborg University Press.

Aoki, M. (1989), 'Global competition, firm organisation and TFP', paper presented at the OECD International seminar on Science, Technology and Economic Growth, Paris, OECD, mimeo.

Archibugi, D. (1985), 'Paradigms and revolutions: From science to technology?', paper presented at the UNESCO-CNR Workshop on Long-Term Scientific Forecasting, Venice, December.

Arrow, K.J. (1971), 'Political and economic evaluation of social effects and externalities', in Intrilligator, M. (ed.), *Frontiers of Quantitative Economics*, North-Holland.

Arrow, K.J. (1974), *The Limits of Organisation*, New York, W.W. Norton & Company.

Arthur, B. (1988), 'Competing technologies: an overview', in Dosi, G., Freeman, C., Nelson, R., Silverberg, G. and Soete, L. (eds), *Technology and Economic Theory*, London, Pinter Publishers.

Axelrod, R. (1984), *The Evolution of Cooperation*, New York, Basic Books.

Chesnais, F. (1988), 'Multinational enterprises and the international diffusion of technology', in Dosi, G., Freeman, C., Nelson, R., Silverberg, G. and Soete, L. (eds), *Technology and Economic Theory*, London, Pinter Publishers.

Dertouzos, M.L., Lester R.K. and Solow, R.M. (1989), *Made in America*, Cambridge, Mass., MIT Press.

Dosi, G. (1982), 'Technological paradigms and technological trajectories: A suggested interpretation of the determinants and directions of technical change', *Research Policy*, vol. 11, no. 3.

Dore, R. (1986), *Flexible Rigidities*, Stanford, Stanford University Press.

Edquist, C. and Lundvall, B.-Å. (1989), 'Comparing Small National Systems of Innovation', contribution to the Nelson Project on National Systems Supporting Technical Change, mimeo.

Elster, J. (1989), 'Social norms and economic theory', *The Journal of Economic Perspectives*, vol. 3, no. 4.

Freeman, C. (1987), *Technology and Economic Performance: Lessons from Japan*, London, Pinter Publishers.

Freeman, C. and Perez, C. (1988), 'Structural crises of adjustment, business cycles and investment behaviour', in Dosi, G., Freeman, C., Nelson, R., Silverberg, G. and Soete, L. (eds), *Technology and Economic Theory*, London, Pinter Publishers.

Giddens, A. (1984), *The Constitution of Society*, Polity Press, Cambridge.

Habermas, J. (1984), *The Theory of Communicative Action, Volume I*, Boston, Beacon Press.

Hallén, L., Johanson, J. and Nazeem, S.M. (1987), 'Relationship Strength and Stability in International and Domestic Industrial Marketing', *Industrial Marketing & Purchasing*, vol. 2, no. 3.

Hirschman, A.O. (1970), *Exit, Voice and Loyalty: Responses to Decline in Firms, Organizations and States*, Cambridge, Mass., Harvard University Press.

Håkansson, H. (ed.) (1987), *Industrial Technological Development, a Network Approach*, London, Croom Helm.

Håkansson, H. (1989), *Corporate Technological Behaviour*, London, Routledge.

Klamer, A. (1989), 'An accountant among economists: Conversations with Sir John Hicks', *The Journal of Economic Perspectives*, vol. 3, no. 4.

Kornai, J. (1971), *Anti-equilibrium*, Amsterdam, North-Holland Publishing Company.

Lundvall, B.-Å. (1985), *Product Innovation and User–producer Interaction*, Aalborg University Press, Aalborg.

Lundvall, B.-Å. (1988), 'Innovation as an Interactive Process – from User–Producer Interaction to National Systems of Innovation', in Dosi, G., Freeman, C., Nelson, R., Silverberg, G. and Soete, L. (eds), *Technology and Economic Theory*, London, Pinter Publishers.

Lundvall, B.-Å. (1989), 'Innovation, the Organised Market and the Productivity Slow-down', paper presented at OECD's International Seminar on Science, Technology and Economic Growth, Paris, 6–9 June, mimeo.

Mowery, D. and Rosenberg, N. (1979), 'The influence of market demand upon innovation: a critical review of some recent empirical studies', *Research Policy*, no. 8.

Nelson, R. (1988), 'Institutions supporting technical change in the United States', in Dosi, G., Freeman, C., Nelson, R., Silverberg, G. and Soete, L. (eds), *Technology and Economic Theory*, London, Pinter Publishers.

Pavitt, K. (1984), 'Sectoral patterns of technical change: towards a theory and a taxonomy', *Research Policy*, vol. 13.

Simon, H.A. (1986), 'On the behavioral and rational foundations of economic dynamics', in Day, R.H. and Eliasson, G. (eds), *The Dynamics of Market Economics*, Elsevier, Holland.

Williamson, O.E. (1975), *Markets and Hierarchies: Analysis and Anti-trust Implications*, New York, Free Press.

Williamson, O.E. (1985), *The Economic Institutions of Capitalism: Firms, Markets, Relational Contracting*, New York, Præger.

13 Variation among nations in the logic of manufacturing sectors and international competitiveness

Rogers Hollingsworth

Introduction

This chapter has several purposes. First, it offers a typology of the various institutional arrangements for coordinating and controlling sectors of advanced capitalist societies. Second, it demonstrates that one kind of institutional arrangement can be quite effective for coordinating industrial sectors when product markets are quite large and stable, technology is slow to change and not very complex, and labor is not very highly skilled or broadly complex. It also demonstrates that a different type of coordination is needed when markets are unstable, technology changes rapidly and is very complex, and the labor force is highly skilled and broadly trained. Third, the chapter suggests that a number of manufacturing sectors in advanced industrial societies are moving to the second form of governance and that the society which has difficulty in shifting to the second form of governance will not compete effectively in the international economy. Fourth, to demonstrate these points, the empirical part of the chapter focuses on the US economy within a comparative framework during the period since 1950.

A typology of economic governance

A critical concern of much recent social science scholarship has been to discover the logic by which manufacturing sectors in advanced capitalist societies are governed (i.e. coordinated). In recent years, a number of scholars have been very concerned with identifying the institutional arrangements with which economic activity is coordinated and understanding why one form of governance or coordination is chosen rather than another in different sectors of the economy (Chandler, 1977; Williamson, 1975, 1985; Hollingsworth and Lindberg, 1985; Lindberg, Campbell, Hollingsworth, 1989; Schmitter, 1989; Schneider, 1989: 1–7), and this chapter expands on that literature. At a rather general level, *governance is the process by which activities and conflicts among various actors are coordinated and managed*. Thus, a major goal of the chapter is to focus on the cooperative, antagonistic, and transacting interrelationships among actors involved in manufacturing, and to understand how they have been managed. The following actors influence almost every industry in manufacturing: (1) suppliers of materials (e.g. producers of raw materials, suppliers of components for final assembly, and researchers who provide basic and applied knowledge), (2) processors of materials (e.g. manufacturing firms), (3) distributors (e.g. wholesalers, retailers), (4) suppliers of capital (e.g. investment houses, banks, private investors, the state, etc.), (5) transport firms (e.g. railroads, airlines, etc.), and (6) labor.

The typology[1] in Figure 13.1 consists of six types of governance arrangements for coordinating relations among these various actors, and it is the basis for the following discussion about the governance of manufacturing sectors. The vertical dimension is frequently found in the literature on forms of economic coordination and control (Williamson, 1975, 1985; Eccles, 1981; Granovetter, 1985; Stinchcombe and Heimer, 1986), and embodies the distinction between formal and informal types of institutional arrangements. The horizontal dimension distinguishes between individual action (including that of individual organizations) and collective action, where there are processes to enable organizations to arrive at definitions of common interest.

The literature on markets, networks, and hierarchies, represented along the left side of the vertical axis in Figure 13.1, has generally ignored collective forms of coordination (Schneiberg and Hollingsworth, 1989). However, forms of collective action play an important role in economic governance, and if actors are embedded in an environment with firmly institutionalized forms of collective action, this will have important implications for whether markets, corporate hierarchies, or obligational networks are chosen as instruments of governance and for how they perform. The three types of governance arrangements on the left side of Figure 13.1 coordinate relationships among the various actors. But these particular arrangements may or may not be embedded in an environment in which the arrangements on the right side of the figure are highly institutionalized. A major concern of the following discussion is the relationship between the various collective forms of governance on the right side of Figure 13.1 with those on the left.

If one's goal is to explain the emergence of institutional arrangements for the governance of industrial sectors, one must focus on a host of contextual variables. Some variables are sectorally specific: the degree of competitiveness and capital intensity of specific industries; the size of the sector measured by the number of firms and employees; and the degree to which consumer tastes are diversified or homogeneous. Variations in governance arrangements are also influenced by product variables: technological complexity and the rate of technological change, as well as by variables which influence the costs of transactions. Of course, the age of sectors influences variability in governance arrangements. New industries generally have greater capacities to undergo structural change than older industries which have highly institutionalized arrangements.

Very importantly, there is the industrial relations system of a sector, a matter of considerable importance in the discussion which follows. It not only reflects the way that relations between capital and labor are constrained within an industry, but it may influence the way that relations among other actors are coordinated. In addition, there are also country specific characteristics: the size of the country (space and population); complexity of the national economy; the degree to which firms are embedded either in a rich or impoverished set of

Figure 13.1 A typology of governance mechanisms

Form of coordination and control: degree of formal integration		Range of Action: Individualistic: coordination based on manipulation of agents' individual incentives	Collective: coordination based on appeals to solidaristic values & collective goods
Low	Imputed coordination and control through dispersed Interaction of units: No discrete organizational structure	Cell one *Markets* – self-liquidating sales – spot market contracts – auctions	Cell four *Monitoring networks* – extensive corporate interlocks for information and control – price leadership arrangements
Med.	Network coordination and control among linked autonomous but interdependent actors	Cell two *Obligational networks* – inside contracting – subcontracting – joint ventures and strategic alliances – relational contracting – franchises – cooperative contracting	Cell five *Promotional networks* – joint development projects – coalitions – action sets – interorganizational clans – industry training institutes
High	Bureaucratic or administrative control structure	Cell three *Hierarchies* – vertical and horizontal integration – conglomerate – job control union contracts	Cell six *Associations* – cartels – trade associations – employer associations – producer cooperatives – unions – research associations

Source: Modified from Lindberg, Campbell and Hollingsworth (1989)

institutional arrangements (e.g. highly developed trade associations, capital markets, training institutions for the development of broad skills for the work force); and the society's belief system, which gives rise to rules and normative constraints for the governance of particular sectors. And finally – but perhaps most importantly of all – there is the role of the state. The state defines property rights and the conditions under which each of the six types of governance arrangements may exist.

There is no universal and immutable logic involved in the governance of capitalist societies. Firms make decisions in response to a large number of variables which are interactive with one another. And while nothing is deterministic, there are in the histories of individual countries decision making styles and strategic policy choices which emerge again and again. The logic of decision-making implies that certain options tend to be chosen in specific time frames in various countries and others are rarely chosen. It is the distinctive interaction of a large number of interdependent elements that forms what Max Weber (1978) labeled as historical individuals (Scharpf, 1989; Sorge and Streeck, 1988). This chapter is in this style of scholarship, and it is an effort to discover the complex pattern by which numerous complex elements interacted at the sectoral level in the United States since 1950.

The coordination of the US economy since 1950 within a comparative framework

Corporate hierarchies and mass standardized production

Through the 1950s, the model of mass production became the undisputed means of enhancing industrial efficiency in numerous manufacturing sectors of the American economy. For many years, American industrial economists assumed that the most efficient means of reducing costs was by employing economies of scale and a standardized system of production. Mass production became the basic strategy for expanding markets, and expanding markets became the means of minimizing costs.

American firms engaged in mass production followed a distinctive logic. They employed a particular form of industrial relations, used specific types of machinery, and related in particular ways to other firms in the manufacturing process. Mass producers took seriously Adam Smith's prescription that the most efficient way of organizing a factory was to routinize and differentiate workers' tasks down to the smallest detail. The key to breaking down manufacturing into even more detailed operations was to employ specific purpose machinery for each manufacturing task along an assembly line. In much neo-liberal thinking about mass production, employment was viewed as an impersonal economic exchange relationship, and machines (when profitable), could easily be substituted for workers. Whatever labor was needed to work on assembly lines could be hired or dismissed on short notice. As machinery became more and more specialized, the skill and autonomy of individual workers often declined – though the process of 'de-skilling' varied from industry to industry. As employees became increasingly 'de-skilled', one worker could easily be exchanged for another. Management had little incentive to engage in long-term contracts with workers or to invest in the skills of employees.

Prior to 1950, American mass production strategies were dominant among: (1) mass producers of low-priced, semi-perishable packaged products, relying on large-batch and continuous-process technology – e.g. cigarettes, breakfast cereals, canned foods, and soaps; (2) processors of perishable products for regional and national markets – e.g. meat packing and processing firms; (3) manufacturers of mass-produced consumer durables which used continuous process technology – e.g. sewing machines, automobiles, office equipment and farm implements; (4) makers of high-margin production goods that were technologically complex but standardized – e.g. elevators and pumps; (5) other industries which were capital intensive, high energy consuming and which relied on continuous large-batch production technology for mass markets – e.g. chemicals, oil refining, glass, paper, and rubber products.

By 1950, numerous manufacturing sectors of the American economy were tightly integrated into a system of mass production, in which the whole was greater than the sum of the parts. This form of production was dependent on stable and relatively defined but very large markets for products which were usually low in their technological complexity and relatively slow in their rate of technological change. Hierarchical governance structures (Cell three) were particularly well suited for mass production and distribution. When the transaction costs of working with external suppliers and distributors became high, firms frequently resorted to vertically integrated structures and performed diverse functions in-house. Such a system was complemented and supported by public sector mass education which provided a labor force with the basic training in reading, writing, and discipline to work on assembly lines. The capital markets were sufficiently well developed to provide the capital the standardized system of production required. Because the American model of coordinating mass production with hierarchical institutional arrangements was widely thought to be the undisputed means of enhancing industrial efficiency, many observers believed through the 1950s that the hierarchical form of mass production was the direction in which manufacturing sectors were converging both within and across countries.

Many of America's industrial firms had succeeded with a hierarchical form of coordination because the barriers to entry were too high for effective competition from other firms – both domestic and foreign – in the same industry. But the day of reckoning was to come. Following the Second World War, various European and Japanese manufacturers adopted the latest technology in industry after industry at a time when transportation costs were declining and markets for high quality consumer goods – as distinct from standardized products – were expanding. Moreover, manufacturers in Japan, Germany, and several other countries had never become as committed to the hierarchical form of standardized mass production as had manufacturers in many sectors of the American economy. Indeed, those countries had a very different form of coordinating manufacturing sectors – coordination geared to flexible forms of production, such as networks based on subcontracting coalitions, strategic alliances (e.g. Cell two or obligational networks) rather than hierarchies based on vertical integration, and collective forms of governance. In Japan, Germany, and other countries where industrialization occurred later and where markets were smaller, forms of coordination which were less hierarchical but more network in nature had long been common. These forms of coordination were more effective in coping with their environment than hierarchical type arrangements once markets became unstable and consumers increasingly demanded products based on technologies which were highly complex and changing very rapidly. In other words, hierarchical forms (Cell three) of coordinating industrial production are quite effective when markets are quite stable, consumer tastes are relatively homogeneous, and the technology is not highly complex and is slow to change. But when markets are less stable, consumer tastes are more heterogeneous, and the technology of products is complex and fast changing, various types of networks tend to be more appropriate for coordinating various types of economic relationships. During the 1950s and 1960s, in newer American industries in which the technology was highly complex and changing very rapidly – pharmaceuticals and other biotech industries, aerospace and computer-related industries, etc. – various types of obligational networks (Cell two) became a dominant form of coordination.

Following the first world oil crisis of the 1970s, many American manufacturing firms which had been engaged in producing standardized products found themselves with products for which there was little demand, and yet their rigid system of production meant that they had little capacity to produce the products which were in demand. Once again, the teachings of Adam Smith were instructive: as markets were shrinking they became increasingly saturated, volatile, and/or unstable.

The historic strengths of vertical production in the auto, aircraft, steel, consumer electronics, and many other industries are increasingly perceived to have enormous costs in a world in which markets are volatile and the technology is increasingly complex and changing. The older, more hierarchically vertical integrated American firms have increasingly become saddled with structural inertia, slow response time, inability to develop in-house components with complex technologies and high production costs.

Many American firms engaged in producing standardized products responded to saturated markets and the decline in profits simply by trying to reduce costs. Some froze or rolled back their employees' wages. Others took advantage of the declining costs of transportation and communication to shift production to low wage areas at home (e.g. the meat processing industry) and/or abroad (as in the auto industry). Most reduced their labor force and introduced new forms of automated equipment. Many firms and their trade associations – especially in the shoe, textile, steel, and automobile industries – pressured Washington for protection against foreign competitors. However, these were only temporary expedients, for it became increasingly evident that standardized systems of production were incompatible with volatile and unstable markets. Even when the market demand for certain goods remained relatively stable, less developed countries with lower wage rates were able to copy standardized products and sell them in the United States at lower prices, so that American mass producers faced severe price competition.

Eventually, it became increasingly obvious that a different coordinating strategy was needed in many manufacturing sectors. And the question was whether industries in the United States, historically coordinated predominantly by hierarchical strategies, could shift to the more flexible and less hierarchical coordinating strategies employed by their foreign competitors. The need for different forms of coordination did not, of course, extend to all industries. Standardized production strategies associated with hierarchical forms of coordination continued to be effective in industries where the technology was not very complex and rapidly changing, consumer tastes were relatively homogeneous, and markets were relatively stable.

The coordinating form which slowly emerged in numerous manufacturing sectors was one involving flexible forms of production, with coordination based on a variety of networks. This was not an entirely new paradigm in the United States, but it had long been subordinate to the mass production perspective. However, a flexible system of production involves different labor management relations, levels of skills and work tasks, and relations among producers, processors, and distributors. Whereas standardized production was characterized by hierarchically imposed division of labor, flexible systems of production require work forces with broad levels of skills, employees who have 'learned to learn' about new technologies and can easily shift from one work task to another, and who can work closely and cooperatively with other employees and management. Labor relations must be structured so that people of various ranks engage in meaningful consultation with one another, rather than constantly responding to commands. Whereas many mass production firms tended to pursue hierarchical strategies of vertical integration to lower transaction costs (Chandler, 1977; Williamson, 1975, 1985), manufacturing firms producing products with rapidly changing technologies reduce transaction costs by relying on long-term stable networks with their suppliers and distributors.

How widespread have flexible systems of production become across manufacturing sectors of the American economy and why have flexible forms of production become more pervasive in some sectors than in others? Firms in some sectors have adopted flexible methods of production in response to the following variables: the degree of and nature of international competitiveness; the level of technological complexity and rate of technological change of the products; the degree to which other types of production arrangements had already become highly institutionalized within the sector; whether the industry has a long or a short history; and the potential of firms in an industry to engage in collective behavior.

For purposes of this discussion, a flexible system of production is simply the inverse of mass production. It is the production of goods by means of general-purpose resources rather than vice-versa, a system of production which can quickly adapt to different market demands. Thus, firms coordinated by obligational networks with flexible strategies of production have enhanced potential to be competitive in volatile environments, for they can make an ever-changing range of goods to appeal to specialized tastes with customized designs (Sabel and Zeitlin, 1985; Streeck, 1987a,b; Piore and Sabel, 1984; Kristensen, 1986).

Markets for many products are changing with great speed, and in such sectors, it is less appropriate for firms to invest in product-specific machines and workers with a capability of doing only one thing. Production systems are closely linked to and conditioned by technology. For example, the emergence of microelectronic circuitry in numerous industries has done much to revolutionize systems of production. Microelectronic circuitry permits firms to produce a variety of products in production runs of variable sizes. Because employees and general purpose machines can be used for many different purposes in a flexible production system, manufacturing must be coordinated in ways that permit its various parts to be combined and re-combined; and the extent of flexibility is measured by the ease with which suppliers, machinery and workers can be rearranged for different tasks (Kristensen, 1986: 38). However, flexible producers require a work force with high levels of skills, workers who can make changes on their own, with less hierarchical work supervision than under a mass production system. Because of the need to shift production strategies quickly, management must be able to depend on employees to assume initiative, to integrate conception of tasks with execution, and to make specific deductions from general directives. Moreover, firms engaged in flexible systems of production tend to be less vertically integrated than firms engaged in mass production, and to be in close technical contact with other firms. Even as producers have improved the quality and diversity of their products and reduced the size of their production runs, many craft-oriented firms have been able to increase their production volume without sacrificing their quality standards. And in the process, such large and small firms increasingly compete in terms of quality as well as price. (Sorge and Streeck, 1988).[2]

As flexible production systems become more pervasive, there is a tendency for mass markets to be broken up into specialized segments, for consumers in more affluent environments to demand more customized and/or diversified products, and for firms to be smaller. Even in capital intensive industries, product diversification and small batch production are becoming very common. Flexible mini-mills are taking over an increasing share of the steel industry from the integrated mills. Large chemical companies have found it necessary to become increasingly flexible in order to produce more specialty products in small volume. Manufacturing firms with flexible production systems develop a wide range of options for coping with demand shifts, material shortages, foreign competition, and other types of disruptions. Whereas firms engaged in mass production historically had a tendency to respond to market disruptions by lowering prices, reducing the scale of operations, and discharging employees and reducing wages, firms with flexible systems of production have tended to respond by developing new market niches and new products. And the emphasis on new products tends to contribute to more expert skills among management, workers, and subcontractors (Sabel, 1987a,b; Lazerson, 1988).

There are several strategies of coordinating flexible systems of production which are becoming quite common. Consistent with the typology above, the coordinating form which is becoming increasingly pervasive in advanced capitalist societies is the obligational network.

Obligational networks (Cell two)

As the typology has indicated, obligational networks assume autonomy of linked firms working in an interdependent fashion. There are several forms of relationships subsumed under the term obligational networks: (i) subcontracting among firms; (ii) cooperative contracting among small firms, and (iii) strategic alliances and joint ventures among large and small firms, at both the nation-state and global level. *While it is possible analytically to differentiate these various forms of coordination, they often overlap and are integrated in practice.* All of these forms of obligational networks involve actors in pursuit of the interests of individual firms. This form of coordination is not part of an effort to govern collectively an entire industry, though obligational relationships function most effectively when they are embedded in a rich set of promotional networks.

(i) Subcontracting among firms (Cell two)

Increasingly, highly capital intensive manufacturing firms producing products which have technologies which are complex and rapidly changing are engaging in long-term stable relationships with other firms. One type of relationship which is becoming quite common can be labeled as subcontracting. In contrast to vertical integration strategies for mass, standardized production, large firms in many manufacturing sectors are increasingly attempting to base their production process on many smaller-scale suppliers. Examples are found in the automobile, consumer and durable electronic goods, metal products, aircraft, aerospace, and computer industries. Because the American automobile industry has historically been more vertically integrated than many of its foreign competitors (e.g. the Japanese), it has been somewhat slow to 'dis-integrate vertically', but even in the American automobile industry, the trend is towards long-term stable relationships among subcontractors rather than the construction of components in-house.

Historically in the United States, in most of these industries, there was usually a hierarchy of firms – the very large corporation at the top and a number of smaller firms under its influence. By relying on subcontracting, larger firms are able to reduce the amount of capital which they must invest, to change quickly the technology upon which their production is based, and to produce a variety of goods in small and medium batches using many kinds of materials and components. In general, the more complex the firm's technology and the better skilled the firm's labor force, the more autonomous the subcontracting firm.

The industrial relations system of industries is increasingly the key to a diversified quality system of production, to obligatory network types of coordination. To attain diversification, and high-quality production, firms must have a labor force with very flexible and broad job skills. But in order to maintain such a labor force, medium- and long-term manpower planning and training are necessary.

In order to stay at the cutting edge of technology, large and small firms are becoming increasingly interdependent. Just as large firms increasingly have broadly trained workers in order to remain competitive in world markets, so also have subcontractors had to train their workers according to flexible principles. This has become feasible as relationships among firms have become long term and stable. Arms length contracting based on short-term market pricing is not conducive to investing in the long-term training of a highly skilled work force.

Increasingly, all around the world large firms in the auto, aircraft, aerospace, computer and other industries must provide capital, marketing, and research services to their suppliers. As products become more complex and change rapidly, many large firms no longer know exactly how to produce their products. To keep up with changes in the market, large and small firms must learn from one another about new markets and new technologies. To facilitate this, middle-level management increasingly moves back and forth between suppliers and final assemblers. Meantime, price considerations have become less important in shaping the relationships among suppliers and customers (Hyman and Streeck, 1988; Sabel, 1987a, 1987b).

The key to high quality, product diversification is technology based on a highly and broadly trained labor force using microelectronic machinery. This strategy has increasingly shifted competition away from a focus on prices to greater concern with product quality and other considerations of work force capability to meet rapidly changing demands.

As subcontractors work with complex technology, they increasingly attempt to develop long-term stable relationships with several companies, reducing their dependence on a single large firm. In some sectors, smaller firms not only have increased the number of buyers to which they sell, but also have diversified into several product lines. Through diversification, firms protect themselves in the event of a downturn in the market for a single product, and they become immune from 'strong arm' tactics that could be exerted by a single customer. It is not uncommon for firms to make dies and molds for auto assemblers, parts for aircraft assemblers, tractor parts, and specialty machinery for electronics and computer industries.

In American firms – in contrast to those in Japan and Germany – the paradigm of mass production is declining more rapidly than a new paradigm of flexible, diversified quality production is being institutionalized. Ideally, before a production system based on an obligatory form of networks can be highly effective, firms must be intricately embedded into an institutionalized infrastructure which provides for training of labor, extensive links with research institutes and business associations, cooperation with local governments and various types of credit institutions – in short in a system of promotional networks. Moreover, firms must have a willingness to cooperate and have trusting relationships with their competitors, suppliers, and customers. However, the degree to which firms are embedded into these two types of networks varies greatly across industries (Sorge and Streeck, 1988; Piore and Sabel, 1984; Herrigel, 1989).

(ii) Cooperative contracting among small firms (Cell two)

In this institutional form, firms are small and skill intensive. Each small shop specializes in one phase in the chain of production, but is engaged in long-term stable relationships with other firms in the same community or region. These relationships are open-ended and continuous, with relationships being periodically redefined. This form of production is especially common in the industrial areas of Italy, in Smaland in southern Sweden, in the western part of Denmark, and in different regions of Japan. In general, the firms rely heavily on microelectronic tools and extremely well-trained labor to produce high-quality shoes, silverware and cutlery, ceramics, textiles, apparel, agricultural implements, special machines, electronic musical instruments, and well-designed furniture. Each shop tends to engage in only one task in a complex chain of production. Thus in the shoe industry, one firm makes the sole and another the heel, another cuts and sews the leather, etc. Cooperative contracting is based on high interdependence, with each actor dependent on every other for survival. This type of production is labeled 'cooperative contracting' rather than 'subcontracting', though the final product of all of these firms may ultimately go to a large distributor as some form of subcontracting (Friedman, 1988; Sabel, 1987a,b).

Cooperative contracting tends to flourish in those regions which historically had strong artisan traditions and where there had also been a tradition of small firms engaging in strong collective action – e.g. where highly institutionalized promotional networks exist. However, the United States lacks a strong artisan tradition in these industries. Moreover, anti-trust legislation, strong ethno-cultural cleavages in American society, the large size of the country, and heterogeneity of interests among firms in the same industry have placed severe constraints on the ability of firms in these industries to develop institutionalized networks for sustaining effective collective actions. In other words, cooperative contracting among small firms can be effective only if there is a vigorous craft tradition and they are embedded in an environment with highly institutionalized promotional networks. With cooperative contracting, small firms in these industries may even join together to reduce all kinds of overhead costs – rent, insurance, accounting and it is not uncommon that they borrow machinery from one another. Through negotiations conducted by their business associations, they often receive special subsidies from local, regional and national governments. Moreover, they are embedded in a set of long-term stable networks with specialized banks, credit associations and government lending agencies. Efficiencies are achieved by rapid responses to market needs because of the flexible organizational structure of the entire system (Lazerson, 1988; Sabel, 1987a,b).

The survival of this kind of relationship depends on the intensity of group solidarity among firms. The stronger their sense of collectivity, the more successful they are in regulating the relations between owners of firms and their employees. Strong group solidarity permits small firms to resist high wage demands and profit pressures from large distributors. Otherwise, firms might lay off workers, reduce wages, move elsewhere, or close down.

(iii) Strategic alliances and joint ventures among firms (Cell two)

Firms in a number of industries which historically conducted transactions with either a corporate strategy of vertical integration or in market-type transactions are presently conducting these transactions through obligational networks, specific forms of which are here labeled as strategic alliances or joint ventures. These are also forms of obligational networks and are becoming increasingly common in advanced capitalist societies.

These relationships are of a different order from the cooperative contracting and subcontracting discussed above. In the language of agency theory, firms in this type of relationship are both principals and agents: risk-takers who allocate tasks and share in the gains or losses of the final product (Powell, 1988). These relationships have the potential to be long-term and durable, much more flexible than the type of hierarchical arrangements which historically characterized firms specializing in standardized production.

Historically, some companies developed special long-term relationships with firms in other countries in order to gain market access or to transfer technology through licensing agreements. In more recent years, however, strategic alliances and joint ventures among firms have become more varied in goal and scope – e.g. the search for economies of scale, the need for market access, the sharing of risks, the need to have access to technology and to pool know-how. Moreover, the frequency of these relationships has increased – especially in industries involving high-cost products with complex and rapidly changing technologies. Like mergers, strategic alliances and joint ventures offer benefits quickly, but without the necessity of giving up much of one's independence. Unfortunately, our knowledge about the conditions under which strategic alliances emerge – at both the empirical and theoretical level – is still quite limited. In general, joint ventures and strategic alliances function best when the various actors are embedded in a highly institutionalized set of promotional networks which facilitate trust and cooperation. However, the American political economy has a strong neo-liberal tradition with networks designed to promote cooperation among actors in the same industry being poorly developed – especially in more traditional industries. But as the section below will argue, cooperative and collective forms of networks have been better developed in newer industries where the technology is more complex and changing more rapidly.

Institutional environment: the importance of collective action (Cells five and six)

Markets and hierarchies can work best when embedded in an impoverished institutional environment. But for obligational networks to operate efficiently, they must be embedded in an environment with institutional arrangements which promote cooperation between firms and suppliers and among competitors, an environment which facilitates the exchange of information among competitors as well as among suppliers and producers. Firms must be embedded in an environment which develops rich, broad and flexible skills for the work force and provides financing on a long-term basis. Advanced capitalist societies require cooperative relations and collective resources far in excess to what is needed for markets and hierarchies to function effectively and in excess of what single firms can develop for themselves (Streeck, 1987a,b). These collective arrangements are displayed in the right-hand column of Figure 13.1.

Highly effective sharing of information and cooperation among various actors requires all of the following: collective action on the part of competitors, rich training centers – whether within vocational schools, universities, or other institutions – and institutions willing to provide financing on a long-term basis. In other words, obligational networks function best when they are embedded in an institutional environment with rich multilateral or collective dimensions which provide these conditions.

Of course, a number of American industries have long been coordinated by obligational networks which have also been firmly embedded in an institutional environment of promotional networks (Cell five). In the American case, some of the most effective promotional networks have involved cooperative relations among university-based firms, commercial firms in the same industry, and the state and federal governments. For example, American agriculture owes much of its twentieth-century success to the way that agricultural producers have been embedded in a rich institutional environment which has provided cooperative activity among producers, the dissemination by the state of university-based knowledge to agricultural scientists, and financial assistance from a number of public and quasi-public institutions. Another sector with a long history of being embedded in promotional networks is the American chemical industry. Since the turn of the century, chemical firms have been extensively involved in promotional networks which have consisted of university-based scientists and from time to time the federal government.

The key actor in promoting multilateral activity among producers in American society has generally been the American state. Significantly, the American state has rarely taken the lead in developing promotional networks among manufacturing firms which historically were coordinated primarily by markets and/or hierarchies. Rather, the American state has been most active during the past half century in developing promotional networks for manufacturing firms which have addressed the military and health-related needs of the society. Some of the products and technologies derived from firms coordinated by obligational networks firmly embedded in an environment (e.g. promotional networks) involving cooperative relations with university-based scientists and engineers, the state (especially the military), and other firms (both suppliers and competitors) are commercial aircraft, semiconductors, integrated circuits, computers, nuclear power, microwave telecommunications, new materials such as high strength steel alloys, fiber-reinforced plastics, titanium, and new methods of fabricating metals such as numerical-controlled machine tools. Without obligational networks embedded in multilateral relations with universities, various agencies and departments of the federal government, and numerous business firms, these technologies and products could not have occurred in the United States at the time they did (Landau and Rosenberg, 1986; Nelson, 1982).

The importance of obligational networks being embedded in an environment with rich promotional networks is that it links together organizations having different knowledge bases. This kind of coordination is not possible within a single hierarchy, as no firm has had the capacity to incorporate all the knowledge and resources to develop any one of the technologies and products listed above. Nor could a single firm linked with its suppliers in an obligational network have developed these products. They could be developed only because of the institutional environment promoting the exchange of information, in which obligational networks were embedded. The Agency for Science and Research reported in 1985 that 20% of the firms it surveyed were linked with other firms, universities, and government laboratories through networks involving joint research projects, commissioned research, or other forms of information exchange, and it predicted that in 1990, there would be 26% (Aoki, 1988).

Because of the flexibility of obligational and promotional networks as well as the flexibility of the external labor market, and because venture capital markets have been quite well developed in the United States, it has been relatively easy for entrepreneurial-oriented American researchers and engineers to develop their own firms in order to commercialize new products. This has been especially common in information-based industries in and around the Silicon Valley, California as well as in the biomedical, biochemical, and artificial intelligence-related industries throughout the United States. It is apparent that this type of institutional arrangement has performed extremely well in American society in advancing knowledge at the frontiers of science and in the development of new products derived from basic science. However, the American R&D system has been less successful in improving upon older and new products for commercial markets. The Japanese system of research and development has quite different orientations and effects. Though the Japanese are much weaker in basic science, have not succeeded in developing radically new products, and are somewhat deficient in entrepreneurial leadership, they have concentrated on establishing very close communication among researchers and engineers with production and marketing personnel involved in existing product technologies. Japanese firms are also increasingly embedded in networks involving university-based scientists and engineers, with these networks closely linked with established production facilities. In Japan, R&D and production personnel frequently move back and forth among production sites, research laboratories, and engineering departments. In the United States, manufacturing networks tend not to involve production personnel – or else the production activity is only at the periphery of the network. Americans are socialized in their educational system to make scientific discoveries and to develop new products. In contrast, the Japanese are socialized to be highly attentive to detail and to improve upon existing products. Whereas the Americans excel in horizontal communication across organizations in the development of new products, the Japanese excel in establishing horizontal feedback types of communication 'from marketing to production and production to redesign' (Aoki, 1988: 247). In other words, the Japanese emphasis is clearly on the production phases of the industrial process, whereas the Americans have tended to be less creative in this area. And this difference in where networks focus their energies in the two countries does much to explain why the Japanese have been so successful in commercially producing and marketing products which the Americans first developed.

Similarly in Europe, producers of more traditional products (automobiles, ceramics, textiles, furniture, machine tools, shoes) are embedded in a rich infrastructure consisting of training institutes, business associations, and local and regional governments. And from this cooperative environment diversified, high-quality production has emerged. One reason why American firms in these industries have not been highly competitive in recent decades – at least internationally – is because of the weakness of obligational networks embedded in a rich institutional environment of promotional networks.

Finally, there is one other aspect of American obligational networks in high-tech industries which deserves mention. Because of the flexible external labor market in the United States, it is very difficult for American firms to keep knowledge proprietary. The moving of personnel from one organization to another undoubtedly facilitates communication, creativity, and the development of new products. But it also tends to limit the ability of individual firms to focus all of their talents on the long-term development of particular products once they come into existence – whereas the rigid external labor market in Japan and Germany permits firms in these countries to focus enormous energy on the improvement and refinement of products.

This brief comparison of networking in the Japanese, European and American environments suggests the varied types and settings of obligational networks and the importance of obligational networks being firmly grounded in a set of promotional networks. All of this variation in national arrangements leads to different performance consequences.

Concluding observations

Historically, there has been much greater variability in the coordination of manufacturing sectors in the United States than in Japan, Germany, or the smaller democracies of Europe. The greater variability in the American case results from several factors – chief among them: (1) the United States has a very complex economy, much more complex than any other advanced capitalist society, with many more manufacturing sectors to be coordinated. (2) The United States is much larger – both in space and population – than any other advanced capitalist country, with very heterogeneous sectors. And because there is considerable heterogeneity of interests within sectors, it has historically been more difficult for firms to engage in collective action, to develop effective associative structures for the governance of their sectors. If there are large numbers of firms in industries but weak associative structures – as in the United States – there is more variability in the coordination of industries than in countries which have extremely well developed systems of associations. (3) Because the United States became a highly industrialized society much earlier than most other capitalist societies, it has a mixture of earlier forms of coordination and new forms of coordination which have emerged since the Second World War. (4) In the United States, more than in most countries, there are sectors at many different stages of the product life cycle – and variation in product life cycle leads to differences in coordinating styles. Some sectors are always growing or declining more rapidly than others. Because the United States industrialized relatively early and over a long period of time, because of the American tendency to be inventive in the development of new industries, and because of the complexity of the American economy, there tend to be more industries at different stages of the life cycle than is the case in most countries. In American history, there have been institutional forms for the coordination and governance of multiple economic sectors at different points in time. Thus, batches of sectors – branches of the economy – have similar competitive and cooperative arrangements. But some of the variability in governance of different branches of the economy occurs because industries developed and coordinating arrangements emerged at different time points. In other words, part of the variation in the coordination of the same industries across countries results from the fact that institutional arrangements for coordinating specific industries developed at different time points.

Taking a long-term perspective on the governance mechanisms used in American manufacturing, one sees truly effective forms of coordination of a collective nature for only brief periods. Monitoring as a form of coordination has been relatively uncommon, and promotional networks have been even less conspicuous, but most prominent in manufacturing industries which address the military and health needs of the society and which involve products which are very costly and with complex and rapidly changing technologies.

In contrast to most advanced capitalist countries, the distinctive feature of collective governance in the American case is its weakness of associative institutional arrangements (Schmitter and Streeck, 1981; Hawley, 1966; Galambos, 1966; Hollingsworth and Lindberg, 1985; Schneiberg and Hollingsworth, 1989). A variable of considerable importance in shaping the associative structure of an industry is the degree to which the country has historically been dependent on foreign trade. Political elites in countries heavily dependent on foreign trade encourage firms to develop strong associational structures in order to cooperate in competing internationally, while countries less dependent on foreign trade tend to be less tolerant of associations, seeing them as institutions that retard change and grant monopolistic status to firms. It is in this context that anti-trust legislation emerged more strongly in the United States than in most other capitalist societies. And in recent years as foreign trade has become more important in the overall American economy, there is evidence that the rigidity of anti-trust enforcement is being moderated.

Countries with strong associative structures tend to have firms which cooperate not only with competitors but also with suppliers and distributors in long-term stable networks (Schneiberg and Hollingsworth, 1989). On the other hand, where associations are more weakly structured – as in the United States – coordination through markets and hierarchies has historically been more pervasive. In the past 40 years, however, there has been very substantial decline in the importance of the hierarchical form of coordination. Obligational networks, in numerous varieties, have become more prominent. Through joint ventures and strategic alliances, they may offer the prospect of economic coordination at a global level, especially when markets are volatile and technology is unstable (Porter, 1986).

In assessing the relative roles of corporate hierarchies and networks in the United States, several points should be noted. First, since stable, homogeneous markets remain for many products which have technologies with low levels of complexity, hierarchical forms of coordination remain in many industries. 'Dis-integration' may be occurring, but hierarchical forms of coordination are still widespread in American manufacturing. Paper products, breakfast cereals, soft drinks, bug sprays, floor wax, deodorants, soaps, shaving cream, and hundreds of other products remain, symbolic of the familiar hierarchical form of corporate America. The mass markets to which these products are directed seem quite stable and unsaturated, ready for products manufactured by semi-skilled workers and distributed by general purpose firms.

Second, the ability of the United States to move rapidly towards more network forms of coordination is drastically limited by prevailing practices of industrial relations. The type of industrial relations which facilitates diversified and high-quality production strategies is one in which workers have broad levels of skills and some form of assurance that they will not be dismissed from their jobs. Indeed, job security or other arrangements which assure long-term employment tend to be necessary for employers to have sufficient incentives to make long-term investments in developing the skills of their workers. And this type of incentive and skill system has become much more widespread in Japan, West Germany and Sweden than in the United States (Hyman and Streeck, 1988; Aoki, 1988). In the United States, manufacturing employment has tended to be much more job-specific, workers have been less broadly trained, internal labor markets have been more rigid, and employers have had much less incentive to invest in their workers' skill development. Because the United States has a very flexible external job market, it is much easier for American workers to leave jobs for other firms than is the case in countries where workers have long-term job security. This also provides disincentives for American employers to invest in worker training. Firms with semi-skilled work forces have little capacity to be substantially involved in flexible production arrangements coordinated by obligational networks.

Third, the associational system of a country influences both its industrial relations system and its ability to engage in obligational networks. Where there is a well-developed associational system, firms have greater capacity to enter into collective agreements with their competitors not to poach one another's workers, in effect creating more rigid external labor markets. But this in turn encourages firms to develop more flexible internal labor markets, to invest in the skills of their employees, to develop more flexible forms of production and more pervasive forms of obligational networks, and to produce higher-quality products. Thus, countries with firms tightly integrated into highly institutionalized systems of business associations (e.g. Japan, Germany), have rather rigid external labor markets but flexible internal labor markets, while countries (the United States) with weak associative structures tend to have more flexible external labor markets but more rigid internal labor markets.

Fourth, the capital markets in the United States have placed constraints on the development of broad employee skills. The kind of capital markets which have emerged in the United States have encouraged firms to engage in short-term maximization of profits. On the other hand, large firms in Japan and Germany have relied more on bank loans and/or cross-firm ownership as a major source of capital because of the lack of a well-developed bond and stock market such as that which developed in the United States (Aoki, 1988; Zysman, 1983). The short-term profit horizon of many American corporations has been very much due to their high dependency on very liquid equity markets, combined with the fact that American banks have been disinclined to provide long-term, low-interest loans. This kind of emphasis on a short-term horizon has placed constraints on the development of long-term stable relations between employers and their employees – a prerequisite for a highly skilled and broadly trained work force.

Fifth, as technology in sectors of advanced capitalist societies becomes increasingly complex, changes more rapidly and becomes more expensive, actors in these sectors are increasingly finding that obligational networks are an effective form of coordinating and transacting with each other. But to maximize their effectiveness as governance arrangements, obligational networks need to be embedded in a rich institutional environment involving various forms of collective behavior. However, American firms with obligational networks as a major form of coordination are very weakly embedded in institutional arrangements of a collective nature. On the other hand, their Japanese and German competitors not only rely on obligational networks as a dominant form of coordination but they are embedded in a rich institutional environment (Aoki, 1988; Hyman and Streeck, 1988). And it is for this reason that they are likely to continue outperforming their American competitors.

As a result of the above discussion, the following conclusions appear to be warranted. A social system of production operating in volatile markets with rapidly changing and highly complex technologies requires a different type of work organization, work skills, system of control, relations with suppliers, and competitors from a system where markets are stable and technologies are less complex and relatively unchanging. Systems capable of producing diversified, high-quality products require high levels of social peace and redundant investments in the skills of employees which are likely to occur only if the labor force has high levels of job security. In short, high levels of growth and productivity in manufacturing sectors with high-quality products and volatile markets will not be enhanced by an environment which has a low level of institutional development in unions, business associations, and promotional networks (Cell five).

To compete effectively in international markets, a diversified quality system of production requires cooperative relations with one's competitors, one with a highly developed association system; well-developed networks; communitarian obligations and collective resources that individual actors cannot normally generate, even if they recognize their importance. Rather, such systems of production exist only by evolving over a long period of time from a complex set of institutions, each having come into existence to achieve specific purposes quite different from the functions which these arrangements presently perform. Social systems of production whether involved in producing mass standardized products or diversified quality products are historically conditioned and historically contingent and are incomprehensible except by means of historical and comparative understanding (Streeck, 1987a,b, and personal conversations with Wolfgang Streeck).

Acknowledgements

In writing this chapter, I very much appreciate the assistance of Leon Lindberg and John Campbell. We have discussed many of the ideas contained in it in connection with a project entitled 'The Governance of the American Economy'. In addition, I have very much profited from stimulating conversations with the following: Masahiko Aoki, Ken Bickers, Alfred D. Chandler, Jurgen Feick, Philip Genschel, Wyn Grant, Jerald Hage, Robert Hanneman, Rob Kennedy, Diane Lindstrom, Tom McCraw, Renate Mayntz, Richard Nelson, Yoshi Okada, Charles Sabel, Fritz Scharpf, Christoph Scherer, Philippe Schmitter, Volker Schneider, Marc Schneiberg, David Teece, Raymund Werle, Oliver Williamson, and Erik Wright. But the two people who did more than anyone else to sharpen my views about the issues developed herein are Ellen Jane Hollingsworth and Wolfgang Streeck.

Notes

1. The reader who wishes to read more about the typology should consult Lindberg, Campbell and Hollingsworth (1989). Each of the six types of governance has its own procedures for detecting deviations from rules and for enforcing compliance, and each involves an unequal set of norms and ideologies that helps to reduce the costs of enforcement.
2. In writing this chapter, I am very indebted to the stimulating essays by Charles Sabel (cited below). However, I do have profound differences with Sabel. Whereas he sees the introduction of microelectronic technology as introducing a new industrial divide in which small producers will become increasingly dominant, my perception of capitalist development is that very large firms in many industries will also adopt microelectronic circuitry technology and will over the long term compete very successfully against small firms.

References

Aoki, Masahiko (1988), *Information, Incentives and Bargaining in the Japanese Economy*, Cambridge: Cambridge University Press.

Chandler, Alfred D. (1977), *The Visible Hand: The Managerial Revolution in American Business*, Cambridge, Mass.: Harvard University Press.

Chandler, Alfred D. (1962), *Strategy and Structure*, Cambridge, Mass.: MIT Press.

Eccles, Robert, (1981), 'The quasifirm in the construction industry', *Journal of Economic Behavior and Organization* 2, December, pp. 355–7.

Galambos, Louis (1966), *Competition and Cooperation: The Emergence of a National Trade Association*, Baltimore: Johns Hopkins Press.

Granovetter, Mark (1985), 'Economic action and social structure: the problem of embeddedness', *American Journal of Sociology* 91, pp. 481–510.

Hawley, Ellis W. (1966), *The New Deal and the Problem of Monopoly*, Princeton: Princeton University Press.

Herrigel, Gary (1989), 'Industrial order in the machine tool industry: a comparison of the United States and Germany', paper presented at Conference on Comparing Capitalist Economies, Bellagio, Italy, 29 May–2 June.

Hollingsworth, J. Rogers and Lindberg, Leon (1985), 'The role of markets, clans, hierarchies, and associative behavior', pp. 221–54 in Philippe C. Schmitter and Wolfgang Streeck (eds), *Private Interest Government: Beyond Market and State*, London and Beverly Hills: Sage Publications.

Hyman, Richard and Streeck, Wolfgang (eds) (1988), *New Technology and Industrial Relations*, New York and Oxford: Basil Blackwell.

Kristensen, Peer Hull (1986), *Industrial Models in the Melting Pot of History and Technological Projects and Organizational Changes*, Roskilde, The Netherlands: Institut for Samfun dsokonomi.

Landau, Ralph and Rosenberg, Nathan (eds) (1986), *The Positive Sum Strategy: Harnessing Technology for Economic Growth*, Washington, DC: National Academy Press.

Lazerson, Mark H. (1988), 'Organizational growth of small firms: an outcome of markets and hierarchies', *American Sociological Review* 53, pp. 330–42.

Lindberg, Leon N., Campbell, John L. and Hollingsworth, J. Rogers (1989), *Economic Governance and the Analysis of Structural Change in the American Economy*. Cambridge, Mass.: Harvard University, Center for Research on Politics and Social Organization, Working Papers Series. Also to appear in Campbell, Hollingsworth, and Lindberg (eds), *The Governance of the American Economy* (New York and Cambridge: Cambridge University Press, forthcoming).

Nelson, Richard R. (ed.) (1982), *Government and Technical Progress. A Cross Industry Analysis*, New York: Pergamon Press.

Piore, Michael J. and Sabel, Charles F. (1984), *The Second Industrial Divide: Possibilities for Prosperity*. New York: Basic Books.

Porter, Michael (1986), *Competition in Global Industries*, Boston: Harvard Business School Press.

Powell, Walter W. (1985), *Getting Into Print: The Decision-making Process in Scholarly Publishing*, Chicago: University of Chicago.

Sabel, Charles F. (1987a), 'Changing models of economic efficiency and their implications for industrialization in the Third World', in Alejandro Foxley *et al.* (eds), *Development, Democracy, and the Art of Trespassing*, Notre Dame, IN: University of Notre Dame Press.

Sabel, Charles F. (1987b), 'The reemergence of regional economies: changes in the scale of production', unpublished paper prepared for SSRC volume, *Experimenting with Scale*.

Sabel, Charles F. and Zeitlin, Jonathan (1985), 'Historical alternatives to mass production: politics, markets, and technology in nineteenth century industrialization', *Past and Present* 108, August, pp. 133–76.

Scharpf, Fritz W. (1989), 'Decision rules, decision styles and policy choices', *Journal of Theoretical Politics* 1, pp. 149–76.

Schmitter, Philippe C. (1989), 'Sectors in modern capitalism: modes of governance and variations in performance', in J. Rogers Hollingsworth, Wolfgang Streeck, Philippe Schmitter (eds), *The Governance of Capitalist Economies*, forthcoming.

Schmitter, Philippe C. and Streeck, Wolfgang (1981), 'The organization of business in advanced industrial societies of Western Europe', Berlin: International Institute of Management. Discussion Paper no. 81–13.

Schneiberg, Marc and Hollingsworth, J. Rogers (1989), 'Transaction cost economics and trade associations', in Oliver Williamson and Masahiko Aoki (eds), *The Firm As A Nexus of Treaties*, London and Beverly Hills: Sage Publications.

Schneider, Volker (1989), 'The governance of large technical systems: the case of telecommunications', unpublished paper presented for the Conference Series on the Dynamics of Large Technical Systems, University of California, Berkeley, 17–21 October.

Sorge, Arndt and Streeck, Wolfgang (1988), 'Industrial relations and technical change: the case for an extended perspective', pp. 19–47 in Richard Hyman and Wolfgang Streeck (eds), *New Technology and Industrial Relations*, New York and Oxford: Basil Blackwell.

Stinchcombe, Arthur and Heimer, C. (1986), *Organization Theory and Project Management*, Oslo: Norwegian University Press.

Streeck, Wolfgang (1987a), 'Industrial relations and industrial change in the motor industry: an international view', *Economic and Industrial Democracy* 8, pp. 437–62.

Streeck, Wolfgang (1987b), 'The uncertainties of management in the management of uncertainty: employers, labor relations and industrial adjustment in the 1980s', *Work, Employment and Society* 1, pp. 281–308.

Teece, David (1988), 'Technological change and the nature of the firm', in Giovanni Dosi *et al.* (eds), *Technical Change and Economic Theory*, London and New York: Pinter Publishers.

Weber, Max (1978), *Economy and Society*, edited by Guenther Roth and Claus Wittich. Berkeley and Los Angeles: University of California Press.

Williamson, Oliver E. (1985), *The Economic Institutions of Capitalism*, New York: Free Press.

Williamson, Oliver E. (1975), *Markets and Hierarchies: Analysis and Antitrust Implications*, New York: Free Press.

Zysman, John (1983), *Governments, Markets, Growth: Financial Systems and the Politics of Industrial Change*, Ithaca, NY: Cornell University Press.

14 The revival of prosperity in industrial economies: technological trajectories, organizational structure, competivity

Michael J. Piore

The last decade has witnessed a remarkable transformation in our views about the world economy. At the beginning of the decade, industrial economies were generally perceived to be in a state of crisis. Unemployment and inflation had both reached levels widely viewed as disastrous; economic growth was stagnating, a major catastrophe seemed imminent. As the decade proceeded, economic analysists held their breath, seeing that catastrophe first in the Third World debt crisis, then in the US trade deficit and the federal government budget deficit. Even countries which actually were doing relatively well economically – Japan, the newly developed countries in Southeast Asia – saw their success as temporary and anticipated reversals.

In the last two years, the cloud of pessimism has miraculously dissipated. The Third World debt and the US deficits persist but expressions of concern have become ritualistic. New sources of trouble have emerged; the collapse of the savings and loan industry in the United States, an inflationary monetary union in Germany, heretofore the icon of world monetary stability. But nobody foresees a calamity. If economic epochs can be measured by the sensibilities of the actors in the system, the crisis has clearly passed and we have emerged into a new period of expansion and prosperity.

This constitutes a real test of *The Second Industrial Divide*, (Piore and Sabel, 1984) for that book is an argument about the prospects for just such a recovery, and this chapter is part of a larger effort to review the argument in those terms. There is only time and space here to do a small part of that job and I want to focus on certain microeconomic (or organizational) aspects of that argument. I will pay much less attention to the broader macroeconomic dimensions of the revival.

The Second Industrial Divide is built around a dichotomy between two approaches to technology: *mass production* and *flexible specialization*. These approaches are associated with distinct forms of economic organizations at the micro and the macro level. Mass production with the large, vertically integrated corporation with a strict functional division of labor and hierarchical lines of authority; flexible specialization with network organizations of loosely integrated units embedded in industrial districts of dynamic interdependent small firms or larger 'federal' corporate structures. The broad contours of the regime which has emerged in the 1980s appear to be consistent with this dichotomy. Among the important indications in this regard are trends in the size distribution of establishments: the progression towards larger units which characterized the condition of industrial economies prior to the mid-1970s has been halted or reversed. There also appears to have been an increase in the demands for skilled labor as the shift toward the craft-like work of flexible specialization would imply: the United States is the most prominent example. After a period of decreasing wage differentials in the 1960s, the relative wage of skilled and/or highly educated labor began to increase in the 1970s, and has continued to rise in the 1980s despite a decrease in the relative supply of unskilled labor. Wage payment systems have also begun to shift in the 1980s: the conventional formulae based upon cost-of-living adjustments, plus the average long-run increase in productivity associated with mass production have been abandoned and are being replaced with profit-sharing, bonuses, and other flexible payment systems involving labor risk-sharing. The business place is replete with reports of dynamic industrial districts and organizational reforms in large companies which seem to support this story as well.

But new case study material, collected by ourselves and others, indicates that the dichotomy is too sharp to capture the full range of organizational experience and that the revival and expansion is probably based upon a hybrid regime. Whether such a hybrid regime can survive over the long run is an open question.

These themes are developed in the following pages in three sections: the first section reviews and summarizes the orginal argument. The second section discusses the wider range of organizational forms which have emerged in the last decade. The final section discusses the relationship between these forms and the macroeconomy.

Mass production vs. flexible specializations

The argument originally developed in *The Second Industrial Divided* focused on the way in which the pursuit of economic growth through technological change affected economic organizations generally and the organization of work in particular. In *mass production*, technological change was associated with the specialization of productive resources, and led to an organization of the production process which involved specialized capital equipment and narrowly trained workers, dedicated to specific products. Because resources were dedicated, they were unable to shift to other uses in response to variations in demand or supply for the final products or in cooperating productive inputs. As a result, this approach to production placed a premium upon stability, continuity and predictability, and fostered organizational structures which were capable of creating and maintaining these background conditions. We traced the emergence of the corporate enterprise, industrial unions, and the key aspects of the welfare state to the effort to develop institutions of this kind. The corporate enterprise operated externally to create large, growing and stable markets which would ensure that its resources could become increasingly specialized over time, and could be employed continuously in these specialities. Internally, the hierarchical structure of the corporation ensured the reintegration of divisions and work stations, each of which was so narrowly focused that it could not see the larger organizational design. Industrial unions operated internally to secure labor peace and maintain the continuity of the production process. Externally, unions worked with the Keynesian welfare state to maintain aggregate demand, validating on the national level the policies which corporations pursued in their own markets to maintain the expansion of stability and profitability. It should be emphasized that, as we developed the argument, the driving force in the enterprise of mass production was its capacity to generate technological change and economic growth, but the mechanisms through which it did so were never fully understood either by the practitioners who promoted its development or by outside analysts.

The alternative to mass production in the argument developed in *The Second Industrial Divide* was flexible specialization. We presented this as the modern analogue to craft productive organizations, which had predominated in the early nineteenth century prior to the emergence of mass production. Craft techniques, we argued, had never disappeared under mass production regimes but they had been relegated to particular functions in the economy and, most important in our argument, had lost their capacity to generate the kind of technological change which could sustain economic growth. Flexible specialization, thus, represented not so much a re-emergence of craft production as a revival of its technological dynamic.

The major difference between mass and craft production is the flexibility of productive resources. Craft resources, both labor and capital, are capable of shifting across a variety of different uses and hence of accommodating a range of variations in economic conditions. In terms of technological dynamicism, however, the difference appears to be that craft production, or flexible specialization, generates growth by 'stretching' existing resources so that they can do more, or different, things. A critical research problem is to specify how this stretching is related to the increase in output from a given set of resources which we associate with technological change. At any rate, it is this difference between static flexibility and growth through stretching that distinguishes craft production in an age of mass production from flexible specialization.

The shift from craft to mass production in the early part of the twentieth century led to a *deskilling* of work as jobs become more narrowly defined and the number and range of tasks which they required was narrowed. More importantly, there appears to have been a qualitative change in the *nature* of skill as well. The work of many semi-skilled mass production jobs involved such a narrow range of tasks and the content of each was so limited that the jobs could be learned by rote. Work, in other words, became a routine set of operations which could be memorized. No theoretical or conceptual understanding of the product or the production process was required at all. And indeed in many cases, the jobs were so narrow and permitted such a partial view of the broader process of production to which they were related that no other understanding of the work process was possible. This was fundamentally different from the understanding of craft work, which frequently required the worker to figure out what to do and, hence, demanded a broader analytical or conceptual framework. Conceptual tasks could not, of course, be eliminated in mass production: somebody had to design and organize the work process; to adjust it when the product changed and to handle the unexpected when the process broke down. But these conceptual tasks were increasingly isolated from execution and assigned to employees who learned and understood their work in a very different way. Thus, one could speak of two kinds of learning: rote learning and conceptual learning. Deskilling proceeded both qualitatively, by eliminating the amount of conceptual learning, and quantitatively, by reducing the number of concepts which had to be mastered and the number of tasks which had to be memorized.

It may be useful to make a further distinction *among* jobs which are basically conceptual. Some of these jobs involve static concepts. Others involve expanding the concepts themselves in a way which extends the range of things which can be done. The organization of work in mass production seems to have tended to encourage a distinction between these two types of conceptualization. Work which stretched existing conceptual categories and/or encouraged a critical appraisal of received concepts was increasingly separated from work involving the application of existing knowledge and assigned to research and development departments or to corporate planning and strategy. This segregation is probably associated with the decline in the technological dynamicism of craft work, since the latter involved the application of existing concepts. A revival of craft-type work in recent years thus implies a reversal of these trends, and an increase in skilled relative to unskilled labor.

New organizational forms

The first signs of the renaissance of craft production was the emergence of *industrial districts*, technologically dynamic communities of interdependent small firms. It was from such districts in central Italy that the model of flexible specialization was first abstracted. But the sense that there is a fundamental shift in the organizational landscape is generated by the fact that even the large corporate enterprises most closely associated with classic mass production are abandoning their traditional structures in favor of forms which they characterize as 'flexible'. They generally describe these adjustments as a response to the heightened flux and uncertainty of the business environment. But it is not clear that they have abandoned mass production or that the new alternatives, however flexible, correspond in any meaningful way to craft production or flexible specialization.

The Japanese company

The organizational structure of the large Japanese corporations, is a case in point. Actually, Japanese practice is undoubtedly much more various than it is made to appear in contemporary literature, but for present purposes actual practice is less important than the model which Japan provides for corporate reform throughout the industrial world. The Japanese organization is an image which is guiding the adjustment process. In this regard, it is significant that Japanese practice grows out of the attempt to imitate the organization of mass production in the United States, and one of the signs of the new institutional regime which has emerged in the 1980s is the recognition, in Japan no less than abroad, that Japanese practice really is a different model. As a model, it involves a blurring of the traditional distinctions among functional roles and jobs. The strict hierarchical structure of the classic US corporation is compromised or eliminated in favor of much closer collaboration and shared responsibilities: between workers and managers; among engineering, design, production and sales; between the company and its suppliers. This is reflected institutionally in quality circles, company unions, and worker representation on company boards; in profit-sharing and bonus systems which reflect company performance; in greater use in design and production of multidisciplinary teams encompassing a broad ranging membership which extends to representatives of outside parts producers. The Japanese model also typically involves much broader training of the rank-and-file labor force than in classic mass production and more extensive employment security. Another salient feature, widely noted and copied abroad, is the attempt to minimize in-process inventories which, in classic mass production, insulate the various stages of the production process within the company and between the company and its suppliers. All of this has at least a family resemblance to craft production and is what one might imagine flexible specialization would look like in a large firm. In other contexts, both Sabel and I have tried to interpret it in that way. But certain features of actual practice are somewhat disquieting for that interpretation.

One of these is the way the system is used by Japanese corporations themselves, at least in the United States. John Paul McDuffy, who in collaboration with H. Shimada has been looking at this question, reports that the Japanese do not see this system as confined to a market which is unstable or unpredictable. They use essentially the same organizational structures to produce long runs of standardized products as they do for short runs or batches. Thus, they seem to abrogate the very distinction between mass and craft production around which the argument of *The Second Industrial Divide* is built. The advantage of this approach, McDuffy claims, is that even in long runs of standardized products there is the possibility of progressive learning and continuous improvement and that a broadly trained labor force working in close collaboration on the line is able to realize these possibilities. But this seems implicitly to misrepresent the practice of classic mass production in the United States: that also involved a commitment to continuous improvement. The difference was that the improvements were to be achieved by managers and engineers, not by the workers themselves. In fact, one interpretation would attribute the difference in productivity between mass and batch production in the United States solely to the fact that batch production occurs in time spans which are too short for these improvements to take place.

US adaptation of the Japanese model

If Japanese organizations appear to be producing standardized products with approaches which the typology of *The Second Industrial Divide* assigns to flexible specialization, American producers appear to be groping towards the opposite solution, a kind of *flexible mass production*. Their notion appears to be to extend the repertoire of products which they can produce with a given resource base. Whereas they once sought to dedicate a shop to a single model with, at most, cosmetic adjustments, they are now prepared to produce several distinct variants of the product on the same production line. Moreover, in many cases, they are seeking to expand the number of variants over time. Their ability to do this is greatly enhanced by computer aided manufacturing since software can now be used to achieve standards of performance on general purpose equipment which were previously possible only with dedicated machinery. Workers are being cross-trained to work multiple products as well. Other reforms borrowed from Japan such as just-in-time delivery make it easier to switch from one product to another. In most cases, these are not one-shot reforms. Firms are looking to expand the number of variants they can produce progressively over time.

But flexibility, which these reforms, even when fully realized, entail, remains limited. Large American companies are not seeking to build a productive apparatus which is capable of responding to the market on an *ad hoc* basis. Market response is preprogrammed into the resource base.

The organizational reforms in flexible mass production bear a family resemblance to the Japanese model but the appearance may be deceptive. For example, workers are broadly trained in the sense that they can work on the several variants which the system is programmed to produce. But they need not have a real conceptual understanding of the work; breadth simply means an expanded repertoire of memorized tasks. The conceptual understanding is still confined to managers, engineers and programmers who lay out the work. In flexible specialization, where the product is not preprogrammed, the job cannot be memorized because the worker, alone or in collaboration with others, has to be able to figure it out. One might think that as mass production systems extend the number of product variants, the dichotomy with flexible specialization breaks down and the difference becomes more of a continuum. But, to the extent that (as this above example suggests) the distinction between flexible specialization and mass production also embodies this distinction between conceptual understanding and rote learning, the dichotomy is fundamental, and the flexible specialization will probably not emerge spontaneously from flexible mass production as the number of product variants grows.

Systems

For some products, however, it *is* possible to think of mass production and flexible specialization as defining the end point of a continuum. A number of industries essentially sell *systems* as their final product: computer systems, audio-visual systems, lighting, heating, telecommunication systems. These systems are made up of a series of manufactured components, and, because the components can be varied and combined in different ways, the possible systems are much more varied than the number of manufactured parts. Sometimes the parts and the systems are produced by different companies, in which case the industry typically divides into 'manufacturing' and 'service' organizations, but, in most such industries, at least a few companies are integrated and do both. Among these integrated companies one observes a range of practices. Some companies mass produce the parts and customize the systems. In these companies, systems are only made up out of standardized components and there is a strict line, organizationally and conceptually, between the manufacturing and sales divisions. Other companies will actually change the components to adapt the system to customer requirements. It is more unusual to find companies where the systems are mass produced, but in certain cases there has been a movement to apply artificial intelligence to systems configuration and, in these areas, systems design is becoming increasingly standardized.

One can see the distinction between mass and craft production played out within companies adopting these diverse strategies. In open companies, where even components can be adapted to customer needs, both engineers and production workers move horizontally across the organization. I have even seen the R&D lab call on the production facility to provide a craftsman to build a prototype and, in one such company, the R&D engineers typically bring the new product into production, managing the product as it moves from the lab to the production facility. In companies which mass produce components by contrast, there are distinct product engineers and process engineers, and production workers are too narrowly trained to serve as R&D technicians. In almost all companies, sales personnel are trained engineers and configure the system, but where the components are mass produced, the skill requirements of the sales department are lower and real barriers to mobility among different divisions of the company arise.

The multiplicity of arrangements in companies of this type in the United States should forestall any simple generalization about the proclivities of US firms for mass-production organizations. On the other hand, once one lays out the problem in this way, it is clear that in a certain sense every industry creates systems. Even the automobile is a system of components, and all of the automobile companies have design and engineering departments that 'configure' the system. These departments with the prototype facilities attached to them are enclaves of craft production within these paradigmatically mass-production firms. It is an ominous sign that none of these companies have even tried to use these internal divisions as models for organizational reform, or even, so far as I know, for ideas about how to adopt more flexible structures and practices.

Simple cost cutting as an alternative strategy

The greatest limitation of the analytical structure of *The Second Industrial Divide* in capturing the organizational response of American industry is its inability to capture the importance of Draconian cost-cutting as a way of responding to competitive pressure.

A very influential current in American managerial thinking and among economists and other public policy analysts attributes the problem of US business in the last decade to restrictions upon managerial authority imposed by union contracts, initial bureaucratic procedures and government regulation. People who subscribe to this view also use the term 'flexibility' to characterize their adjustment strategies, but for them, flexibility means the freedom of management to deploy labor at will and the capacity to set wages and other payment in direct response to short-run market pressures and competitive opportunities. In its crudest form, this approach involves the attempt to reduce the cost of production simply by cutting wages and reducing the prices paid to parts manufacturing while preserving the basic organizational structure with its highly articulated division of labor and strict hierarchical lines of authority. Typically, this approach has also involved the attempt to increase managerial authority and control by weakening secondarily countervailing power, particularly trade unions but also subcontractors, franchises, etc. Even when the organization itself never becomes formally committed to this approach, the model which underlies it tends to undermine efforts to realize reforms in the structures of classic mass production. Thus, for example in many companies seeking to institute the Japanese Kanban system of just-in-time delivery, the cooperative relationship with suppliers is sabotaged by factions within the company who see the reform as basically an instrument for shifting the burden of inventory costs to suppliers. Collaborative approaches to labor–management relations have been similarly undermined by particular managers who see collaborative institutions as a way of breaking the union and new compensation systems and the revision of rules governing work assignments as a way of increasing work effort. In the United States, public policy has enhanced the attraction of the cost-cutting model by weakening legal protection for union organization and reducing the level and scope of the minimum wage and of unemployment insurance payments.

Organizational structures and macroeconomic regimes

The implications of these different organizational adjustments depend ultimately upon the macroeconomic foundations of the current 'recovery' and upon the relationship between the technological trajectories associated with these organizational structures and economic growth. The latter is most important but it is also the area about which I think we know least about.

The fundamental hypothesis of *The Second Industrial Divide* is that mass production was the engine of growth from the late nineteenth century to the early 1970s, but that in the 1970s it stagnated. At the same time, flexible specialization which had been basically stagnant in the period of mass production's hegemony, revived a dynamic which it had lost in the mid-nineteenth century. The argument as developed in the book implied that a revival would be based on one or the other of these trajectories but not both. In point of fact, however, the current revival does appear to involve both kinds of growth. Organizations devoted to flexible specialization have done well; but mass-production organizations have done much better in the late 1980s than in the earlier period as well.

There are two plausible hypotheses about how this revival of mass production has been achieved. They reflect the distinction which economics has conventionally made between efficiency and growth. The efficiency hypothesis is that in the 1970s, mass production was over extended. Given the changes in the business environment, firms found themselves using mass-production technologies in markets which were too small and too unstable to sustain this form of production; for markets, in other words, in which mass production was inefficient. Over time, these firms retreated into sections of the economy which were more hospitable to mass production, and mass-production capacity was reduced; as a result, there is now a better fit between mass production and market structure. The economy is more efficiently organized. The increase in efficiency appears as a revival of economic growth. Since it has not occurred all at once, but is a process spread out over time, the growth appears to be sustained. But, in fact, it is a one-shot deal attributable to a realignment of organizational forms and market structures.

The alternative hypothesis is that the technological dynamic of mass production has in fact been revived, and that firms committed to this approach to production will be able to generate progressive increases of output out of the same resource base in the future as they did prior to the crisis of the 1970s. To pin down this effect, one would need more of a theory of the cognitive foundations of mass production than we can develop here. However, most theories of how mass production generates growth require that, as Adam Smith argued, the size of the market hospitable to it grows over time. Has this occurred? And if so, is it likely to continue? Is there, in other words, a second effect; once mass production retreated into those markets structurally suited to it, did the expansion of these markets resume? I think it is plausible to argue that that is indeed the case. In the late 1980s, the movement towards 1992 in Europe, the US – Canadian free-trade pact, the general expansion of international trade among industrial countries, and the general resumption of economic growth in the industrial world, have all led to an extension of the market in Adam Smith's sense of that term. One can argue, in addition, that flexible manufacturing and the shift from classic to flexible mass production, which it permits, has the additional effect of enlarging that portion of the total economy which is hospitable to mass production: classic mass production is efficient only in those markets where demand is stable for the one product produced; flexible mass production is efficient if demand is stable for an *aggregate* of the products which the equipment has been programmed to produce. That aggregate, moreover, is growing over time due to the perfection of the new technology.

There is a real question, however, about how long these effects can be maintained. Two factors suggest that they may be limited. First, the institutional changes which have led to an expanding mass market involve one-shot reforms. Some of these reforms, such as the integration of Europe, may be drawn out over time, but as currently planned, they are self-limiting. It is possible to envisage subsequent changes which would extend the effects – in the case of the European Community events in Eastern Europe have opened a whole new range of possibilities – but they would have to be worked out.

A second problem is that the new dynamism of flexible specialization makes the current expansion very different from that of the earlier post-war period. Because flexible specialization is now technologically dynamic, it has the potential to cut into mass markets from the side, and the future of mass production then depends, in a critical way, on how fast the technological dynamic of flexible specialization cuts the cost differential between standard and specialized products. For mass production to prosper, the size of the market must not only grow rapidly enough so that costs fall progressively over time; it must grow so rapidly that costs fall as fast as the costs of specialized products are falling.

If the current revival is aborted by these effects, then national economic prosperity will depend upon the capacity of organizational structures still committed to forms of mass production to shift towards flexible specialization. If Japanese companies actually do use an organizational form for mass production that can produce small batches as well as long runs, Japan will have an easier time weathering such a transition than the United States. One can argue that, nonetheless, flexible mass production is better adapted to the American environment. We are a much more heterogeneous country and have been less successful in reaching the uniformly high levels of educational achievement which seem required to sustain the conceptual understanding of the rank-and-file work force which flexible specialization entails. Flexible mass production could thus prove an important half-way house for our industry, buying us time to adjust our educational system to the requirements of a genuinely flexible system but, at the same time, moving our organizations in the right direction, laying the ground work for further adjustments in the future. The real threat to long-run adjustment of American organizations appears to be the attraction of the simple cost-cutting model. That model seems to freeze all of our old organization habits. It does not provide the basis for further adjustment; indeed, as noted above, it seems to be undermining the ability of many organizations to work out intermediate organizational forms. Here, public policy may have an important role to play. It appears that our system for the provision of social security actually widens the cost differential, forcing high-wage employers to subsidize low-wage industry. This is particularly true of our system of private medical insurance. High-wage employers typically provide coverage not only for their own workers but for members of those workers' families. The fee structures covered by that insurance is further inflated to absorb the care of the medically indigent. The medically indigent and secondary earners in covered families then work in uncovered low-wage employment. American automobile companies claim this creates a $500.00 differential between Canada and the United States. A shift in medical care alone would thus substantially change the attraction of the low-wage strategy. The reversal of public policies adopted under Reagan encouraging union-free approaches to industrial relations and lowering the relative level of the minimum wage could also have a favorable effect in positioning American industry for the adjustments which are likely to be required in the future.

Reference

Piore, Michael J. and Sabel, Charles F. (1984), *The Second Industrial Divide: Possibilities for Prosperity*, New York, Basic Books.

15 Studied trust: building new forms of cooperation in a volatile economy*

Charles F. Sabel

If trust is the mutual confidence that no party to an exchange will exploit the vulnerabilities of the other, then in many parts of the world today it is almost a commonplace that economic advantage increasingly depends on persons trusting one another more than they do. It is also almost a commonplace that trust can be found, but hardly ever created, and that its creation is much more likely to be the unintended consequence of historical events than the result of a deliberate strategy. Thus realization of the growing importance of trust in large corporations, in their relations to their suppliers, and in the regional networks of small and medium-sized firms of which these suppliers are a part typically goes hand in hand with despair of ever creating it by any combination of private and public action. I concur with the first commonplace and will not defend it here against its critics, although their criticisms are not negligible. I disagree with the second, and the aim of this chapter is to show that despite its superficial theoretical and empirical plausibility, it is false. Properly understood, I will argue, trust is a precondition of social life. Hence the proper question is not how trust can be created, but rather how in any situation it can be redefined, in, to take the immediately relevant example, an economically expeditious way. I support this argument with an account of current efforts to revitalize four traditional industries in several areas of the US state of Pennsylvania.

1. Trust in a volatile economy

Trust is widely regarded as a precondition for competitiveness because of current world-market conditions. As markets become more volatile and fragmented, technological change more rapid, and product life cycles correspondingly shorter, it is too costly and time consuming to perfect the design of new products and translate those designs into simply executed steps. Those formerly charged with the execution of plans – technicians, blue-collar workers, outside suppliers – must now elaborate indicative instructions, transforming the final design in the very act of executing it. But in a world of half-

* Research for this chapter was funded in part by the International Institute for Labor Studies of the International Labor Organization. In writing it I profited greatly from discussions with Joshua Cohen, Robert W. Coy, and, as usual, Michael J. Piore. The usual exculpations apply.

formed plans to collaborate in the production of highly specialized goods or services, any party can hold up the others – most ruthlessly by simply enticing a collaborator into dedicating resources to a joint project, and then refusing to dedicate the necessary complementary resources until the terms of trade are renegotiated in its favor. Absent trust, no one will risk moving first, and all will sacrifice the gains of cooperation to the safe, if less remunerative, autonomous pursuit of self-interest.

But it is frequently concluded in theory and noted in practice that collective appreciation of this dilemma and its potential costs are insufficient to resolve it. The theoretical arguments ring the changes on the hold-up theme. So long as there is a sufficiently high probability of a sufficiently costly breach of trust in the last of many exchanges, the argument runs, the economic agents will, by a process of reasoning aptly called backward induction, refrain from engaging in any of the mutually beneficial prior ones, and thereby exclude experiences which might dispose them to be more trusting and trustworthy.

From this perspective cooperation is likely in two contrary and unusual circumstances. First, when the exchanges are many and the gains from future dealings highly valued in relation to current ones, then it can well be more advantageous to risk betrayal in the end than to forego the profits to be made in the meantime. Second, for reasons rooted in a common history – belief in the same god, dedication to the same political ends, or a common ethnic or cultural heritage – the parties may come to see themselves as members of a community of fate whose implicit (and sometimes explicit) conditions of membership exclude exploitation of the economic vulnerabilities of their fellows.

Neither circumstance, of course, justifies the hope that trust can be created where it is needed. In the first case, in fact, it would be wrong to associate cooperation with trust at all, because cooperation results from continuous calculation of self-interest rather than a mutually recognized suspension, however circumscribed, of such calculation. Here it would surely be more accurate to speak of a *modus vivendi* than trust. In the second case, trust is a by-product of events which, to the extent they are planned at all, did not have the creation of trust as their goal. Seen this way, trust is one of those states, like drowsiness, spontaneity, or – archetypically – having no thoughts, which cannot be produced directly by willing them, and hence at first blush are inaccessible to individual or collective acts of volition. Surely, in this view, it is as self-defeating to try to cease continuous calculation of economic advantage because of the (calculating) conviction that it would be beneficial to do so as to concentrate on the thought of thinking nothing.

These theoretical conclusions are echoed by the economic agents' own perception of their situation. In declining or threatened economic areas firm owners and trade union officials very often agree that labor and capital would both benefit if firms cooperated more with one another and with labor, yet despair of ever doing so. In each case, the outsider is given to understand, 'our' industry has developed an unalterable tradition of universal suspicion. This tradition, exacerbated by the struggle to survive in hard times, precludes the cooperation which is necessary in principle to reverse developments. In those areas which do prosper because of the collaboration which trust makes possible, on the other hand, these same actors regard their confidence in one another as a natural fact, as much an expression of their way of life as their language, and no less resistant to political direction. Here I am thinking in particular of the many small and medium-sized firm industrial districts in Western Europe and Japan. Each tells a different story of how trust became for it a fact of life; and the only common feature of these stories is that none could be reproduced elsewhere precisely because each depended on an improbable chain of fortuitous local circumstances. What better confirmation of the theoretical discouragement than the actors' fatalism?

This pessimism is, I think, unwarranted. In what follows I will argue that the theoretical views discussed so far mischaracterize human nature and, as a correlate, trust, so as to suggest a gross underestimation of the possibilities of creating the latter (section 2). I will claim further that the construction of trust depends in part on the actors' reinterpreting their collective past, and especially their conflicts, in such a way that trusting cooperation comes to seem a natural feature, at once accidental and ineluctable, of their common heritage (section 3). But if these two claims are sustained, the *prima facie* case for the futility of creating trust collapses; and in order to test and deepen the alternative view it becomes worthwhile to examine in some detail a case where economic agents and public officials are struggling to improve economic performance in mistrustful environments, through introduction of forms of cooperation which ultimately depend on trust. Thus (in section 4) I take up the case, alluded to a moment ago, of current and arguably successful efforts to revitalize segments of the garment, foundry, injection-molding and machine-tool industries in certain areas of the Commonwealth of Pennsylvania. A brief conclusion (section 5) reaffirms the central theme of the chapter while cautioning against making too much of those arguable successes.

2. Vigilant trust and negotiated loyalty

Given what has already been said, the heading of this section joins two oxymorons, because trust precludes vigilance (a classic expression of distrust) as much as loyalty excludes negotiation (a sign that the parties do not spontaneously share the very standards of fairness in all circumstances which the notion of loyalty supposes). Nonetheless, I want to argue that the oxymorons express the practical truth that the line between trust and mistrust is much more blurry and easier to cross than our theories suggest. This practical truth finds partial expression in such ambiguous phrases as 'blind trust' and 'undying loyalty', which frequently suggest abdication of responsible judgement, not a breach of duty. We (particularly 'we' in the Anglo-American world) often fail to register this disjuncture between our idea of trust and our practice of it, because of deep-seated ideas of personhood and human motivation. It is to these, their defects, and a communitarian understanding of individuality as an alternative to them that I now turn.

The notions of trust discussed so far rest on two complementary and widespread views of human nature. The first is the idea, familiar from neoclassical economics and rational-choice social theory, that individuals act to maximize satisfaction of their current desires, where these desires are given as the results of an unexplained and theoretically irrelevant process of individual development. Where this is so, cooperation is the result of an accidental complementarity of maximizing strategies, and its culmination can only be a *modus vivendi* and never trust. The second is that under rare circumstances, notably those associated with archaic or pre-modern societies, or modern communities which have retained key features of these latter, many persons have compatible preferences. Hence they have compatible motives, know that they do, know that they know they do, and know that in their simple worlds the cost of obtaining all this knowledge and verifying its implications is not prohibitive. Consequently they trust one another. On this view a miss is again as good as a mile, because anything short of a synchronization of preferences and motives so perfect as to be self-evident (and thus easily assimilated to the preconditions of collective life) sets in motion the backward inductions that transform the second case into the first.

To this view of personhood and human nature I want to counterpoise a lawyer's list of alternatives, all of which blur the distinction between trust and mistrust and suggest passages from one to the other in ways that I shall spell out in a moment. By a lawyer's list I mean a catalogue of arguments which could be simultaneously true, although it is unlikely that they are, and any one of which, if true, would discredit the claim to which all are opposed. A familiar American illustration would be the lawyer who defends a client accused of stealing a bucket by arguing that the accuser is not the rightful owner, that the bucket was in any case borrowed with his or her consent, worthless because it leaked, and returned in improved condition. I prefer this form of criticism of the liberal or neo-classical view to exposition of a single alternative for two reasons. First, those readers who already have doubts about the liberal picture are likely to have different reasons for doing so; and within the wide limits of the catalogue I am about to supply it is immaterial to the subsequent argument which these are. Following Rawls' principle of avoidance, I do not want to pick any fights to whose outcome I am for present purposes indifferent. Second, the variety of alternatives may itself sow some doubt where none currently exists, although after more than a century of explicit and inconclusive debate on these matters, I am aware that this seed takes root only under extraordinary and unpredictable circumstances. For ease of exposition I will initially draw a crude distinction between criticisms that have grown out of sociological theories whose suppositions are antithetic to the methodological individualism of liberalism, and views which have emerged as the result of liberalism's own critical self-examination. This distinction carries less and less weight in current debate, and ideas from one source are more and more often combined with those from the other. Subsequently I will treat conclusions drawn from both sometimes as complementary, sometimes as equivalent formulations.

The core of the sociological or communitarian view of personhood is simply the idea that persons are constituted or can constitute themselves only in society. Individuality can only be expressed and appreciated as individual application of a complex body of common norms which define a shared universe of meanings and expectations: a community. Because it supposes this community, the only spontaneity there is, is coordinated, the only originality collectivized. Just as acquisition of a style or even creation of a new one (which after all only changes some of the prior constraints) simultaneously subordinates the artist to a common culture and enables expression of his or her otherwise ineffable individuality, so specialization requires individuals to define themselves by means that in many ways reinforce the common culture within which and by whose lights they seek distinction.

In this perspective, of course, trust in the sense of shared expectations (and confidence that the expectations are and will continue to be shared) is the constitutive fact of social life. What needs to be explained here is how the boundaries of a particular community are drawn, or, collaterally, how mistrust arises. One way would be as the result of disputes that begin as disagreements (prompted, say, by novel experiences) over the interpretations of common norms, and end as the articulation of irreconcilable views of the world. Another would be as the result of the clashes of different cultural worlds which were, so to speak, irreconcilable from the first. But whatever the answer to this question, so long as social life is possible the realm of trusting behavior can always in principle be enlarged by extending to new realms the shared understandings which make sociability possible in the first place. The contentious interpretations can be reinterpreted to re-establish consensus; two alien cultures can discover – each in its own way but prompted and provoked by the other – the grounds for a common self-definition. Put another way, every *modus vivendi* can be made a trust relation simply by assimilating it to the kinds of exchanges regarded as just by the common culture – a culture whose very existence is denied in the liberal characterization of the modern world.

There are, moreover, variants of this view which connect the social constitution of individuality to specific spheres of social life, and in the spirit of lawyerly listing I mention two of these. One is the idea, central to certain strands of Marxism, that work is the collective activity – the artistry of material survival – which simultaneously subordinates the individual or groups of individuals to common constraints, yet thereby provides the material means for self-development upon which the emancipation of humankind from these and subsequent constraints depends. Another is the notion, central to much of late twentieth-century philosophy, that we constitute ourselves and our humanity by trying to make ourselves intelligible and understanding others' efforts to do the same: in short, by speech acts. With all its ambiguities and lacunae constantly revealed by application in social situations (whose central issue is precisely the struggle to impose or agree on meanings), language here becomes the medium of socialization and individuation.

Because language is so imperfect, and mutual intelligibility so dependent on hard-fought collaboration, shared understandings and hence trust are always extensible in this view too. Unless individuals constantly assumed that others, like themselves, were seeking common meanings – the assumption which Donald Davidson calls the 'act of charity' necessary and sufficient to produce mutual intelligibility – there could be no communication at all. But these same acts of charity constantly create (without necessarily realizing) the possibility for the substantive agreements which are the foundations of trust. A world in which all know that no one can think his or her own thoughts without conversing with the others is a world which is constantly reminded how much it has to lose from the selfish exploitation of ambiguity. In this world it is literally impossible to discuss how trust is possible, because if trust as the willingness to make mutually intelligible sense of the ambiguous were not a primitive characteristic of human nature, it would not be possible to talk about anything at all. Rather, the question here, too, is how boundaries between communities are formed, and under what conditions they may be revised.

Conclusions that overlap these can be reached through criticism of liberalism's core idea. The characteristic claim of this critique is that the notion of an individual maximizing satisfaction of an arbitrary bundle of current desires simply cannot capture crucial aspects of our intuitions of what it is to be a person, especially our capacity for prudential action and even our capacity to understand and draw moral conclusions from what it means to be a person.

Take first the problem of prudence. To act prudently is to do something for which there is no current motivation, but which one will have to want to have done at some time in the future. But unless the notion of current desires is implausibly stretched to include the desire to do things that one would regret not having done in the future, it is unclear why the desires of my future self should influence my current behavior at all. Indeed, why should I care for the satisfaction of the bundle of desires which by accident will define my future self any more than the satisfaction of the desires which define any one of my immediate contemporaries? If persons are capable of, and in part defined by, the capacity for prudential action, as I take them to be, then they must have a view of their personalities as extended in time. Because of this extension, present desires lose their absolute motivational priority; and now is a particular time, one among many in the life of a being aware of its continuous existence, and not the locus where identity forms.

An analogous argument connects the ability to grasp others as persons – and hence to achieve some conception of personality which is not an extrapolation of one's own unique experiences – with the need to view the self as one among many: a someone. To understand someone else as a person, it is necessary to fix what they have in common with you. But to do that, it is necessary to ask how the other, posing the same question, would view you. Thus to imagine someone else as a person it is necessary to view oneself impersonally as a someone. Just as the capacity for prudential action supposes the existence of a personality that encompasses the present in an extended lifetime, so the capacity to grasp the very idea of a person supposes a personality which can understand itself as one self among many. And by demonstrating that the self is not hostage to its own selfish present, and indeed by its very nature entertains thoughts about what would be good for persons (of which it is one) in general or itself in the future, this criticism of liberalism creates the preconditions for a psychology or even metaphysics of the person in which the question, what would happen if everyone did *that*?, can always be asked, and hence trust is always an arguable proposition.

These critical arguments directly suggest the common theme of all the entries in my lawyer's list of objections to the neo-classical picture of the person and the pessimism about the possibility of trust which follows from it. The theme is that the self is always a virtual or reflexive self. Its defining capacity is the ability to choose through reflection which possible self will actually motivate action. The choices express and elaborate a personal identity which is shaped by consideration of the future and other, and which can in principle be more and more shaped by them. Writers in the tradition of self-critical liberalism will refer to this self-creation as the ordering of preference orderings, or, as Amartya K. Sen puts it, 'rankings of rankings'. Writers in the sociological tradition may prefer to speak of self-creation hermeneutically, as the definition of an individual personality through the struggle to make sense of one's self in relation to one's own (constantly redefined) community of shared but ambiguous norms.

Either way we have here a conception of personality and, by extension community, which escapes the liberal's dismaying oscillation between a conception of the self and the group which can perceive the benefits of trust but cannot act to obtain them (because of a paralyzing belief that present preferences inevitably rule), and a conception of the self and the group which benefits from trust, but cannot conceive how others might similarly benefit (because, as an accident of history, they share values whose commonality is taken to be irreproducibly accidental). The reflexive or communitarian self, which on this account is the one we actually have, can entertain and act on the idea of creating or extending common values regarding loyalty and forbearance in the face of vulnerability precisely because it knows that other selves can entertain and act on the same idea. Whether and under what conditions such a change is likely to occur is an empirical question, whose answer depends, among other things, on the prevailing economic conditions and their history, as well as the agents' skill in reinterpreting these. The crucial point is that there is in this view nothing in principle mysterious about the creation of trust in economic affairs. Mutual dependence is the precondition of both individuality and sociability, and it is in some sense known to be such. What precise bearing this mutual dependence has for economic exchanges is another open question, to which one might expect a wide range of answers, depending on circumstances. The answer one would *not* expect is precisely the one liberalism, in the sense intended here, has taught us to count on: that by our nature we cannot discuss the question as though the discussion could be consequential.

Notice, finally, that from this viewpoint trust in economic affairs becomes much less of an all or nothing psychological gamble on the reliability and verifiability of guesses about the harmony of our motives and others'. A community of reflexive selves is by definition both prudent and other-regarding. It can imagine a trusting world and imagine others imagining the same. It can also devise stratagems for testing and encouraging these beliefs. These stratagems, moreover, will appear to be part of the continuous and inevitable process of individual and collective self-definition in a mutually dependent world, rather than leaps of faith hedged by various fall-back strategies. In this world loyalty could be negotiated and trust vigilant without fomenting disloyalty and mistrust precisely because the elaboration of norms of collective behavior as a concomitant and precondition of the elaboration of one's own or group personality is precisely what the exercise of autonomy is all about. Blind trust and undying loyalty are here regarded as deformations, at once admirable and deplorable, because they bespeak a suspicious renunciation of the self's or the group's powers of self-creation. By the same token, it would be possible for individuals and groups to outwit themselves and achieve by indirect means changes which could neither be willed directly nor made accessible to a single act of collective volition. In the next section, therefore, I want to return to the broad-brush characterization of the agents' view of trust to see whether on closer examination their experience does not lend support to the views arrived at here.

3. Reconciliation and genesis amnesia: the politics of trust

The views of the economic agents, it seemed, confirmed neo-classical expectations on two central points: those who suffered the costs of mistrust could imagine the benefits of trust, but despaired of obtaining them given the motivations of persons such as themselves; those who profited from trust thought it an historical accident that they and the (few) others in similar situations did so. In this section I want to focus on this second perception, although I will touch at the end briefly on the first before discussing it at length in connection with the analysis of the case of Pennsylvania.

In re-examining the reports of trust as historical good luck I do not, of course, mean to challenge the existence of such reports or the good faith of those who repeat them. I myself have heard or read different variants of this tale, each emphasizing the particularities of local circumstance, at least in a baker's dozen industrial districts in Italy, West Germany, Austria, France, and Denmark. For what it is worth, all strike me as sincere and guileless in the sense of concentrating on themes and events which everyone in a particular locale regards as important to their collective self-definition. Indeed, once you have come across one of these stories about how misery in a particular place was turned to prosperity through the cooperative exploitation of folk ingenuity galvanized by the genius of a few widely travelled native sons (daughters in these tellings tend to stick to their knitting, quite literally; but that is another story), you can be sure of two things: so long as you keep the conversation on that place, you will hear almost exactly the same story again and again, and it will be different in many details – for instance, whether the setting is in the city or the country – from similar stories about similar places. But it is precisely these two regularities – that accounts of the same place are always the same and those of different places always different – which, I think, ought to arouse the suspicion that the story you are getting is not the whole story.

What is suspicious or at least curious about the similar stories is that they sound similar enough to have been if not rehearsed, then at least repeated with proverbial frequency. But rehearsal or repeated for whom, and why? Surely they do not circulate for the benefit of outsiders. As a rule the stories themselves explain why the local society is different from, morally superior to (because capable of greater solidarity), and hence in some profound sense inaccessible to outsiders. On this point the story does seem to be the whole story. There is nothing put on, so far as I can tell, in either the self-effacement or the pride contained in the remark periodically heard in the Italian metal-working district of Emilia-Romagna: 'America is here!'

But why should the insiders, the members of the local society, tell themselves these homilies so often that they sound rehearsed? In the most austerely liberal account of psychology in these societies, everyone simply has the same or compatible motivations, and there is no need for anyone to advertise that fact. In less austere versions there might be room for a little sociological folklore in which the common knowledge is ritually publicized to reassure everyone that their assumptions about shared expectations still hold. Even taking account of that possibility, however, it is hard to see why people who do rely on one another, and allegedly believe that it is a matter of (cheerfully accepted) historical inevitability that they do, spend so much time getting straight what they all are supposed to take for granted.

The obvious alternative explanation in the light of the preceding discussion of the reflexive construction of self and community turns the liberal account on its head. It is that the stories are articulated when persons or groups which once had incompatible stories agree on a common history which resolves or renders irrelevant those differences: the stories continue to circulate because they set bounds on subsequent disagreements. Instead of expressing a consensus, the stories in this view are part of the process of creating it: they create a past in which the prior conflicts resulted from mistakes and misunderstandings rather than fundamental differences, and suggest a future in which all subsequent conflicts will be limited in virtue of being defined in advance as family fights. Seen this way it is not us outsiders but themselves that the tellers of these stories aim to fool; and it is no fault but our own if we allow their efforts to induce a kind of genesis amnesia in themselves to lead us to believe that their history was actually without conflict.

A deliberately shocking example from the practices of the Ilongot headhunters of Northern Luzon in the Philippines will illustrate the process I have in mind. The Ilongot bear grudges, and bear them hard. Offenses to certain classes of their kin produce in the young males of the society a resentful anger that can only be relieved by severing the head of someone appropriately related to the offender and hurling it joyously in the air. In this way feuds start and ramify, with the result that in time the very survival of the society is jeopardized. When things reach this point, the feuding groups meet and – here is the crucial point – reinterpret kin relations in such a way that the original offence need not have obligated a revenge killing, nor, because of the understandable error, the revenge killing a feud. If this sounds far fetched as a description of a dispute-resolution mechanism that might apply in any way to advanced industrial societies, consider how insistent formerly hostile nations are that, as an indispensable part of their reconciliation, the textbook history of their hostility be rewritten in a mutually recognizable way as a series of tragic misunderstandings. A nineteenth-century example is the United States and Great Britain; twentieth-century examples include France and West Germany, the United States and Japan, and currently Japan and China and South Korea. Letting bygones be bygones requires a collective act of self-redefinition, not simply forgetting.

As I do not want to break stride to introduce the historical evidence that would buttress my argument about the creation of trust relations in particular economic settings, I will merely mention a few *prima facie* cases in favor of my interpretation and indicate several well-documented instances of the 'discovery' of historical conflict in industrial districts that once did or do today tell stories about themselves which exclude such possibilities. The *prima facie* cases are the twentieth-century experiences of countries like West Germany, Italy, Austria, and Japan. All of these came to the brink of or actually fought civil wars in the 1920s or 30s. Yet all today enjoy, reasonably, in the light of their economic performance, a reputation for putting to productive use a national culture of cooperation which is itself a source of national pride and identity. The power of current success to blind nationals and foreigners alike to the obvious fact that these 'cooperative' cultures did not exclude the possibility of fratricide cannot be overestimated. I recall, for instance, once being told by a group of AFL-CIO trade union leaders that certain West German institutions of labor–management cooperation were unsuited to US conditions because German employers could never *think* of doing to German unions what American managers did to organized labor here.

The historical 'discoveries' I have in mind concern industrial districts such as the modern woollens center at Prato, near Florence; the Sheffield cutlery industry which saw its heyday in the late nineteenth century; the center of injection molding in Oyonnax, near Lyon; or the nineteenth-century Swiss watch-making region in the Jura. One common feature of all these districts – and every other one of which I have knowledge – is a history of complex struggles between shifting alliances of merchant factors or converters, high-volume producers of semi-finished goods, artisans making highly differentiated finished products and working independently or in large factories, and less-skilled workers employed directly or as nominally independent subcontractors by the independent artisans or the large firms. A second common feature is that whenever the parties to these conflicts regulate their disputes through arbitration boards or councils which police quality or setting and monitor wage schedules, the districts flourish; when not, then not. The third common feature, and the reason 'discoveries' is set in quotation marks, is that the preceding two are immediately obvious in historical retrospect, but almost completely invisible (because taken for granted and obscured by tales of a cooperative culture) to contemporaries so long as the districts are prospering. A visitor to Prato today, for example, will hear lots about how trade unions and employers' associations are working (as they apparently always have) to solve the problems of industrial adjustment, but nothing about the fact that for almost a decade after a wave of decentralization in the late 1940s, the unions and manufacturers were unable to sign a single collective bargaining agreement.

This view that conflict and an almost mythologically extolled consensus coexist in high-trust systems is, moreover, supported by the few ethnologically sophisticated studies of such systems in action. 'Negotiated loyalty' and 'vigilant trust' are in fact characterizations drawn from, respectively, an oral history of labor–management relations in a US shoe firm in the 1920s and subcontracting in the modern Japanese textile industry. Surely these phrases evoke worlds that mistrust blind trust without being for all that mistrustful.

Consider next the diversity of (unique) stories about the history of particular districts. A few diverse histories support the liberal assumption that trust is a rare accident. Many diverse histories, on the other hand, suggest that there are many paths that end in the creation of trust relations, or, to connect the point to the earlier discussion of the reflexive self and community, that many groups of producers can reinterpret themselves and their history in such a way as to make trust the natural outcome of their common experiences. Whether there were few or many, or – above all – an increasing number of industrial districts and other trust-based economic systems is, of course, a matter of considerable debate. To caricature the positions: those who see many claim any instance of cooperation as a case of trust. Those who claim few claim any instance of conflict as a proof that it does and cannot exist. From the foregoing it should be clear that this way of framing the question makes it impossible to answer. Cooperation that is not more or less institutionalized in a system which enlists public sanctions in the defense of shared values may be no more than a *modus vivendi*. But it should also be clear that conflicts can lead to the creation of trust and persist, transmogrified and limited, after it has been established. How many production systems have crossed or are crossing the *modus vivendi* threshold and institutionalizing conflict in a way which does not obstruct flexibility I cannot say. But my strong suspicion, watching developments in Western Europe, Japan, and the United States, is that the number is growing. If that turns out to be true, then the history of the contemporary period will treat the switch to high-trust systems much as history has treated the whole process of industrialization beginning with Great Britain: apparently impossible for any country to do until it does it, and dependent on some short list of rare preconditions which is successively extended as new cohorts of nations or regions prove that it is in fact possible to industrialize under conditions different from those of their predecessors.

Finally I return to the first of the two confirmatory reports of economic experience: the complaints of those in declining or threatened regions that they are incapable of the trust they require. Again it would be foolish to deny the prevalence of such complaints or impugn their sincerity. Taken at face value, they suggest that the agents themselves recognize the force of the theoretical conclusion that it is impossible to create trust by an act of will. And in this, the theoreticians and the actors may be right. But the interesting question, we saw, is not whether trust can be created at will. If the reflexive view of self and society is correct, then the real problem is how trust can be built in particular circumstances through a circuitous redefinition of collective values. In the next section I provide an example of one way this can be done by reporting on recent developments in the Commonwealth of Pennsylvania, a large industrial state just south of New York whose declining mass-production industries and efforts at economic revitalization mirror the dilemmas and possibilities of industrial America as a whole.

4. Studied consensus: learning to cooperate in fragmented industries

This section tells a tale-within-a-tale about the revitalization of the foundry industry near Pittsburgh in the south-west part of the state, plastics firms in the north-west on the Lake Erie shore, apparel firms in the Lehigh Valley in the north-east, and four scattered clusters of tool-and-die firms. The larger narrative concerns the reorientation of economic development policy both nationally and in Pennsylvania. In the last 15 years, that policy has shifted from a strategy aimed at providing individual firms with the services needed to increase their innovative capacity to one aimed at helping the actors in particular industries and locales – firms, trade associations, trade unions, educational institutions, and local governments – to define collectively which services they need severally and collectively. Put in a way which resonates with the language of the preceding discussion, the consensus is drifting from the view that individual actors know their interests, and that government's role is to remove obstacles to realizing them, to the view that it is only by recognizing their mutual dependence that the actors can define their distinct interests, and that government's role is to encourage the recognition of a collectivity and the definition of particularity.

The smaller, interior story, and the one on which I concentrate, concerns this two-fold process of identity formation. The aim here is to show how, in coming to a common, and generally surprising view of an economic situation which each thought it had understood fully, mutually suspicious groups can redefine their relations and (prudently) begin to construct communities of interest – yet another practicable oxymoron – where none had seemed possible. These beginnings are no more than that and I will underscore their fragility later. They are nonetheless worth discussing because they have already gone beyond the limits of what the conventional liberal view of the wellsprings of cooperation suggest is probable; and should they in fact revitalize their respective local industries, the projects will have demonstrated that it is possible to create vigilant trust in settings – declining industries in the heart of mass-production, unrepentantly individualistic America – where theory and practice would rule it out entirely. If economic community can be discovered in Erie or the Lehigh Valley or the environs of Pittsburgh, then it can certainly be discovered in many other places. Not least for this reason, as we shall see, those concerned with economic development in the United States are following developments in Pennsylvania with increasing attention.

In retrospect, of course, it is easy to trace the changing direction of economic development policy in the United States. For ease of exposition, I will distinguish three successive models, although they in fact overlap, and the seeds of the later ones are contained in the earlier experiences.

By the second half of the 1970s, as traditional mass-production industries in the leading manufacturing states came under increasing domestic and, particularly, foreign competition. As the federal government proved less able and willing to cushion, much less reverse the changes, state governments were forced to articulate new strategies of economic development. To Republicans on the right and Democrats on the left two points seemed clear. First, 'smokestack chasing' – the use of tax incentives to attract branch plants of large, typically mass-production corporations – did not work. Study after study showed that firms' locational decisions were rarely influenced by the universal (and hence self-defeating) offers of tax breaks; and when they were, the kinds of firms that were attracted by such means tended to close up shop at the first sign of economic trouble. It was therefore, second, necessary to foster the state's existing economy, particularly by encouraging the foundation and growth of firms in new industries such as computers, semiconductors, or test equipment, where competition depended on the ability to innovate rather than reduce manufacturing costs. Both convictions seemed well-founded in the product-life cycle theory of industrial development: the view that new products and industries are born in the most sophisticated and technologically advanced economies, and then diffuse to less-advanced and lower cost settings as the products and production processes become standardized and hence easier to master and transfer. From this perspective the only plausible strategy was to stop chasing smokestacks – the 'mature' industries – and encourage the transition to the youthful, technologically dynamic ones.

The upshot was the first new strategy of growth, which might, for want of a generally accepted name, be called the development-bank model. Given the assumptions about the self-directing powers of the market and its close relation to technological development on which the product-life cycle view rested, the problem for policy was simply to reduce the barriers to innovation for firms. These were seen to be of two kinds. One was access to capital: banks were very reluctant to accept new ideas as collateral for loans, and hence it was difficult to finance their translation into commercial products. The other was access to technology: somehow the ideas developed in university laboratories had to be brought within the reach of the economy. State-funded development banks in various guises could solve both problems. By providing grants or guaranteeing or matching private-sector loans to firms with innovative projects, the banks could address the capital problems. Use of similar instruments would encourage joint industry–university research efforts in areas of benefit first to individual firms, but by extension to whole sectors of the state economy. Massachusetts, pushed by the decline of its textile and shoe industries and pulled by the prospect of distributing the treasure trove of ideas accumulated (and in part already commercialized) at MIT and Harvard, led the way. But it was followed in the early 1980s by Pennsylvania and Michigan, among others. The idea was attractive because, beyond its resonance with deep-seated conceptions of economic growth, it promised a lot for a little and appealed to a wide spectrum of political interests. As was widely noted, the development banks operated 'wholesale' not 'retail': by demonstrating the viability of certain kinds of investments, they drew private and semi-private capital such as pension funds into previously avoided areas, thereby vastly multiplying the effect of the state's own lending activity. For those on the free-market right of American politics, moreover, all this could be seen as a remedy for clearly limited market failures, and hence a way of reinforcing the broader appeal of a market economy. For those on the New Deal left, it could be seen as a form of French-style indicative planning or a publicly controlled variant of the coordination of private firms exercised through long-term loans by German universal banks. The ambiguity of these viewpoints and the coalitions they permitted recalls the progressives' efforts to establish social welfare programs in the United States in the first decades of this century by creating public insurance schemes which were meant to serve as both an alternative and a compelling model to private insurers.

By the mid-1980s, a second, extension-service model began to take root in the undergrowth of the first one. One cause was the discovery that even firms in the new, high-tech sectors of the economy needed a longer list of services than originally imagined. If access to capital and technology were problems, so too were access to managerial expertise in areas ranging from marketing to manufacturing, as well as training of technicians and workers. A second cause was the realization that often, indeed typically, the solution to industry's problems lay not in abandoning current markets for new ones (the recommendation of the product-life cycle theory), but rather in introducing high technologies into the products and production processes of the 'mature' industries (the strategy at firms in such advanced economies as West Germany, Japan, Italy, Sweden, and Switzerland which were outcompeting their US counterparts). To do that as well, new services were needed, particularly for those small and medium-sized firms. Their modernization was indispensable to the success of their larger clients because the reintegration of conception and execution requires increasing reliance on collaboration with sophisticated suppliers. But the latter had been so weakened by the crisis that they could not afford the necessary help. In the United States, the Agricultural Extension Service of the Department of Agriculture had successfully provided general consulting services to family farms since the Great Depression. It was therefore a convenient model – or, rather, analogy, as few agricultural institutions were actually transferred to the industrial sector – for state governments moving from provision of resources which firms combined as they saw fit to provision, in the form of consulting services, of knowledge about how best to combine resources. The Technology Deployment Service in Michigan (later part of the Michigan Modernization Service), which serves primarily the automotive parts suppliers, the Cooperative Regional Industrial Laboratories in Massachusetts, which serve the Springfield machine-tool industry and the New Bedford-Fall River garment industry, as well as the nine industrial Resource Centers spread through Pennsylvania are all examples.

In this case too the model was in part palatable because it was consistent with politically divergent interpretations of its underlying motivations. For the right, the extension services could be seen as an elaboration of the preceding, market-perfecting activities: marginal interventions needed to make further, and consequential, state intervention unnecessary. For the left, the new activities could be seen as part of a comprehensive effort to reshape markets through politics: to establish, by new institutional means, new boundary conditions appropriate to the current conditions of competition, as the New Deal had done by the then appropriate use of regulatory agencies and the like.

Once again, however, a new model has begun to grow in the underbrush of the old. This third, incipient model of economic development, which I will provisionally call associative or cooperative, also has twin roots. One is the discovery by those immediately involved with economic development that what makes the successful extension services and development banks work is that their operation creates more or less informal networks of business persons, trade unionists, local government officials, bankers and educators who together discover ways to bring resources to bear efficiently to the problems at hand. These resources include, but are hardly limited to, those provided by the state or local authority. Indeed, the original program, redefined to serve ends only distantly connected to the intent of its originators, often comes to play a subordinate part in a package of resources which could not have been assembled without the efforts of the network – whose own formation would in turn have been unlikely but for authorization of the now marginal program. Thus the programs 'worked' only at one remove, and for reasons not anticipated at the time of their conception. The second root was the growing realization that the systems of West German, Italian, or Swedish firms whose success had played an important role in casting doubt on the product-life cycle theory and suggesting the plausibility of the extension-service model themselves depended on (often highly formalized) networks of a similar type.

In short, the more carefully US observers studied domestic or foreign industrial structures well adapted to current economic conditions, the clearer it became that what mattered was the social *system* by which packages of programs were defined and administered, rather than the precise definition of any single program or service. But if that were so, then instead of trying to define programs directly – extend the extension services – state government should try to design programs which encourage the actors to define their own needs. Insofar as the definition of needs depends on the creation of certain types of local cooperative networks, this means programs which encourage creation of the appropriate forms of cooperation among the actors in particular industries in particular locales.

One of the first, and certainly the most comprehensive programs of this sort, and a learning process in how to create one, is the Manufacturing Innovation Networks (MAIN) project announced by the Commonwealth of Pennsylvania in the spring of 1989. The program's sponsors, all in their thirties and early forties, held responsible staff or line positions in the Commonwealth's Departments of Commerce and Labor and Industry, or worked directly for the governor. All had come of age in government in the late 1980s, just as ideas of economic development were changing most rapidly. One, Jacques Koppel, had helped found and currently supervises the Commonwealth's industrial-extension service, the Industrial Resource Centers. Another, Robert Coy, who was chiefly responsible for organizing MAIN and currently directs the program, had served as executive director of a committee to foster labor, management, and government cooperation on a range of statewide economic development issues. All were familiar with the related efforts of other states, and Coy was particularly knowledgeable about the situation in Western Europe. All were aware of the shortcomings of discrete services, no matter how appropriately designed, and sensed the possibility of improving the performance of all the pieces of the already extensive economic development program by integrating them first locally, and perhaps later at the state level as well. No one, however, had a clear idea of how to do this, and the MAIN program was in large measure an experiment to see whether, as current fashion had it, the actors themselves could play a decisive role in solving the problem.

The program itself was straightforward and fiscally modest enough to be almost negligible. In the spring of 1989 the Commonwealth invited groups of firms in a particular locale and serving common markets to submit, together with trade associations, unions, and any relevant public entities, plans for assessing the strategic situation of their industry and the utility, if any, of more cooperative arrangements among themselves. Small- and medium-sized firms were to have preference for reasons having to do with the general shift in policy towards the weakest part of the traditional industrial base. But plans contemplating increased cooperation between large and small firms were also encouraged. The proposals also had to stipulate the membership of a steering committee, composed of representatives of all these groups, whose task would be to oversee administration of the project. The applications were due less than two months after the Commonwealth's request for proposals was announced, and the four submissions judged best by a committee of the

program's originators were to be funded for one year in amounts of not more than $100,000.

By giving the applicants little time, the Commonwealth could be sure that only industrial agglomerations already toying with the idea of new forms of association would be likely to enter the competition. The kind of collective 'strategic audit' (a consultants' term of the art for an assessment of a firm's competitive possibilities) which the request for proposals called for would only seem worthwhile to assemblies of distinct groups which had begun to contemplate themselves as collectivities. By making the grants small, the Commonwealth insured that no group of firms already cooperating in the way envisioned would bother submitting a proposal: the bother was simply more trouble than it was worth for a going concern. By forcing the applicants to present a single steering committee of the 'relevant' groups, the Commonwealth brought to the surface submerged conflicts which might, had they remained hidden, have thwarted subsequent efforts. In fact, in several cases, different groups claiming to represent the same industry in the same area did submit or consider submitting competing proposals; and the differences could only be composed through Coy's mediation.

Given these constraints, then, it was no surprise that the projects selected came from industrial agglomerations whose situation makes it nearly impossible, on the liberal view, to establish trust. On the one hand, all had long traditions of entrepreneurial independence; and competition among firms, generally considered a fact of business life, had turned cutthroat in the late 1970s and early 1980s. The Pittsburgh-area foundries, to begin with, went into crisis with their traditional customers, the steel industry. The Erie plastics firms came on hard times as their customers in the automobile industry were battered by imports, and as foreign injection-molding firms began to compete in the United States. For firms in Erie, the only way to survive was often to shoot and ship plastic cups and other low value-added products, where relentless price competition increasingly pitted local firms against one another. The situation was similar for the apparel firms in the Lehigh Valley. In the 1950s and 60s, these firms had prospered as subcontractors (or simply contractors, in the industry's language) for the large New York manufacturers who essentially dictated fashions and provided the corresponding piece goods to the giant department store chains which dominate retailing in the United States. In the 1970s and 80s, however, most of the traditional manufacturers were pushed aside as the retailers began to design garments themselves and subcontract production to new, usually foreign firms (private labels), or to a new more flexible kind of manufacturer able to keep abreast of more rapidly changing fashions in large measure by also using foreign production sites (Liz Claiborne). Men's wear, women's wear, and sportswear firms in the Lehigh Valley were left chasing the business for reorders of those items which proved more successful than anticipated. Given the relatively high price they commanded, these reorders could be resupplied on short notice by relatively high-cost local producers. For the tool-and-die producers from Pittsburgh, Philadelphia, Erie, and the Pennsylvania Dutch area in the center of the Commonwealth, the decline of domestic industry and the rise of imports had also apparently led to an intensification of local competition. The Erie plastics industry was the only case in which employment grew through the 1980s, but this was widely perceived to be the result of an explosive growth in demand for plastic goods of all kinds, rather than a sign of the technological prowess of the local firms. During this period other firms grew in number without strengthening their competitive position.

On the other hand, all four industry groups had had in the more or less recent past some experience of inter-firm or labor–management cooperation which, together with their increasing awareness of the organizational basis of their foreign competitors' success, made it plausible to think that each could benefit from rethinking its current forms of association. The most extreme case was apparel, where unions had long played a role in stabilizing the industry. Unions put a floor under wages by controlling the flow of work from unionized manufacturers to unionized contractors; they even provided industrial engineering services to small shops to help the latter operate efficiently enough to pay the higher union-scale wages. But Erie, which billed itself in the 1920s as the 'plastics capital of the world', was evidently for a time an industrial district quite similar to Oyonnax, near Lyons. The plastics industry there grew out of the activities of two resistor manufacturers (which accumulated experience with the new materials) and a zipper maker (which trained several generations of highly skilled tool-and-die makers, who turned their knowledge to mold-making). Most firms in the area still trace their lineage back to one of these three companies. Awareness of this common heritage, and many of the friendships and associations on which it is based, seems to have survived the increased competition of the 1970s and 80s. In the foundry and tool-and-die industries, an analogous role is played by craft pride and the cognate sense of membership in a community where everyone has something to learn from the others. In sum, then, the four project groups were balanced, as the liberal view suggests they would be, between the view that efforts to organize cooperation or create trust would founder on calculations of current self-interest, and a longing for the benefits of cooperation which would prove futile precisely because it was too imprecise to evaluate.

The ability of the MAIN project to tip the balance in favor of trust depended, as almost all the participants seemed to realize, on the utility of a simple and transparent trick. By 'studying' their industries jointly, it was hoped, the parties would at best discover new sources of vitality which could serve as models for collective reorganization. At worst, it was further hoped, they would discover a reality different enough from the one they expected to force reconsideration of their traditional assumptions. This reconsideration would, in turn, lead to formulation of new ideas of best practice. Just as insomniacs can be led to forget their insomniacal thoughts, and fall asleep, by the request that they write down the symptoms of their insomnia, so the industry groups were invited, or invited themselves, to connive in a form of self-distraction which would allow them to catch sight of new possibilities. I will call the kind of consensus and the associated forms of economic transactions which theoretically result from such a process studied trust.

The 'trick' seems to be producing such studied trust because the groups are capable of redefining their community interests, as the previous discussion has suggested they could, and because they have reason to do so given the results of their investigations. Most generally, all have found that they overestimated the costs and underestimated the benefits of cooperation. This was because the estimations were typically based on the experience of the crisis years of the early 1980s, when the firms were truly competing head-to-head and cooperation would have been meaningless. But the firms that survived that period did so largely by specializing in mastery of a particular production process or manufacture of a particular product. Hence, many had become more and more dependent on the provision of complementary products or processes by local companies, without, however, grasping the cumulative significance of the incremental changes. Even as it was becoming more pervasive and mutually beneficial cooperation among specialists was still generally regarded as an exception to the old rule of cutthroat competition, and in any case, assimilated to traditional forms of neighborliness or craft solidarity.

These findings were not self-evident. Entrenched views and interests of the trade associations, firms, and unions drew attention away from those situations which might have alerted them to emergent changes in the environment. The Commonwealth's role was to bring these limiting assumptions to light, either by bringing in consultants to suggest alternative ways of looking at the situation or by using the surprises encountered by one project to jog the others. It was this process, of course, which created the possibility for redefining collective identities and cleared the way for studied trust. Had it been otherwise, the groups would have solved their coordination problems long before they were identified as such.

Some examples will clarify the sorts of barriers which obstructed joint efforts and how the groups have begun to dismantle them in the name of a new collective self-definition. The Lehigh Valley project was initially conceived as an effort to improve relations between the unions and the unionized contractors. The initial participants were aware that there was a growing non-union sector of recent Vietnamese, Lebanese, and Syrian immigrants. But to the Italians and Jews who dominated the firms and unions in the organized sector, the new firms were essentially sweatshops, and the problem – to be addressed in part by the project – was how to make operations there conform to union standards, whether the workers themselves were organized or not. From the point of view of the new immigrants, a number of whom had adopted best-practice technologies and were paying wages well over minimum union rates, the established groups and the unions in particular were simply taxing them: firms which subcontracted work to non-union shops are forced to pay an amount into the union social welfare funds equal to the contribution which a unionized subcontractor would have made, and this regardless of the level of wages and benefits actually paid by the non-union firm. To the extent the non-union firms absorbed the penalties paid by their unionized clients, they were subsidizing union members at the expense of their own employees and profits. But through meetings and shop visits, employers from unionized shops and union officials became more aware of the dynamism of parts of the non-union sector. Conversely, representatives of the latter began to similarly distinguish a dynamic part of the unionized sector which was prepared to trade services with them and a stagnant sector which counted on enforcement of old rules and practices to assure its flow of work. As this happened the problem became less one of regularizing sweatshops (the original 'union' view) or outwitting an outdated but still powerful monopolist (the new-immigrant view) and more a question of rethinking the collective needs of the new specialist firms in the Valley, their relation to one another, and ultimately their relation to the manufacturers and retailers in New York.

The Pittsburgh foundries provide a second example. Their trade association is the Pennsylvania Foundrymen's Association, a for-profit organization which sells its services to individual foundries. The principal service is workmen's compensation insurance, which the Association, through one of its entities, can provide at advantageous rates because it has identified and taught its members to eliminate the chief hazards in foundries. Because the Association is run by persons who do not come out of the industry and think of themselves as providing discrete services, they do not have a detailed grasp of changes in the foundry technology or the structure of the industry: for example, the customers' growing emphasis that foundries provide design services and furnish products which require little if any further machining. Nor did they learn of these changes from their current members, because these latter tend to be the oldest, most traditional, and often economically most desperate of the surviving firms. Again shop visits, organized by outside consultants with little prior knowledge of the industry, turned up facts that surprised the more traditional shop owners and trade associations alike.

The first was that many foundries were in fact selling foundry-related services, or rather providing customers with services which went far beyond the simple pouring of metal according to the client's specifications. Of the 42 (of 45) foundries in the area which responded to a questionnaire, 20 provided just-in-time delivery, 29 machining services, 20 welding services, 20 drafting services, and 15 engineering services. Plant visits revealed, moreover, that the firms had begun to cooperate systematically in the provision of complementary services. Some went so far as to bid jointly on jobs; many were enthusiastic about advertising the region as a foundry center. The industry, finally, was attracting a surprising amount of investment. Eight of the firms, or nearly 20%, had either been purchased by new owners, sold to management, merged into joint ventures, or started from scratch. All of those, domestic or foreign, who entered the business were foundrymen, and all brought knowledge of the latest technologies.

The strategic audits also uncovered less surprising surprises: situations which confirmed widely held views about common problems, but which cast those views in a new light and led to unexpected conclusions regarding their implications for the organization of particular industries. Training is the chief example. It is banal to remark that US workers have not been trained for the tasks which the new competitive conditions require of them. All four project groups observed in their applications to the Commonwealth that collective provision of training was a crucial area in which they expected gains from cooperation, and none were disappointed. What was surprising was the precise relation in each case of training to other problems of organization. In apparel, for example, training was crucial to the flexible use of new garment-routing technologies: unless sewing operators mastered several tasks, there was no point in introducing conveyance systems which allowed managers rapidly to reconfigure the shop for new products. In the tool-and-die industry, shop-owners and managers were compensating for skill deficits by programing machines and organizing work themselves. As a result they had little time to manage, and considerations of long-term strategy were sacrificed to the daily struggle for survival. Increased training in this industry is thus literally the precondition to thinking through other changes. As foundries expand into foundry services, to take a final example, they discover the need for broad training in metal working which overlaps with the needs of the mold-makers in plastics and the tool-and-die industry. Hence cooperation in training soon came to include the idea of cooperation across industry lines, and raised the possibility of collaboration well beyond what had been imagined by each group in isolation.

All of these discoveries, surprising or less so, reacted in turn on the organization of the parties to each project and ultimately on the Commonwealth as well. The dynamic forms are gently emarginating the more traditional ones in the Foundrymen's Association, and the Association's leaders see the possibilities of leading an expanding sector, rather than administering the decline of a moribund one. Non-union firms are playing an increasing role in the apparel project, particularly in the important committee which has been established to promote the full range of new technologies appropriate to producers in Lehigh Valley. These changes have just begun, but everyone concerned in the projects remarks on the new, and more numerous faces which appear at meetings; and this in itself is a strong hint that the institutional identities of these industries are in flux.

The Commonwealth's role has changed in at least two ways. First, whatever the outcome of these projects, the state is now part of the social system within which firms in these industries decide their future. In this sense it has become an unobtrusive participant rather than an intrusive spectator. Information about the economy flows to the state because the state has helped create the system by which that information is generated. The better informed the state is, the better it is, presumably, at making decisions of all kinds regarding the economy.

Second, as the projects learn to coordinate the efforts of their member groups and the latter define the services they need, the state has begun to learn how to define and coordinate provision of public services. Making the extension-service Industrial Resource Centers and the community colleges respond to the needs of the MAIN groups means changing the administrative structure of both. Just what that will require is still unclear, but that it is required is not. In sum, the state's successes in reshaping the economic actors' identity demand that it rethink its own.

5. Some pessimism of the intellect

At the outset I said I would conclude by reaffirming the central theme of the essay and cautioning against overly enthusiastic conclusions, and I intend to do just that. I have argued that liberal pessimism about the possibility of creating trust is theoretically untenable, and that the actors' echo of it can be reconciled with the alternative view that trust is a constituative – hence in principle extensible – feature of social life. The example of Pennsylvania, I think, supports this skepticism about neo-classical claims. But it does not prove that there is even one sure – Pennsylvanian – way to actually extend trust when the actors believe it is in their interest to do so. Many economic-development projects, as everyone knows, are either well-intentioned failures or publicity-minded frauds. This one is indubitably neither. But it is only a year old, and it would be foolishly premature to celebrate its success. The most that can be said, and I think it enough, is that it has done well enough to strengthen confidence that something like it will succeed if it does not – and that such success will further loosen the hold of ideas that undermine trust where trust is possible.

16 Work organization, technical progress and culture

Susumu Watanabe

This chapter is confined to a specific aspect of work organization in Japan – 'small-group activities (*shôshûdan katsudô*)'. A flood of occidental literature continues on the topic. There are, however, some gaps between the actual philosophy and mode of operation of Japanese 'small-group activities' and their Western interpretation. Workers' motivation is also misunderstood. In the present chapter I would like to make these points clear. It is not my intention to discuss technicalities of these activities and to survey the literature on their effects. Those who are interested are advised to consult works such as Imai (1986), Suzaki (1987) and Onglatco (1988), which contain many company cases.

As will become clear below, different companies call their 'small-group activity' differently, although its content and mode of operation are much the same. 'QC circle' is the most common term, and it is often used to cover all types of small group activities for the sake of simplicity. I will do the same. These small groups formed to undertake quality control and other *kaizen* activities (defined below) should not be confused with small groups or teams of workers engaged in actual production activities, such as the semi-autonomous groups at Volvo's Kalmar plant and the more recent Udevalla works in Sweden. Work in Japanese industries is team work almost by definition.[1] Accordingly, there is no special Japanese expression for team work in actual production processes.

Before proceeding to the discussion on our main subject, it will be relevant to question what is 'competitiveness' and how competitive Japanese industries are. Section 1 is concerned with these issues. I will then briefly discuss, in section 2, what 'QC circle' as practised in Japan is and how it differs from its counterparts in Western countries. In sections 3 and 4 its diffusion and effects on Japanese industries' competitiveness will be dealt with, followed by a discussion on Japanese workers' motivation in such activities in section 5.

1. The competitiveness of Japanese industries

Labour productivity in Japanese manufacturing industries, defined as the sectoral gross domestic product per hour worked, was the lowest among ten major OECD member countries in 1970. Even in 1987 it was behind the United States. Why then do people talk about the competitiveness of Japanese industries? One possible explanation rests with the high rate of productivity improvement. Another possible explanation is that Japan's labour productivity has surpassed that of other countries in some key industries.

The competitiveness of a firm or an industry, however, accrues only partially from the cost or price advantage. It also depends on the quality of the product and services including reliable and timely delivery, marketing and after-sales services, and Japanese practices appear to be attracting increasing attention in this connection. The authors of *Made in America*, for example, argue:

> Multifunctional design teams and orientation toward simplicity and quality have been a feature of Japanese industry for a good deal longer and have contributed to Japanese advantages in quality and productivity. . . . The Japanese engineers had incorporated quality-enhancing features into the design itself. Building quality in at the design stage is a lot more efficient than applying quality controls retroactively. . . .[2]

The main source of the international competitiveness of Japanese industries has apparently been shifting from low cost/price to superior quality. In order to offer better quality products and services at marketable prices, however, a great effort at productivity improvement is usually required, because better quality involves more work. This may be illustrated by the following analogy. To some extent, the quality of a technical paper is a function of the amount of revision work. Quantitatively, however, the output remains the same, just one paper, no matter how much revision has been made. My contention here is that significant portions of the productivity (i.e. the physical amount of work executed per unit of labour input) improvements in Japanese industries has been absorbed in qualitative changes rather than price reductions.[3]

Improvement of labour productivity and product quality does not always require a large amount of capital investment. Dertouzos *et al.* (1989) argue that the competitiveness of Japanese industries accrues to a large extent from continuous minor improvements and modifications to the existing processes and products, rather than major technological breakthroughs.[4] This assessment is consistent with my own findings in the automobile industry, which will be discussed below. Japanese firms use simple devices such as loaders, unloaders, chutes and stoppers[5] for the purpose of 'low-cost automation'. Application of these devices often presupposes adjustments to product design, plant lay-out and work procedures.[6] Such small but continuous improvements in manufacturing processes and products, including the development and preparation of simple labour-saving devices, usually originate from QC circles.

2. The QC circle in Japanese industries

Scientific quality control methods were introduced in Japan before the Second World War by the National Railway Corporation, navy arsenals and pioneering private sector companies such as Toshiba.[7] Serious efforts to master and apply statistical quality control techniques started after the war. In order to improve the reliability of the Japanese telecommunication system, the US Occupation Authorities provided guidance on quality control techniques for the Japan Department of Communication (today's Nippon Telegraph and

Telephone Corporation (NTT)) and other companies. From them, the know-how was transferred to other firms such as Nippon Electric Corporation (NEC). When Japan adopted an export-oriented growth strategy, her industries had to reduce costs and at the same time get rid of the 'cheap and low-quality' image of their products. The Industrial Standardisation Law of 1949 provided for the inspection not only of manufactured products but also of facilities and methods of design, manufacturing, analysis and testing. To encourage and assist companies in their efforts, organizations such as the Japanese Union of Scientists and Engineers (JUSE) and the Japanese Standards Association began to publish journals and organize seminars on quality control. Companies responded enthusiastically. Serious efforts at workers' education and training in quality control started at major companies around 1951. In 1958 a study team brought back from the United States Feigenbaum's concept of TQC (Total Quality Control). This was developed into a 'Japanese-style TQC' or 'Company-wide Quality Control (CWQC)'. Today, the term 'TQC' is often used in Japan interchangeably with 'CWQC'. In April 1962 JUSE launched its quarterly *QC Circle Magazine* (*QCCM*, hereafter) and urged industries to organize workers into small study groups ('circles') using the magazine as a textbook and to register them with QC Circle Headquarters established within the JUSE. The members of the registered circles were to share their experiences, e.g. by reporting on their achievements at local, regional or national conferences which JUSE would organize regularly. This is the origin of the QC circle.

The Japanese Industrial Standard (JIS) regulation defines quality control as a system of means of economically producing goods or services of the quality meeting the buyer's needs. CWQC aims at achieving this goal efficiently through the participation and cooperation of everybody in the company from top management to section managers, supervisors and ordinary workers in all domains of corporate activity: marketing, R&D, product development, design, pre-production processes, material and component purchase, manufacturing, testing, sales, after-sales services, finance, personnel, workers' training and education, etc. Here, the term 'quality' is interpreted in the broadest sense to mean 'anything that can be improved', and in this context 'quality is associated not only with products and services but also with the way people work, the way machines are operated, and the way systems and procedures are dealt with. It includes all aspects of human behaviour.'[8] Hence, Imai associates quality control not simply with quality or productivity improvements but with improvements in business activities in general (*kaizen*).

While quality control in the West is focused on inspection of the product, the emphasis in the Japanese system is on the non-production of defective products. Quality control in the West is far more centralized and specialized in the hands of, e.g. quality control engineers. In Japan, the role of technical specialists is to provide technical advice when needed and to train other workers.[9]

Probably a more basic difference is that QC circle activity in Japan aims not just at the satisfactory execution of a given task, but even more at the *improvement* of existing products, methods and systems. This is possible, probably because in Japan job descriptions and job classification systems are not narrowly defined. Task allocation is loose, and a task is shared by many workers.[10] Under the Western system, where each worker's responsibility and authority are clearly defined, going beyond one's own area of responsibility, even with a good intention, may be interpreted as an invasion of another person's territory. Citing a college professor and a president of a management consultant company she interviewed in the United States, Onglatco reports that many American managers not only fail to recognize the need for a QC circle but feel their authority would be threatened if their subordinates suggested modifying work methods and standards.[11] Recent occidental literature on Japanese managerial practices abounds with similar accounts.[12]

It seems also true that small group activities in the West tend to be more formal, even at Japanese-owned plants.[13] Japanese workers often talk about their circle activities during a tea break, at lunch time or at an after-work social gathering. This is rare elsewhere. (It is well known that there is a similar difference between Japan and Western countries with regard to the management–labour consultation system.) During a visit to Mexico in March 1990 I was told that a US car assembler had made it obligatory for his local component suppliers to set aside a room in their plants for their employees' small group activities.

Finally, J.M. Juran – one of the two experts, along with E.W. Deming, whom Japanese QC experts respect as their great teacher – argues that while in Japan the QC circle has been regarded as an extension of the continuing corporate efforts at progress, in Europe and North America it has been treated as something more or less independent of the overall corporate strategy. In Japan it forms a part of the overall corporate organization and operation. Elsewhere it is somewhat detatched and, more often than not, left to external consultants to look after. In the late 1970s the QC circle came into vogue in Western countries, in his view, largely because its worker participation aspect caught on with the media. Because companies did not try to promote the movement as an integral part of the corporate management, however, workers became disillusioned and the 'fever' quickly died down.[14] This assessment is consistent with Oliver and Wilkinson's report from the United Kingdom that the feverish interest in QC circles which gripped managers in the late 1970s and early 1980s has eased off and that among all the Japanese-style practices it is now the least favourably rated, although there are exceptional success stories such as Rolls-Royce (aero-engines), Wedgwood Potteries and more recently Jaguar Cars.[15]

Juran, however, forgot to mention what could be an even more basic constraint on the spread of QC circles in the West. In the United Kingdom, it has been suggested that the concept of TQC does not work partly because of ordinary workers' low educational standards, notably lack of knowledge of simple maths.[16] British Ford had to upgrade its recruitment standards in order to ensure that employees were capable of using statistical process control techniques. About 60% of school leavers were ruled out and the company's manpower requirement could not always be met. Moreover, some of the existing staff resisted the new techniques 'mainly due to a lack of confidence with sums'.[17] The situation appears to be no better in the United States. Local speakers at a conference I attended in Stanford in March 1990 stressed that a fundamental problem facing US industries was the high proportion of illiterates and drug addicts in their workforce.

Table 16.1 Sources of advice or guidance for the QC circle, 1983 and 1987 (% of replying promotors)

Source	Important		Fair		Not very important	
	1983	1987	1983	1987	1983	1987
Manager	44.9	41.3	47.4	52.5	6.3	6.0
Promotor	25.6	20.8	67.5	70.4	6.1	8.6
Leader	34.3	29.1	53.8	53.9	11.1	16.5

Note: 'Manager' means department or division chief. The replies do not add up to 100 because of a small number of 'other' responses
Source: *QCCM* Study Subcommittee, 1987: 28

As the QC circle is organized within the overall framework of the corporate strategy, about two-thirds of the existing circles are operated as part of the TQC as defined above.[18] The theme or target of their activity is usually closely linked with the current main concerns of top management. Circle members receive advice and guidance more often from the unit manager or from the office or persons designated as 'QC circle promotor (*suishinsha*)' than from the actual circle leader who is selected from among the workers (Table 16.1). All the workers at the unit which is chosen or volunteers[19] for the activity are supposed to participate. This was the case at 86.7% of the responding establishments in the third survey on QC circle activities which the *QCCM* Study Subcommittee conducted in 1987. Forty per cent of the department or division chiefs in this survey regularly, and just over 50% occasionally, provided guidance or 'indicated their wishes' to the circle members. Results of a more recent survey by the Ministry of Labour are consistent with the *QCCM*'s finding: at 83% of the sample establishments all the workers belonging to the unit concerned were supposed to participate.[20]

From the above it is clear that participation in the circle activity is usually part of the worker's duties. The operation of the circle and the choice of its themes are left to the participants' *jishusei* (spontaneity), within the broad framework of corporate policy. There seems to be some misunderstanding or confusion among Western observers on this point. The problem arises, in my view, from the erroneous translation of the term '*Jishusei*' or '*Jishuteki*'. In this context it should be translated as 'spontaneity' or 'spontaneous', not 'voluntary'.

3. The spread of the QC circle

Major companies in Japan began to organize QC circles in the 1960s. In part, this was meant to be a solution to the mounting shortage of engineers and technicians in very rapidly expanding industries. It was not, however, until towards the end of the 1970s that the spread of small group activities gained momentum (Table 16.2). During the low growth period after the oil shocks the business performance of some companies with a CWQC programme was

Table 16.2 Growth of the numbers of QC circles and participants registered at QC Circle Headquarters, FY 1962–90

Year	QC circles	Participants
1962	23	n.a.
1965	4,930	70,920
1970	33,499	388,543
1975	72,475	723,201
1980	115,254	1,062,759
1985	223,762	1,831,299
1990	325,681	2,535,289

Note: The figures are related to the end of February of the following year
Source: QC Circle Headquarters, JUSE

outstanding and attracted attention: energy savings and cost reductions at New Nippon Steel Corporation with its JK (*Jishu Kanri* or self-management) programme, for example. In the meantime, the average education level of the Japanese industrial workforce rose significantly, younger workers now having at least senior, instead of junior, high school education and the proportion of college and university graduates rising. Even ordinary production workers could now analyze their problems and think of solutions, and this facilitated the rapid spread of the QC circle.[21] The increasing *tanôkô-ka* (acquisition of multiple skills or diverse technical knowledge) through job rotation[22] was another contributing factor, as knowledge not only of one's own unit but also of the entire flow of work is essential for a successful QC circle.

At the end of August 1990, the numbers of registered circles and participants reached 313,924 and 2,454,635, respectively. The responsible official at QC Circle Headquarters estimates the number of active member circles at about half that figure. Some circles become dormant after a period of active operation, often for lack or exhaustion of a target theme. Starting in the production lines of manufacturing industries, the QC circle has spread to their offices and to non-manufacturing industries such as finance, insurance and trade. As the circle moves away from the production line, however, its target tends to become less concrete and its operation less active.

However, the coverage of the registry is not complete. Only about 20% of the small group activities identified in a recent *QCCM* survey had been registered at QC Circle Headquarters. One-fourth of the total and one-third of the manufacturing establishments had some kind of small group activities, 11.3 circles per establishment on the average. Roughly 60% of them were actually called QC circles, the rest bearing various other names such as JK and ZD (Zero Defect). Three per cent of the sample establishments had discontinued their QC circle programme, while 5.4% intended to start one in the near future. On the basis of these findings, the authors of the report estimated the total numbers of practising establishments and circles at 65,800 and 743,540, respectively. As the number of participants was 84.1 per establishment, the total number of participants in the country was estimated at 5.5 million.[23]

The *QCCM* Third Survey cited earlier provides the most recent information on the registered circles. About 75% of them are actually called QC circles and two-thirds are organized as part of the establishment's TQC programme. Nearly 90% of the circles are composed of five to eight workers, and a large firm can have thousands of them. On average, the participants meet just over twice a month, for about an hour each time. When the activity extended beyond regular working hours, 36% of the responding companies paid overtime allowances and another 34% more modest amounts. During a recent visit to QC Circle Headquarters I was told that major companies try to accommodate small-group activities within regular working hours, e.g. by setting aside a certain amount of time (usually two hours a month) for the purpose, and pay for out-of-working-hours activities, provided they have been approved beforehand. Workers often collect data used by their circles while discharging regular duties. Where the programme is still at an early stage of development, its organization is loose and a lot of time is wasted. In such cases, the activity tends to extend further beyond regular working hours and is more likely to be unpaid.

The QC circles are heavily concentrated in larger companies. According to the already cited June 1989 survey by the Ministry of Labour, 76% of the establishments belonging to companies with 5,000 or more workers had some kind of small-group activities. The proportion declined steadily with firm size, down to 39.9% in the size group with 50 to 99 workers (Table 16.3). Two-thirds of the sample establishments accommodated the QC circle activity within regular working hours or some other time specially set aside for the purpose. The proportion fell with firm size but increased again in the smallest size group, probably because firms in this group were more eager to avoid

Table 16.3 Presence and modes of operation of the QC circle in different firm size groups, 1989 **(%)**

Firm size (no. of workers)	Practising establishments in the sample	Within regular workhours	Outside regular workhours	(Unpaid)	Other
Total	52.1	61.8	33.0	(15.2)	5.2
5,000 +	76.0	67.6	23.2	(9.4)	9.1
1,000–4,999	64.2	62.7	33.4	(15.0)	3.9
300–999	53.6	57.9	37.6	(13.3)	4.5
100–299	49.2	56.9	39.0	(18.5)	4.1
50–99	39.9	63.9	31.7	(14.5)	4.4

Notes: 1. The figures in the third through sixth columns stand for proportions in the practising establishments

2. 'Unpaid' includes all the cases other than those paying overtime allowances, i.e. 'do not pay overtime allowances', 'other' and 'unknown' in the original publication

3. The total sample consisted of about 4,000 establishments in the private sector

Source: Adapted from ML, 1990, tables 19 and 21

Table 16.4 Objectives of the QC circle activity by firm size
(% of practising establishments)

Firm size	Productivity improvement	Quality improvement	Sense of solidarity	Capability development	Safety improvement	Other
Total	62.1	47.2	32.9	21.4	21.0	1.6
5,000 +	61.7	44.7	22.1	29.2	26.7	2.7
1,000–4,999	59.6	49.3	30.3	24.7	17.9	3.4
300–999	65.4	56.3	36.4	19.6	14.2	0.4
100–299	67.5	52.7	31.4	17.4	20.1	1.3
50–99	56.7	36.8	41.7	18.6	23.8	0.8

Source: ML, 1990, table 19

extra expense. The QC circle activity was paid at 85% of the total practising establishments and at over 90% of the practising establishments in the largest size group. The 'unpaid work in Japanese QC circles' is exaggerated in Western literature.

Productivity improvement and quality improvement are the two main objectives of the QC circle programme. There is no systematic difference in their relative importance by firm size (Table 16.4). In some other respects there are fairly significant differences. The larger firms use the programme more to develop workers' capabilities than do the smaller ones. Clearly, this has something to do with the life employment and job rotation systems, which are spread more widely among the larger firms. (I will come back to this subject later.) In contrast, smaller firms resort to the QC circle to promote a sense of solidarity among their employees. This may be interpreted as follows. The life employment system at larger firms tends to unite the workers in corporate efforts as this helps promote their longer-term, if not immediate, welfare. It also seems true that recruits of a prestigious company tend to develop, sooner or later, a sense of 'corporate nationalism'. Smaller firms without such built-in catalysts need to find other means and the QC circle offers one.

4. The QC circle and competitiveness

The implications of the QC circle for competitiveness (or the level of overall corporate efficiency) may be illustrated as in Figure 16.1. The effects of discontinuous dramatic innovations (i.e. introduction of the latest production techniques or managerial concepts) are depicted by the staircase-like solid line, while continuous incremental progress achieved through the QC circle programme is represented by the ascending dotted line. Experience shows that any system tends to start deteriorating as soon as it has been established. Continuous efforts for improvement are required even just for maintaining the

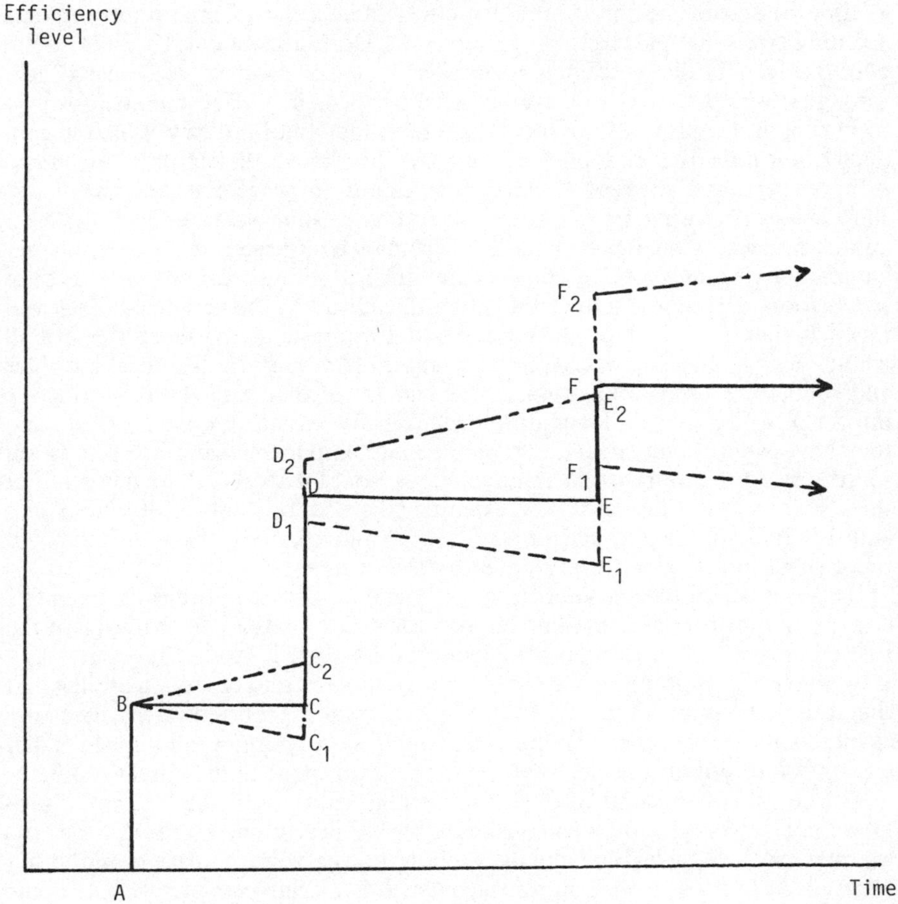

Figure 16.1 Effects of innovations and QC circles

Source: Adapted from Imai, 1986: 25–7.

status quo. Otherwise each innovation will have to start at a level somewhat lower than that attained by the previous one. Let us assume this is the case with Firm I. Then its actual development path will not be the staircase-like line ABCDEF but a saw-toothed dashed line $ABC_1D_1E_1F_1 \ldots$, even if we assume each innovation creates its full impact initially ($C_1D_1 = CD$, $E_1F_1 = EF, \ldots$). If, on the other hand, Firm II with a QC circle programme manages to avoid the deterioration and build on the basis of the innovations through successive incremental improvements and modifications to the newly adopted products and processes, it may be able to follow the development path $ABC_2D_2E_2F_2 \ldots$ Even if the two firms adopt the same innovations at more or less the same time and with more or less the same direct effects ($C_1D_1 = C_2D_2$, $E_1F_1 = E_2F_2, \ldots$), Firm II can outdistance Firm I more and more as time passes due to the cumulative effect of successive small improvements accruing from the QC circle programme. As Dertouzos *et al*. (1989: 74) point out, this can be large enough to overshadow the effect of technical breakthroughs, which seems to be exactly what happens in Japanese industries.

During a field survey in the Japanese automobile industry which I conducted towards the end of 1983, all the five major assemblers and some component manufacturers stressed that minor improvements at the workshop level originating from QC circles and suggestion schemes had a greater cumulative effect on labour productivity than microelectronics-based factory automation technology. To some extent this is because the average level of automation in Japanese industries before the advent of the new technology was much higher than in North America and Europe, and therefore the overall labour-saving effect of microelectronic machinery such as NC machine tools and robots has been much more limited in Japan than elsewhere.[24] Still, the impact of QC circles and suggestion schemes appears to be indisputable, and the above-cited statement by automobile and component manufacturers is not so surprising if one bears in mind the vast scale of workers' participation at these companies. At Toyota, for example, there were some 5,000 QC circles and 1.9 million suggestions, or 38.8 per employee, were made in 1982. Of those suggestions, 95% were adopted by the company.[25]

The most dramatic and most frequently cited estimate of the gains from QC circle activities is related to New Nippon Steel Corporation. In the wake of the first oil crisis of 1984, the company launched its 'Attack 1000' campaign with a view to cutting production costs by 1,000 yen per tonne of steel materials. At the core of the campaign was the JK programme. Reportedly, the company spent about 700 million yen per year on this programme and saved a net amount of 40 billion yen per year. This was equivalent to its annual profit.[26]

In the *QCCM* Second Survey, the average gains from the QC circle programme at 256 responding establishments were nearly one hundred times the average costs (external payments excluding wages and salaries and other internal costs): 231.5 million yen against 2.4 million yen per year.[27] Some company reports give quantitative information on 'returns from the small group activities',[28] but they do not give information on the costs. Even in the cases cited above, it is not at all clear to what extent the reported expenses and gains were strictly attributable to the QC circle programme. The case of New Nippon Steel is no exception. For this very reason the *QCCM* Study Subcommittee has stopped publishing such data.

QC circles may enhance the competitiveness of industries through (1) the development and production of low-cost devices and equipment, (2) more efficient use of existing equipment, e.g. through modifications to plant layout and work procedures, (3) the development of workers' capabilities (training effect) and/or (4) intangible attitudinal and organizational improvements.

As I already mentioned, in Japanese industries labour-saving and quality-enhancing devices such as chutes, loaders and unloaders, and stoppers are developed and produced in-house to meet specific needs of individual workplaces. This may be illustrated with reference to Mazda's experience. In June 1962, this company (Toyo Kogyo until 30 April 1984) launched a special automation programme based on such a strategy. In 1963, simple automatic devices were attached to 8 of a total of 11 machines on a cylinder production line. Two machines became redundant and the number of workers on the line was reduced from 9 to 3. Subsequently, similar changes took place on lines producing camshafts, pistons, rear axles, gears, etc. By the end of 1966, 624 devices had been produced, ranging from loaders to automatic assembly and heat treatment equipment. Processing of about 30 kinds of parts and components and transport (transfer) of some 80 parts and components had been automated.[29]

Mazda's experience is far from exceptional. With the rising educational standard, workers seem to have become capable of more difficult tasks. When I visited Honda's Sayama plant in 1983, they had a 'hand-made automation programme' and even robots were built as part of the QC circle activity.

The use of such simple devices has the important side effect that the firm can use its capital concentratedly in key areas. During my visits to European automobile and component manufacturers in the mid-1980s, I often noticed sophisticated robots were used simply to transfer workpieces. Robots with multiple degrees of freedom were required to perform complicated movements. In such cases Japanese engineers and workers in the QC circle would seek ways to simplify the movements, for instance by changing the plant layout. Thereby the requirements would be reduced to devices like chutes and simple manipulators, which the workers could build by themselves. The capital saved in this way would be spent on sophisticated machines where they are indispensable.

Obviously the difference does not accrue from the QC circle alone. In the West, investment decisions are predominantly influenced by engineers, who tend to pursue technological excellency or sophistication almost for its own sake. In Japan, in contrast, economic calculation – 'cost-performance ratio' – is more decisive.[30] At a major American company I recently visited, I was told that their new plant had been so overloaded with high-tech equipment that some had had to be removed after the plant came into operation. Japanese companies take meticulous care to avoid 'overspec (over-specification)' of their products and production facilities. This may be a matter of philosophy to some extent, but more importantly it is a question of work organization. In Japanese industries the views of production workers and marketing personnel are taken into account more fully in the choice of technology as well as in the development of new products.[31]

I have already referred to the rising quality of the Japanese workforce as an important factor encouraging the spread of the QC circle. The relationship is not unidirectional. In a survey conducted in 1982,[32] 31.3% of about 22,000 technical workers who had received vocational education and training in the past three years referred to the QC circle activity as a mode of training. The proportion rose from 17.1% in the firm size group with 30–99 workers to 29.3% in the group with 100–999 workers and 39.6% in the group with 1,000 or more workers. This is consistent with the result of ML (1990) cited in Table 16.6. The 'extremely large' on-the-job training effect of QC circle activities is also stressed in NIEVR (1986), which argues that they promote *tanôkô-ka* by more or less compelling individual workers to study the production processes before and after their own and to pay attention to supervisors' views and top management's policies. A respondent in this survey reported that the training effect had been particularly remarkable while the education level of the workforce was low (girls from junior high schools), as the ability of numerical data processing and analysis could be acquired through the QC circle programme.[33]

As for the main intangible effects of the QC circle, 36.5% of the respondents in *QCCM*'s Third Survey mentioned increased problem awareness; 24.0% improvements in productivity, quality, working conditions and so forth; 22.9% better human relations; 10.3% improvement of workers' capabilities, knowledge and skills. The order of importance of individual items naturally changes over time because of differences in the industrial composition of the sample, general conditions of the economy, etc. The second item above was ranked first in 1979 and the third item was at the top of the list in 1983. It may well be that as small group activity extends from manufacturing to the service sector its effect becomes more intangible. In a number of other surveys conducted by different organizations respondents invariably mentioned greater problem awareness and better human relations, team work, communication and work motivation as the main benefits.[34]

5. Workers' motivation

Although I have argued above that the Western image of the Japanese QC circle activity as unpaid voluntary work is not right, our data do not entirely invalidate such interpretation. Of the practising establishments in Table 16.3 15% did not pay overtime allowances for QC circle activities that took place outside working hours or paid only modest amounts. The proportion was almost certainly much higher in earlier years when Japanese firms were financially weaker. It is therefore relevant to say something here about the workers' motivation.

Interpreting Japanese workers' 'unremunerated' QC circle activity as a sign of loyalty to the company and devotion to their work, Western observers usually associate their behaviour with their Confucian culture. I believe that this is a mistake. As I will argue later, Japan's socio-cultural tradition may have some relevance, but only indirectly. Basically, the workers' motivation is economic.

A majority of Japanese workers are happy with their QC circle. In a survey of technical workers in the manufacturing sector which was conducted in 1982, for example, nearly 70% of the respondents confirmed this (Table 16.5). Asked about the reason, 56.9% of these respondents replied that the QC circle activity helped improve their capabilities; 51.3% said 'because we know the situation in our unit better than anybody else'; 23.8% felt that the circle provided a stimulus for work; and 20.9% were happy 'because it is good for the company'.

For our later discussion it is important to note that the proportion of 'very happy' and 'happy' workers increases with firm size. In Table 16.5 it rose from 54% in the smallest firm size group to 67% in the middle and to 85% in the largest. More workers at larger firms appreciate the training and educational effect of the QC circle programme than at smaller firms (Table 16.6).

Among the workers who replied either 'can't tell', 'unhappy' or 'very unhappy' (22.8% of the respondents), 36% either did not or could not explain

Table 16.5 Technical industrial workers' attitude to QC circle activities
(% of total respondents)

Firm size (no. of workers)	'Very happy'	'Happy'	'Can't tell'	'Unhappy'	'Very unhappy'	No answer
Total	32.4	36.4	18.8	3.3	0.7	8.4
1,000 or more	44.9	39.9	8.3	2.6	0.5	3.5
100–999	30.8	36.6	20.3	3.7	0.7	8.0
30–99	21.7	32.0	28.2	2.5	0.8	14.8

Notes: 1. The total number of effective respondents was 10,158
2. The gap between the total number of respondents and the sum of the respondents in the three size groups is presumably due to some respondents' failure to indicate the size of their firm

Source: VTRC, 1983: 192

Table 16.6 Reasons for being happy with the QC circle
(% in the total happy workers)

Firm size (no. of workers)	'Helps improve our capabilities'	'We know the workplace best'	'Gives a stimulus to work'	'Good for the firm'
Total	56.9	51.3	23.8	20.9
1,000 or more	61.0	49.1	28.0	17.5
100–999	54.7	51.5	20.3	23.7
30–99	53.2	54.2	22.5	22.2

Note: Marginal proportions (0.5 to 3.4%) of workers gave 'other' reasons, and 0.5 to 0.7% gave no reason

Source: The same as in Table 16.5

why. The percentage rose from 22.6% in the largest firm size group to 43.7% in the smallest. The most common explanations were 'this is a task that should be carried out by managers and/or technicians/engineers' (25.8%) and 'free time is reduced' (24.8%), followed by 'overtime allowances are too small' (11.2%) and 'the worker benefits little' (10.9%).

Clearly, a great majority of Japanese workers are motivated by personal benefits, not by loyalty to the company or devotion to their work. Still, one may argue that Japanese workers are, on the average, more 'loyal' and 'devoted' than their counterparts in Europe and North America and that this helps the QC circle programme. Even this loyalty and devotion are basically motivated by economic considerations.[35]

Long-term incentives

Japanese workers' acceptance of the QC circle and their so-called loyalty and devotion are largely attributable to the same factors: the life employment system, the egalitarian remuneration system and the promotion system that prevail among the larger sectors of Japanese industries.

Although the life employment system is now extending into the smaller sectors,[36] it is still more common among the larger firms, partly because of the generally greater instability and shorter life of small enterprises. So is the QC circle, as Table 16.3 testifies. The relationship between the two is clear.

The life employment system presupposes the worker's versatility and flexibility. He is not recruited for a given job. Transfer from one job to another and from one unit to another is routine. His promotion is closely linked to such job rotation, which entails *tanôkô-ka*. If the worker is incapable of accepting it, he has no chance. This is why both management and workers place so much importance on the training effect of the QC circle, especially at larger enterprises. As there is no narrowly defined job description and as tasks are shared by many workers,[37] moreover, workers are willing to give a hand to their colleagues whenever necessary. They are more company-oriented than job-oriented, as a matter of necessity.

At the same time, the life employment system enlarges Japanese workers' time horizon in their economic calculations. Even where the immediate reward is lacking, they work hard counting on the rewards that they will reap during their subsequent career, notably job security and promotion that can be assured only as long as the company survives and prospers. Severe competition among Japanese firms and among workers within a firm reinforces such attitudes. Competition among workers does not preclude their mutual cooperation, e.g. in the QC circle, because an uncooperative attitude will endanger their annual pay increase, biannual bonuses and promotion through a poor report from their supervisor.

From the viewpoint of workers' morale, the importance of the egalitarian remuneration system cannot be exaggerated. Earning differentials between ordinary workers and those at the top of the company hierarchy are small in Japan. Although the lack of systematic information forbids a precise comparison, one source estimates that the average annual gross earnings of presidents of major Japanese companies were 5.7 times higher than those of all workers (22.3 million yen against 3.9 million yen) in 1985. In the same year, executives of major American companies earned, on the average, 33.5 times more than the average of all workers (US$703,891 against US$20,991).[38] Moreover, Japanese executives and workers share the effects of ups and downs in the company business through their annual increments and biannual bonuses. In fact, companies reduce their losses by the executives' remunerations before those of the ordinary workers. At the same time, the dividend–profit ratio in Japan is low, partly due to the heavy reliance upon bank loans: during the late 1970s and early 1980s the ratio of dividends to gross profits in Japanese industries was roughly half the level in the United States (11.8% against 21.2% in 1982).[39] Consequently, workers have no reason to suspect that they work hard only to enrich the executives and shareholders.

The relevance of the QC circle activity to the worker's long-term welfare should now be clear. I need only add that while the QC circle provides individual workers with opportunities to demonstrate their managerial capabilities and leadership, it also helps develop such abilities. In consequence, active participants tend to have a better chance of promotion.[40]

Shorter-term incentives

These long-term incentives and calculations alone might not keep so many workers sufficiently enthusiastic about the QC circle. Smaller firms, moreover, cannot always assure their employees of life employment. To supplement the long-term incentives and fill the gap where they are lacking, there are at least two short-term incentives.

NIEVR (1986) argues that Japanese firms have succeeded in providing incentives by linking the QC circle programme with the suggestion scheme. Under the latter scheme, if a suggestion made by an individual worker or a circle is adopted, the originator is awarded a prize. In many companies the suggestion scheme existed before the QC circle. For example, in the Ministry of Labour's 1977 survey on management – labour communication, 74.1% of the responding establishments had a suggestion scheme, as compared with 40.4% having QC circles. However, the suggestion scheme had rarely functioned effectively and often fell into decay, until it acquired a group effort element through linkage with the QC circle programme. Most of the awards amount to only a few hundred to a few thousand yen, but for a really significant suggestion the award can be considerable. NIEVR (1986) refers to a precision machinery manufacturer, who usually paid 300 to 10,000 yen per suggestion and 100,000 to one million yen for a suggestion leading to a patent. Workers usually pool small awards to use for their social activities such as parties and excursions.

In the 1990 survey by the Ministry of Labour, the spread of the suggestion scheme was almost exactly the same as that of the QC circle both in extent and firm-size distribution (cf. column 2, Table 16.3). On average, over 9,000 suggestions were made per establishment over the one-year period preceding the survey, the actual number ranging from nearly 15,000 in the largest firm size group to a mere 34 in the smallest group. Most suggestions were concerned with the rationalization of production or office work (81% of the establishments with a scheme), followed by safety and health (53%), workers' welfare (36%), development and marketing of new products (35%), workers' education and training (21%) and organizational improvements (16%).[41]

Another important incentive to the QC circle activity is the competitive presentation of results. Individual companies organize such presentations regularly at department, establishment and enterprise level. The nine regional branches of QC Circle Headquarters organize regional or sub-regional presentations, at which the winner circle is invited to a national convention. These meetings are aimed at sharing the participants' experiences and encouraging efforts at greater achievements. The system provides incentives not only in the form of awards, but also by offering opportunities for travel. Japanese workers seem to take special delight in this.

Socio-cultural factors

I have argued that Japanese workers' motivation is basically economic. This does not mean that socio-cultural factors have no influence. Indeed, one might well say that the life employment system and the egalitarian remuneration system themselves reflect Japan's cultural tradition.

The historical tradition of communal responsibility may have some influence. Under the Tokugawa Shogunate (1603–1867), every five households in rural and urban areas were organized into a unit for administrative and taxation purposes. Land tax and other duties were borne collectively by the member households. In a considerably reduced form this tradition has survived as the *tonarigumi* (neighbours' group) or more recently *jichikai* (self-governing group) for mutual assistance in daily life and for the communication of administrative decisions. In the QC circle one clearly recognizes a trace of this tradition. As one may suspect, the system works partly because each member jealously ensures that the others fulfil their share of collective duties. It is also true that, as is often pointed out, the Japanese have a tendency to conform. Even if they feel that something is irrational, they tend to accept it so long as people around them do not protest. This is why some workers feel uneasy about the QC circle 'as it is not quite "voluntary"'.[42]

My argument above fits perfectly with the popular image of group-oriented Japanese workers held by Western observers, but the Japanese tradition seems to be responsible also for a contrary feature. I have already touched upon the attitudinal difference between Japanese and US managers regarding workers who initiate changes to an existing system. Unless Japanese workers are prepared to take such initiatives, the idea of the QC circle will not work. It is in this context that the following findings of Whitehill and Takezawa (1968: 178–87) are interesting, presenting images of Japanese and American workers which are the opposite to the popular perception.

When the researchers asked what a supervisor should do when work methods were to be changed, 73% of the Japanese workers and 66% of the American workers said that he should 'first ask workers for their suggestions regarding proposed changes, and then decide what to do'. When the workers were asked what they would do if they felt a change in their work methods or assignments was unjustified and yet could not convince their supervisor, 25% of the Japanese, as against 7% of the Americans, replied that they would resist his orders unless they were convinced by their co-workers that they were wrong. Of the Japanese 26% replied that they would 'obey his orders since, as supervisor, he must have the authority to give orders'. In contrast, 50% of the American workers chose this alternative. On communal or social matters, Japanese workers tend to be quiet and more readily accept their superiors' authority. On technical matters closely related to their own work, however, they speak more freely and claim a greater authority, and their superiors accept this. It is tempting to associate this with the high social status traditionally enjoyed by the craftsman.

Conclusions

Japanese industries have managed to catch up with Western competitors by rapidly improving productivity over the last few decades. To this end they have maintained high rates of capital investment and technological innovation. The effect of their efforts has been considerably enhanced by small but continuous improvements resulting from QC circle activities. QC circles develop and produce low-cost devices and equipment, find ways to use existing equipment more efficiently, develop workers' capabilities and contribute to attitudinal and organizational improvements. Such small improvements can cumulatively have a greater effect on labour productivity and product quality than major technological innovation such as robotization. At the same time, major innovations are encouraged by QC circles which permit concentrated use of available capital in key areas.

The Western notion of the Japanese QC circle is mistaken in a number of respects. First, QC circle activity without compensation is rare. Second, in a great majority of cases it is organized by the company as a part of the 'total quality control' programme involving all workers from the top management down to the ordinary production workers. It is therefore not 'voluntary', although the operation of the circle including the choice of its themes is largely left to the circle members' spontaneous decisions. Third, it is also wrong to ascribe the motivation of Japanese workers in QC circle activities, and their 'loyalty' and 'devotion' to their company and work, mainly to their socio-cultural background.

Japanese workers are motivated primarily by economic incentives and calculations. The crucial factors in their acceptance of, and enthusiasm for, the QC circle are the life employment system, the egalitarian remuneration system and the promotion system. Japan's socio-cultural tradition has some influence, but only indirectly through these economic institutions and possibly also in other, less important, ways. Recognition of the essentially economic motivation of Japanese workers is important, because economic institutions are much more malleable than the socio-cultural framework of a society.

As part of their efforts to reduce the competitive pressure from Japan, various industries in the West are now seeking to transplant Japanese methods. The QC circle seems to be one of the last to take root. This is unfortunate in view of its potential for improving industrial efficiency in the West. By correcting their notion of this practice in Japan Western industries may be able to pave the way for its successful transplantation.

Notes

1. Watanabe, 1991.
2. Dertouzos *et al.* 1989, p. 70.
3. For a more detailed discussion on this point, see Watanabe, 1986, pp. 253–5 and *idem*, 1987a, ch. VIII, pp. 188–90.
4. Dertouzos *et al.*, *op. cit.*, pp. 74–6.
5. A stopper is a simple device used for the purpose of interrupting the operation of an automatic machine whenever something starts going wrong.
6. Seering, 1987, p. 27. Recently, an international group of researchers organized at MIT coined the term 'lean production system' to characterize this aspect of Japanese industries. See Womack, Jones and Roos, 1990.
7. NIEVR, 1986, p. 2.
8. Imai, 1986, p. 9.
9. Onglatco, 1988, pp. 13–4. See also Dertouzos *et al.* 1989, pp. 75–6 for a similar argument in a broader context.
10. Cf. Watanabe, 1991.
11. Onglatco, op. cit., p. 109.

12. Shimada and MacDuffie, 1987, cited in Lawrence, 1990, p. 29; Aoyama, 1988, p. 55; and Dertouzos *et al. op. cit.*, p. 76.
13. Ueda, 1980, p. 229.
14. Juran, 1987.
15. Oliver and Wilkinson, 1988, pp. 80 and 95.
16. Aoyama, op. cit., p. 102.
17. Oliver and Wilkinson, 1988, p. 59.
18. *QCCM* Study Subcommittee, 1987, p. 7 and *idem*, 1988, p. 20.
19. There *are* cases where workers take the initiative.
20. ML, 1990, p. 62.
21. NIEVR, 1986, pp. 4–8.
22. Cf. Watanabe, 1991.
23. *QC Circle Magazine*, No. 319, December 1988, pp. 4–27. Underlying these estimates are the 1982 population of establishments with 30 or more workers amounting to 222, 113.
24. Cf. Watanabe, 1986 and idem, 1987a, chs. III and VIII.
25. Toyota, 1984, p. 22. A large majority of the adopted suggestions are, of course, trivial. Companies use the suggestion scheme partly as an incentive to thinking and initiative. At Toyota, it is currently organized under the slogan 'Good Thinking, Good Products'.
26. Nitta, 1978, p. 23.
27. *QCCM*, 1983, p. 23.
28. See Onglatco, 1988, for examples.
29. Toyo Kogyo, 1972, pp. 400–1.
30. Here one recalls Wells' distinction between 'economic man' and 'engineering man' in the choice of production methods. See Wells, 1975.
31. For a discussion on this point in the context of the Japanese NC machine tool industry, see Watanabe, 1983.
32. VTRC, 1983, p. 202.
33. NIEVR, 1986, pp. 15 and 36.
34. Cf. Onglatco, 1988, pp. 44–5.
35. For a more detailed discussion on this subject, see Watanabe, 1987b.
36. EPA, 1986, pp. 138–9.
37. Cf. Watanabe, 1991.
38. An internal document of the Economic Research Department of Japan Federation of Employers' Associations (NIKKEIREN) which is based on data from the following sources: Rôdô Gyôsei Kenkyûjo, *Nenkan Chingin, Shôyo no Jittai* (Survey of annual wages and bonuses), annual; Seikei Kenkyûjo, *Yakuin no Hôshû, Shôyo, Nenshû* (Corporate executives' remunerations, bonuses and annual earnings), annual; US Department of Commerce, *Statistical Abstract of the U.S.A.*, 1987; and *Business Week*, 4 May 1987.
39. Maruo, 1986.
40. NIEVR, 1986, pp. 140–3.
41. ML, 1990, pp. 69–71.
42. MI, 1985, appendix p. 121.

References

Aoyama, Sigeki (1988), 'Nihon Kigyô no Takokuseki-ka to Koyô-Rôdô Mondai: "Nihon-teki Keiei" no Tekiyô Mondai o chûshin ni (Japanese firms' multinationalization, employment and labour, with special reference to the applicability of "Japanese-style management")', Magazine Keizai (ed.), *Nihon Kigyô*

Kaigai Shinshutsu no Jittai (Actual conditions of Japanese firms' overseas investment), Tokyo, Shin Nihon Shuppan.

Dertouzos, M.L., Lester, R.K., Solow, R.M. and MIT Commission on Industrial Productivity (1989), *Made in America: Regaining the Productive Edge*, Cambridge, Mass., MIT Press.

EPA (Economic Planning Agency, Keizai Kikaku-chô) (1986), *Gijutsu Kakushin to Koyô* (Technological innovation and employment), Tokyo, Printing Bureau of the Ministry of Finance.

Imai, Masaaki (1986), *Kaizen: The Key to Japan's Competitive Success*, New York, Random House Business Division.

Juran, J.M. (1987), 'QC circles in the West', a foreword to QC Circle Headquarters, *QC circle Katsudô 25-nen-shi* (A 25-year history of QC circle activities) (Tokyo), the title in English and the text in Japanese.

Lawrence, Robert Z. (1990), 'Foreign-affiliated automakers in the United States: An appraisal', Washington, DC, Brookings Institution, March.

Maruo, N. (1986), 'Nihon-gata Keiei to Seika Haibun – Incentive system to shite no Hyôka (Japanese-style management and profit-sharing – An appraisal as an incentive system)', *Nihon Rôdô Kyokai Zassi*, September.

ML (Rôdô-shô, Ministry of Labour) (1985), *Rôdô Hakusho* (White Paper on Labour), Tokyo.

ML (Rôdô-shô, Ministry of Labour) (1990), *Nihon no Rôshi Communication no Genjô* (A fact-finding survey on the labour-management communication in Japan), 30 July.

NIEVR (National Institute of Employment and Vocational Research (Koyô Sokushin Jigyô-dan, Koyô Shokugyô Sôgô Kenkyôjo) (1986), *Nihon ni okeru Shô-shûdan Katsudô no Jittai to sono Tenkai Jôken ni kansuru Jirei Kenkyû Hôkokusho* (Case studies on the current situation of small group activities and conditions of their development), Tokyo, March.

Nitta, Michio (1978), 'Tekkôgyô no 'Jishu Kanri Katsudô ("Self-management activities" in the iron and steel industry)', *Nihon Rôdô Kyokai Zassi* (Monthly Journal of the Japan Institute of Labour), September.

Oliver, N. and Wilkinson, B. (1988), *The Japanization of British industry*, Oxford, Basil Blackwell.

Onglatco, Mary Lou Uy (1988), *Japanese Quality Control Circles: Features, Effects and Problems*, Tokyo, Asian Productivity Organization.

QC Circle Magazine, QC Circle Katsudô Kenkyû Sho-Iinkai (Study Subcommittee on QC Circle Activities) (ed.) (1983), *QC Circle Katsudô no Jittai* (The second fact-finding survey on QC circle activities), Tokyo, Hinshitsu Gekkan Iinkai.

QC Circle Magazine, QC Circle Katsudô Kenkyû Sho-Iinkai (Study Subcommittee on QC Circle Activities) (ed.) (1987), *Dai-sankai QC circle Katsudô no Jittai Chôsa* (The third fact-finding survey on QC circle activities), *idem.*

QC Circle Magazine, QC Circle Katsudô Kenkyû Sho-Iinkai (Study Subcommittee on QC Circle Activities) (ed.) (1988), 'QC Circle Katsudô no Fukyû Jittai Chôsa Hôkoku (A survey on the diffusion of small group activities)', *QC Circle Magazine*, no. 319, December.

Seering, Warren (1987), 'Robotics, numerical control and the computer', in Watanabe (ed.) (1987a) below.

Shimada, Haruo and MacDuffie, John Paul (1987), 'Industrial relations and "humanware"', briefing paper for the First Policy Forum MIT, mimeo.

Suzaki, Kiyoshi (1987), *The New Manufacturing Challenge: Techniques for Continuous Improvement*, New York, The Free Press.

Toyo Kogyo (Mazda) (1972), *1920–1970 Toyo Kogyo 50-nen-shi* (1920–1970, 50 years of Toyo Kogyo), Hiroshima.

Toyota Motor Corporation (1984), *Outline of Toyota 1984*.

Ueda, Toshio (1980), *Seisansei to Shô-shûdan Katsudô* (Productivity and small group activities), Tokyo, Nihon Seisansei Hombu (JPO).

VTRC (Shokugyô Kunren Kenkyû Center (Vocational Training Research Center)) (1983), *Mechatronics Jidai no Jinzai Kaihatsu* (Human resources development in the age of 'mechatronics'), Tokyo, Printing Office of the Ministry of Finance.

Watanabe, Susumu (1983), 'Market structure, industrial organisation and technological development: The case of the Japanese electronics-based NC-machine tool industry', World Employment Programme working paper WEP 2-22/WP. III, Geneva, ILO, February.

Watanabe, Susumu (1986), 'Labour-saving versus work-amplifying effects of micro-electronics', *International Labour Review*, May–June.

Watanabe, Susumu (ed.) (1987a), *Microelectronics, Automation and Employment in the Automobile Industry*, Chichester, John Wiley.

Watanabe, Susumu (1987b), 'On socio-institutional conditions of Japan's modernisation', *Political Economy* (Rome), vol. 3, no. 2.

Watanabe, Susumu (1991): 'The diffusion of new technology, management styles and work organisation in Japan: A survey of empirical studies', in OECD, *Technical Change as a Social Process: Society, Enterprises and Individual* (Paris).

Wells Jr, Louis T. (1975), 'Economic man and engineering man: Choice of technology in a low-wage country', in Peter Timmer *et al.* (eds), *The Choice of Technology in Developing Countries: Some Cautionary Tales*, Cambridge, Mass., Harvard University Press.

Whitehill Jr, Arthur M. and Shin-ichi Takezawa (1968), *The Other Worker: A Comparative Study of Industrial Relations in the United States and Japan*, Honolulu, East-West Center Press.

Womack, James P., Jones, Daniel T. and Roos, Daniel (1990), *The Machine That Changed the World*, New York, Rawson Associates.

17 Skill flexibility, labour market structure, training systems and competitiveness*

David Marsden

Introduction

The impact of workers' skills on economic performance has long been recognized. As early as the late nineteenth century Alfred Marshall warned of the threat to British economic strength arising from the superiority of the German system of technical and scientific training (Marshall, 1920). Much of the debate about skills and economic performance has concerned the quantitative contribution of education and training and economic growth. However, of equal importance is the contribution of training systems to the adaptation of the skills of a country's work force to new economic circumstances, which is the focus of this chapter. Scope for such adaptation can be seen in the flexibility of the jobs in which skills are used, the adaptability of the content of skills and jobs, and in the adaptability of the institutional frameworks within which skills are formed both outside and inside the enterprise. In recent years, one of the main concerns of such adaptability has been organizational and technical change in response to changing economic demands.

The argument of this chapter is that adaptation of training systems to economic change is constrained by their links with the labour market structures into which they are integrated. These labour market structures contain their own models of flexibility which differ from one type of structure to another so that flexibility in one dimension is often at the expense of flexibility in another. The problem of institutional reform is not that of devising the one best model of skill development for all labour market environments, but rather to identify the constraints and possibilities offered by the labour market structures prevailing in a given sector in a particular country.[1]

The history of training systems provides many examples of attempts to change which have either fallen far short of expectation, or more interestingly, evolved in unexpected ways. In France, attempts to set up a German-style industrial apprenticeship system in the 1930s failed, and current efforts to increase the quality of skills by training young people in state vocational

* This chapter draws on joint work with Paul Ryan (Kings College Cambridge), and with François Eyraud (now at the ILO) and Jean-Jacques Silvestre (Laboratoire d'Economie et de Sociologie du Travail, Aix-en-Provence) some of which was funded by the Rowntree Memorial Trust, and some by the CNRS and the ESRC.

schools have encountered problems of qualified young job seekers unable to find skilled jobs. In Britain, the reform of apprentice training in the 1960s and 1970s resulted in the creation of a high-quality system, but which few employers wanted, or could afford to support because of its cost. The Youth Training schemes have encountered immense difficulties in providing paths to skilled jobs. In contrast to Britain and France, West Germany seems to offer a picture of relative success. There apprenticeships were also reformed in the 1960s and 1970s, notably increasing flexibility by reducing the number of specializations, and by voluntary action by the parties involved to expand the number of training places.

It will be argued that the source of many such problems lies in the link between training systems and labour markets. The chapter starts by setting out the argument concerning the link between training and labour market structure, then explores the adaptability of training and labour markets, and considers their impact on economic performance. In all, the chapter is mainly confined to the industrial sector of the economy.

Vocational training systems

Intermediate-level vocational training falls into three broad categories, each of which can be found in Western Europe: mixed on-the-job and off-the-job training in the form of apprenticeships; off-the-job training in specialized vocational schools; and enterprise based on-the-job training. To these, one should perhaps add the large number of unqualified workers who benefit from none of the above.

Enterprise-based training, and apprenticeship

Apprenticeship training usually involves a number of elements. First, it provides on-the-job training (OJT) standardized across firms providing such training, and second, theoretical instruction, usually dispensed off the job. Third, apprentices usually enjoy a special status within the enterprise recognized by their peers, the unions and the employer. Apprenticeship is widely used in West Germany for both blue- and white-collar workers, and in both industry and the services. It is also the main path to skilled industrial work in Britain, but is confined mainly to manufacturing and construction, to blue-collar jobs, and to males.

In both Britain and West Germany apprenticeship training underwent reform and modernization between the late 1960s and the early 1970s. In both countries, the reforms sought to increase the amount of theoretical training in schools, in the belief that this led to higher quality, and provided more adaptable skills for the future. In West Germany, although employers and unions retained control over the apprenticeship system, the state intervened by boosting the provision of special vocational schools which dispense the theoretical component of training. Despite the poor press of British apprenticeships, similar moves were undertaken in certain industries, notably the

engineering industry,[2] greatly increasing the amount and quality of theoretical instruction to British engineering apprentices (Venning *et al.*, 1980), but not all of the sectoral training boards were as farsighted or as successful (one reason for the abolition of the majority of industrial training boards shortly after Mrs. Thatcher's election).

Apprenticeship in France accounts for a fair volume of training: in France in 1981–2 there were about 220,000 apprentices (Combes, 1988a). The reform of apprenticeship in 1971 treated it as one of the paths by which young workers could obtain their CAP (*Certificat d'Aptitude Professionnelle*), the other being through vocational schools (see below). Consequently, apprenticeship training is only recognized by employers if it has led to successful completion of the CAP exams, and it has tended to be the least favoured route among young people, and the one with the highest failure rate. Subsequent legislation in 1987 has sought to open up higher-level diplomas to those following an apprenticeship (Combes, 1988b). In Italy, although there is a large number of apprentices in the small-firm artisan sector, the duration is generally fairly short, and its quality is generally believed to be mixed (ISFOL, 1982).

School-based vocational training

Intermediate-level vocational training may also be provided outside the enterprise in specialist state or private schools. In this case, the training is usually sanctioned by a diploma which provides a good degree of standardization of the theoretical knowledge imparted. One of the problems encountered by this kind of training is that it often does not contain much practical experience, and that gained by young workers after qualifying is usually much less homogeneous than that included with an apprenticeship.

The predominant school based vocational qualifications in France are the intermediate-level CAP and BEP (*Brevet d'Etudes Professionnelles*), mostly lasting two years, and the various streams of the technical *baccalauréat*, which provides the backbone of technician training. The volume of all these has increased over the last 15 years, that of the CAP and BEP now running at over 150,000 a year (Mouy, 1983).

Reliance upon private-training provision by individual employers

A third method of skill formation is by the accumulation of experience on the job of the kind described by Doeringer and Piore (1971). This may complement the other two methods, or it may, alternatively, stand on its own. Such OJT is distinguished from that of apprenticeships by the fact that it is generally specific to the individual enterprise providing the experience, and the skills thus formed can vary greatly from one firm to another even within the same sector of activity.

The degree of reliance upon individually organized employer training provision has varied considerably between countries. During the rapid industrial growth in France and Italy in the 1950s and 1960s, great reliance was placed upon on-the-job training organized either formally or informally by individual employers. This has been one of the reasons for the strength of internal labour markets in these two countries. In both of them, the problems of industrial restructuring highlighted the problems of non-transferable skills, and some of the public policies of the 1970s and 1980s can be understood as attempts to create more adaptable skills.

In many countries, there are signs that, in recent years, employers have sought to increase the volume of company level training, to rely less upon the old informal methods of on-the-job training, and to be more systematic in training for non-transferable skills. One influence has been the experience of enterprise restructuring during the late 1970s and early 1980s in which retraining was seen as one of the keys to enhanced work force flexibility (Atkinson and Meager, 1986). A second has been the spread of new technology and the accompanying increase in the demand for both quantity and quality of skilled labour. A third has been the spread in the popularity among managers of 'Human Resource Management' which places a greater emphasis upon training as a key to quality in production than in the past (Guest, 1988).

The unqualified

One of the serious failings of many countries' educational and training systems (West Germany is an exception), is the high percentage of young people with neither general educational nor vocational qualifications. In Britain and France, for example, the figure is about 40%. Such people are almost entirely dependent upon what they can learn on the job, which is usually not transferable. In France, the strong position of internal labour markets and the practice of upgrading does at least provide a limited second chance (those with the CAP or BEP have a much higher probability of such upgrading), whereas in Britain, the more limited use of upgrading in many major sectors leaves them with little prospect beyond that of a life of largely unskilled work.

Labour market structures

Sectoral labour markets are dominated broadly speaking by one of three types of labour market, occupational (OLMs), internal (ILMs), and unstructured or secondary, labour markets.[3] OLMs are particularly widespread in West Germany, and, for skilled manual workers, in Britain, whereas internal labour markets provide the predominant model in France and Italy, and in many countries for managerial employees.

The occupational model requires a set of transferable skills and fairly standardized job descriptions or job slots, otherwise there can be no market for transferable skills. They also require a standardized system of training and skill development. Concentrating investment in such skills early in a person's working life will increase the period over which a return can be earned, so that methods of training such as apprenticeships would seem to be particularly appropriate to this kind of labour market structure. In the standard human capital analysis, the trainee should bear the cost of training for skills in the form of a trainee wage below the value of his or her output (Becker, 1975).

This solution is particularly fragile in an environment in which skilled adult workers fear substitution by cheap trainee labour (Ryan, 1987; Marsden and Ryan, 1990). Consequently, in practice, employers frequently bear a large part of the net cost of such training, thus creating scope for free-rider problems which, if unchecked, can undermine the supply of skilled labour and break up the occupational market as employers devise ways of restricting skilled mobility (Marsden, 1986). For this reason, a strong institutional framework is essential to such markets for their long-run viability. Successful apprenticeship systems need a framework which can reduce skilled worker fears of substitution, thus rendering a special low trainee rate of pay more acceptable, and reduce the dangers of free-riding among employers. The difficulty and cost of establishing such a framework explains why OLMs tend to exist for a limited number of occupations, and to cover fairly broad sectors of activity. Thus, apprenticeship involves much more than a simple combination of OJT and theoretical training.

Internal labour markets, in contrast, rest on skills which for technical and institutional reasons are not transferable, and because of the lesser risk of wastage by skilled workers employers have greater freedom in how they organize training and the build-up of work experience. But once workers have invested in such skills they too may fear substitution from outside, and thus will seek to ensure that the rules by which they reached their skilled status apply to subsequent new entrants into the firm.

The third type, that of unstructured labour markets, typically involves little training or skill beyond the general skills provided by basic secondary education. Workers can move fairly easily between firms because neither they nor their employers have invested in developing their skills, and they remain confined to unskilled jobs. In recent years, such labour markets have to some extent been favoured indirectly by public policies seeking to reduce youth unemployment as these have sought to 'price' certain categories of workers back into jobs either by lower pay, or by exclusion from certain levels of employee protection.[4] Although the special training and employment measures for young workers, which have often involved employers taking on young people outside the normal employment relationship, have improved youth access to jobs and training places, at the lower end of the quality spectrum they have also boosted secondary labour market conditions.

The link between training systems and labour market structures may be further explained by a consideration of why general school-based qualifications have not usually given rise to OLMs. In theory, such qualifications should be transferable because the theoretical knowledge dispensed is standardized, and the quality of attainment is certified. Yet many studies attest to the integration of such qualifications into ILMs, for example, engineers in Britain (Mace, 1979), secretaries and computer programmers in the United States (Osterman, 1982), and the experience of young qualified workers with the French CAP and BEP diplomas (Coeffic, 1987; Marsden and Germe, 1991).

Several factors militate against such qualifications being sufficient for creating OLMs. First, they generally do not contain much in the way of work experience, the acquisition of which is costly. Second, so long as the cost falls on individual employers they have little incentive to make the resulting skill transferable. Third, workers already in skilled positions have little incentive to accept that such workers should be on special trainee rates because there is no general framework regulating the work of the inexperienced new recruits. Fourth, unless there is already a suitable OLM, incumbent workers will be working under ILM conditions, and so will be reluctant to accept the employment of qualified workers ahead of them in the queue for promotion.

There is now quite a body of evidence suggesting that one or other of these models prevails over large sectors of employment in various EEC countries. The comparison of a matched sample of manufacturing plants in France and West Germany by Maurice *et al.* (1978 and 1985) showed clearly the influence of internal enterprise norms on skill acquisition and skill organization in France, compared with that of the apprenticeship-based organization of skills and skilled work in the German firms. This work has been extended by plant-level studies to Britain (Maurice, Sorge and Warner, 1979), and using a broad range of aggregate statistical indicators Marsden (1990) has shown that for industrial workers the internal market model predominates in industry in France and Italy, and the occupational model, in West Germany and in Britain.[5] It appears also that some of these differences are longstanding (e.g. Lorenz, 1987).[6]

The extent to which the two models can coexist within the same sector is limited by the fact that using an occupational market for skilled labour effectively truncates upgrading possibilities for unskilled and semi-skilled workers. Moreover, the dependence of occupational markets on a strong institutional framework implies high fixed costs in their establishment, and so limits the number of such markets in an economy.

Flexibility of job structures and deployment

The first type of adaptation concerns the flexibility of the job structures created by training systems and their related labour market structures in response to changing pressures of labour demand.

As the debate about labour force flexibility has evolved it has become clear that one can distinguish two broad types: flexibility of deployment between jobs and functions within the enterprise, and flexibility of employment levels, sometimes referred to respectively as 'functional' and 'numerical' flexibility (e.g. Atkinson and Meager, 1986). These two types of flexibility are to a considerable extent substitutes in that functional flexibility requires a willingness to learn a mixture of skills which is unlikely to be transferable to other firms, implying a risk of loss of earning power to such workers if they are laid off. Equally, workers with transferable skills, such as those with apprentice-trained skills in Britain and West Germany, have less incentive to accept deployment to areas outside their main competence because it will entail the development of a non-transferable component of their skill, and thus weaken one of the prime sources of their individual bargaining power – namely their potential to quit.

Some evidence of different labour turnover patterns between the countries with a predominance of ILM and those with OLM-based skills (respectively France and Italy, and Britain and Germany) can be detected in length-of-service patterns among manual workers (shorter average length of service in Britain and Germany than in France and Italy: Marsden, 1990), labour turnover patterns (higher turnover in Britain than in France: Eyraud, Marsden and Silvestre, 1990), and from employment functions (ECE, 1982; OECD, 1989), all of which indicate a greater sensitivity of employment levels to output in the OLM countries (Britain and Germany) as compared with the ILM countries (France and Italy). As concerns the employment functions, analyses by the ECE (1982: 53) for 1960–79 gave short-term elasticities of 0.197 and 0.103 respectively for France and Italy, and of 0.451 and 0.291 for FR Germany and the United Kingdom. Updated estimates for 1981–7 by the OECD (1989: 41) present a similar picture: 0.04 and 0.08 for France and Italy, and 0.16 and 0.14 for FR Germany and the United Kingdom.

Thus the OLM and ILM models appear to enjoy differing degrees of numerical flexibility, the more flexible employment levels being found in British and West German manufacturing.

More difficult to pin down is the possibility that the two countries in which ILMs predominate benefit from a higher degree of functional flexibility. One piece of evidence may be gleaned from a comparison of flexibility agreements in Britain and France (Marsden and Thompson, 1990; Eyraud, Marsden and Silvestre, 1990). In so far as these may be considered to reveal the negotiated work rules whose change employers believe holds the key to greater productivity, one may use their content as a clue to the nature of existing systems of job regulation. During the 1980s, the main emphasis of such negotiations has been upon job demarcations and changing rules which inhibit the deployment of labour in Britain, and on working time regulation in France. As job demarcation rules control access to skilled work they are closely related to the regulation of OLMs for skilled workers. In France, such preoccupations have been absent from flexibility negotiations, suggesting that negotiated constraints on functional flexibility are fairly limited.

A second may be derived from differences in the ability to adapt to incremental technical change of the kind which prevailed during much of the period of post-war growth. The prolonged rate of productivity growth in France as compared with Britain is consistent with the idea that ILMs, by virtue of the greater ease with which additional knowledge can be added onto existing skills enabling redeployment without loss of skill transferability, have favoured adaptation to incremental technical change of the type that prevailed during the 1960s and 70s. Equally, the threat that such adaptation poses to the demarcation boundaries around skilled work for OLMs appears consistent with the difficulty experienced by UK firms in achieving the rates of productivity growth found in much of continental Europe.

Despite a predominance of OLMs, West German industry appears to have enjoyed high rates of productivity growth, and a higher rate of functional flexibility than its British counterpart. A comparison of labour and technical change in British and West German firms (Jacobs *et al.*, 1978) suggested that West German managers benefited from a higher degree of functional flexibility than their UK counterparts. One contributory factor is that West German skills are more broadly based than those in the United Kingdom, having a stronger theoretical component. Another is that the strong role of the works councils in supervising workplace training provides skilled workers with an effective degree of protection of against the erosion of their skill base. This provides an alternative to regulation of the boundaries of skilled work by reference to detailed job demarcations of the kind found in the United Kingdom which are jealously guarded by work groups.[7] A third factor is that a good deal of further training (*weiterbildung*) has been focused on skilled workers and this would provide a source of functional flexibility by enabling incumbent skilled workers to broaden their skills. The works councils have a good deal of control over such further training, and so may be able better to ensure that skilled workers' interests are not undermined. Where such skills are certified, then they may work as additional elements to the basic apprenticeship, and so pose no threat to skill transferability. Nevertheless, the dangers for skilled workers of accummulating enterprise specific skills and thus losing skill transferability were highlighted by a survey of works councils (Maase, Sengenberger and Weltz, 1975).

Thus, there is some indication that ILMs are better at providing functional flexibility, and OLMs at providing numerical flexibility, but the German case illustrates that in practice the question may not be as black and white as suggested by the theory of labour market structure.

Adaptability to new demands

Technical and economic change require gradual, and sometimes dramatic, changes in the nature and content of skilled jobs. Hence, another way in which skill systems may contribute to relative economic performance is by the greater or lesser ease with which they adapt to such changes. This problem may be posed in two ways: the first, dealt with in this section, concerns the adaptation of job categories and the content of training, and the second, dealt with in the next section, deals with the adaptation of training and labour market institutions.

In an era of rapid technical change, one might expect internal labour markets to prove more responsive in the short run than occupational markets. As mentioned earlier, internal labour markets are essentially private goods owned by the firm and its workforce. Both have a strong interest in rapid adaptation for survival and for competitive advantage. In contrast, OLMs are more akin to public goods, and adaptation of the skill categories they distribute requires coordination between a large number of different users, workers and firms, many of whom may be at different stages in adopting technical innovations. In either case, adaptation has to take account of the distribution of interests among incumbent workers. In the ILM, incumbent workers have a strong interest in employment security in the firm, and in job and earnings progression, all of which may be disturbed by technical and organizational change. In the OLM, incumbent workers may attach less importance to job security, but job boundaries and their adaptation may prove more contentious.

In ILMs in France, recent research on technical change and on flexibility has highlighted that much of the change in the workplace has come about through the adaptation of existing skills and jobs to the new technical requirements rather than by the introduction of radically new categories of skills. In doing so, the firms have sought to adapt existing job structures and job classification systems to the demands of new technology, and have divided up the tasks associated with the new technology between existing categories of workers (Maurice *et al.*, 1988).

Repercussions of new technology can also be seen at the national level in the agreements to revise industry-wide job and pay classification systems to add a new, intermediate, category of highly skilled manual worker (*technicien d'atelier*) to provide promotion opportunities to the existing skilled workforce, but also to compensate for their declining chances of promotion to technician level. Thus, one might see these changes as seeking to adapt while preserving the logic of internal labour markets.

In an OLM environment, in the United Kingdom, one of the main concerns of skilled workers' unions has been to find ways of dividing work associated with microelectronic technology between existing categories of workers. For example, how should elementary programming tasks be divided between semi-skilled workers, existing crafts, and technicians? New technology has also brought into question some of the long-standing criteria for dividing work between different categories of skilled workers (e.g. between electricians, fitters, and mechanics) (e.g. Cross, 1988). One of the striking findings of Atkinson and Meager's (1986) survey was the large number of employers interested in increasing the flexibility of these criteria, and in merging skill categories.

The flexibility agreements mentioned above have been one avenue down which such changes have been pursued, but their slow progress, and their limited coverage attest to the difficulty of obtaining agreement on changes in OLM-related rules. Although it is likely that many employers have sought to introduce change outside such procedures, the evidence of workplace surveys suggests that incumbent workers and their representatives have retained the means to oppose such changes (Marsden and Thompson, 1990), which would impose a limit to the extent to which employers can alter such rules unilaterally. Attempts at industry-wide change have been made, notably in the shipbuilding and the engineering industries, but it has proved difficult to coordinate the large number of interests involved. In shipbuilding, an industry-wide agreement was reached, and then implemented in a series of local agreements, but in engineering, acceptance of a general agreement on flexibility and job demarcations has proved elusive.

Coordination of adaptation of OLM skills has also proved slow in West Germany. Although the law of the reform of industrial training was passed in 1969, by 1975 agreement had been reached only in the construction industry, the other industries taking even longer (Streeck, 1985).

This suggests that in both OLMs and ILMs there has been scope for altering the rules governing job structures in response to economic changes, and that the main constraint on the ILM side has been to maintain promotion opportunities, and on the OLM side, to maintain the pattern of jobs and training which enable transferability.

Institutional adaptability

The third question concerns the adaptability of the institutional structures regulating training and labour markets to changing economic demands. An important element in their adaptation lies in the possibility of revising training norms, and of integrating new knowledge into the system. Two major channels have been used in recent years: the first has been the expansion of school-based theoretical training to boost the level of competence of intermediate-level vocational skills, and the second has been that of adult or further training. The first enables an updating of basic training, and the second an updating of the skills of the current stock of skilled workers.

In the case of internal labour markets, one way of increasing the technical competence of skilled workers is by the provision of school-based training, although its effect is confined mainly to new cohorts of skilled workers. This path has been actively pursued by successive French governments which have sought to improve the quality of skills by expanding school-based vocational training, notably by improving the CAP, and introducing its sister qualification, the BEP. Both of these provide effective channels for learning and for keeping up with the new technical requirements, but it has proved hard to integrate these qualifications into skilled jobs. In practice, the internal labour market logic has prevailed when young people with these qualifications enter the labour market, and they have experienced a period of initial occupational downgrading as they have been taken on in unskilled and semi-skilled positions (Marsden and Germe, 1990). Of the young males who left school in 1983 with the CAP or BEP, the biggest single group entered non-skilled jobs (Coeffic, 1987). Subsequently, many are upgraded to skilled positions by the age of 25 (Pohl and Soleilhavoup, 1981), but the process is not without wastage: some qualified young workers do not make the transition, and for those who do, there is a period of underutilization of their skills. The uncertainties of such initial downgrading would seem to be a severe disincentive to young people thinking of studying for such qualifications. In addition, such qualifications do not guarantee skill transferability as a high proportion of skilled workers who change firms also experience such occupational downgrading (Eyraud, Marsden and Silvestre, 1990).

The reason for resistance to taking on such qualified young workers is not hard to understand. Adult workers who have reached skilled positions themselves by upgrading, many of whom will not have these vocational qualifications, will see their own investment in on-the-job training, or even in simple queuing for the better jobs, as threatened. In addition, the provision of work experience by employers is costly, so integrating its acquisition into their existing ILMs is one way of ensuring that their investment will not prove very mobile.

An earlier example of reform of the French industrial training system is provided by the attempt to establish a German-style apprenticeship system during the 1930s (Maurice, Sellier and Silvestre, 1978). This also foundered on the inability of existing ILMs to integrate the newly trained workers, combined with opposition from the national educational system.

As mentioned earlier, the apprenticeship systems of both Germany and the United Kingdom have experienced a major injection of theoretical training in the 1960s, and this has continued to provide a means by which updated technical knowledge can be fed into existing training provision. In view of the high percentage of apprentice-trained workers who enter skilled jobs on qualifying, this system would appear to be fairly effective.

In Britain, reform of the apprenticeship system during the 1960s included a major injection of theoretical training such that, although apprentices still did not require to pass an exam on completion of the apprenticeship itself, an increasing number acquired other related diplomas during the course of their training (e.g. City and Guilds Diplomas). This injection increased the cost to employers by raising the standards of the supervision they had to provide, and by increasing the apprentices' need for regular time off work to study. Yet the institutional status of apprentices *vis-à-vis* other young workers was left unclear with the result that correspondingly reduced apprentice allowances seemed unattractive. Indeed, during the 1960s and early 1970s, apprentice pay as a percentage of skilled pay rates rose to double that in Germany. As a result, Britain emerged with a high-quality high-cost apprenticeship system against a background in which free-rider pressures were already threatening existing apprentice training.

The injection of greater theoretical training took place in West Germany at about the same time, and was also not without its problems. In a case study of the reform in the construction industry, Streeck (1985) showed that in the late 1960s, the industry was faced with an acute shortfall of apprentice numbers, young people being discouraged by the low allowances, poor employment prospects, and the difficulty of finding a job on account of the large number of fragmented trades. After a protracted negotiation, the federations of artisan and of industrial employers and the unions agreed reforms in 1974. These raised apprentice allowances, and reduced the fragmentation of trades by reducing the number of different trades, and introducing a common core of training in construction skills, thus increasing skill flexibility. One of the difficulties was that the big civil engineering firms had different skill requirements from those of the artisan firms and the reformed apprenticeship system had to satisfy both. The employers also agreed on a levy on themselves to contribute towards training costs. Agreement in other branches was not so easily obtained, and the government introduced, but did not apply, a law which included a levy for apprentice training in 1976. It repealed it in 1980 after taking note of a big increase in the number of apprentices, but the threat of application was enough to stimulate the employers' organizations to act in order to keep the government out of an area they considered to be best left to voluntary arrangements.

The relative success of adaptation of the West German apprenticeship system suggests that adaptation can be achieved within such a system despite the public-good problems of coordination, and thus that OLMs need not be a brake on economic adjustment.

Adaptation of skills among the existing stock of skilled workers as opposed to the new cohorts is the field of adult and of further training. In this case, the apprenticeship-based OLMs appear to be more exclusive than ILMs on account of the low rates of pay and long period of training for apprentice places. Government intervention, for example, to subsidise apprentice places for adults would threaten the delicate consensus on which lower trainee rates of pay rests, and without these it is difficult to see how older workers, with family responsibilities, could afford the investment. In contrast, ILMs do not have special trainee rates except for pay systems which reward length of service. In some respects these provide employers with a *de facto* discount on inexperienced employees, but the rates are considerably above those of apprentices. Thus, ILMs would seem to allow more scope for late entry by people who have retrained, and who are prepared subsequently to await promotion to more skilled jobs. This would explain part of the success of the French continuous training programme (*formation continué*). In West Germany, the fact that most of the further training goes to already skilled workers suggests that it is not used to help semi-skilled workers make up for earlier missed opportunities.

Conclusions

Economic performance, or competitiveness, has not been defined very precisely in this chapter, and the link between this and training systems is only tenuous. The basis of this guesswork is that the capacity to adapt to technical and organizational changes which are driven by product market changes facilitates adjustment by firms, and so helps their market performance. Some recent empirical support for this intuition is provided by the studies of Steedman, Prais and Wagner (notably, Steedman and Wagner, 1987), and by Finegold and Soskice (1988). A fuller treatment of the subject would require consideration of labour supply changes, and also of the moral aspects of the different kinds of exclusion that different labour market systems operate.

Nevertheless, this chapter has aspired to show that training systems can be flexible in different ways, and that this owes much to their links with labour market structure and with the jobs to which trained workers will eventually go. Consideration of the impact of labour market structure reveals the narrow limits within which successful reform of training systems is constrained. The failure of the French attempt to set up a German-style apprenticeship system, and the lack of penetration of the current artisan apprenticeship system into skilled industrial employment illustrate how difficult it is to mix systems. Equally, the experience of 'occupational downgrading' of qualified young labour market entrants in France illustrates the difficulty of grafting school-based training onto a system of ILMs. Apprenticeship systems have perhaps proved more able to incorporate a greater component of theoretical training, but as this increases the cost of apprenticeships to employers, the weak link of reform has proved to be apprentice pay.

The other aspects of flexibility also deserve mention. The greater functional flexibility which appears to be associated with ILMs provides an opportunity for adaptation to localized technical and organizational change. The growth of further training in Germany showed that such flexibility was not confined to ILMs, but, according to some researchers at least, there was a danger that further training could undermine the OLMs on which the basic apprenticeship rested. Greater functional flexibility in OLMs may be more effectively pursued by means of broader-based skills, as were adopted in the German construction industry, than by adding on bits of extra training which can gradually erode skill transferability.

The seriousness of constraints on the reform of training was highlighted by the second type of flexibility discussed, that of job structures. It appears that existing job structures have a profound influence on how the tasks associated with new technology are allocated. This could be seen in the new category of workshop technician introduced in French plants, and in the tensions over the division of new programming tasks between workers of different craft skills, semi-skilled workers and technicians in Britain. These examples show both that change was indeed taking place, but that it was shaped by the existing systems. As such, the new jobs would be likely to remain within the existing patterns of labour flexibility.

Thus the most fruitful paths for developing new skills and for reforming any given training system will be determined by the existing pattern of jobs and labour market structures prevailing in the sectors concerned. International comparisons can be a fruitful source of ideas, but they cannot be transplanted mechanically without taking account of existing institutional and labour market structures in borrowing countries.

Notes

1. Training systems are usually specific to large sectors of individual national economies. This is to be explained partly by the role of the state as a provider and organizer of education and of much training in many countries, and by 'public good' problems which often occur when there is private provision. For this reason the chapter relies heavily on international comparisons (within Western Europe) in order to establish the link between training systems and labour market structure, and then to evaluate some of the contributions they may make to competitiveness.
2. The Engineering Industry Training Board is currently in the process of being wound up (1991), in favour of the new system of employer-led Training and Enterprise Councils (TECs).
3. We take an occupational market to relate to persons endowed with a particular skill or qualification, validated by a diploma, or by the opinion of their peer group. An internal labour market may be said to exist for a particular job when an employer regularly fills vacancies for that job from among existing employees.
4. An example would be the British Young Workers' Scheme which provided a marginal employment subsidy to employers of young unemployed people provided that they were engaged at below a particular level of pay. The philosophy of the scheme was to price young workers back into employment, but this would inevitably be low-wage employment.
5. In Britain, in industry, the strong position of skilled workers in most firms has meant that they could prevent employers upgrading workers who had not served an apprenticeship (for recent evidence of the small amount of upgrading and additional training to skilled jobs see Millward and Stevens (1984) and the Warwick Survey (Brown, 1981)). This limited the rewards employers could offer semi-skilled workers in return for accepting on-the-job training, and so inhibited the growth of this kind of training.
6. As Lorenz's (1987) study of the British and French shipbuilding industries in the first part of this century suggests, reliance on one or other type of labour market structure is long-standing, and given the institutional dependence of OLMs, it would seem that the predominance of one or other type of labour market structure in major sectors and in particular countries is historically contingent. In other words, ILMs have tended to predominate in France and Italy almost by default because of the absence of OLMs. Indeed, in the 1930s, the French government tried to set up a German-style apprenticeship system, but it failed, in part because of the lack of suitable jobs (Maurice, Sellier and Silvestre, 1978).

7. An additional control over the link between vocational training and the jobs done by German workers derives from the close link between the wage groups specified in the industry agreements and qualifications. These carefully set out just how far a worker may progress into higher-wage groups without gaining additional qualifications. Their application is closely supervised by the German works councils.

References

Atkinson, J. and Meager, N. (1986), *Changing Patterns of Work*, National Economic Development Council, London.

Brown, W. (ed.) (1981), *Changing Contours of British Industrial Relations*, Blackwell, Oxford.

Capdeville and Grapin (1975), 'L'insertion professionnelle à la sortie du système scolaire: quelques exemples sur la période récente', *Economie et Statistique*, no. 81–82, pp 57–72.

Coeffic, N. (1987), 'Le devenir des jeunes sortis de l'école', *Données Sociales*, INSEE, Paris.

Combes, M-C. (1988a), 'L'apprentissage en France, *Document de Travail CEREQ*, no. 33, March, CEREQ, Paris.

Combes, M-C. (1988b), 'La loi de 1987 sur l'apprentissage', *Formation Emploi*, no. 22, April–June, pp. 83–100.

Cross, M. (1983), *Changing Requirements for Craftsmen in the Process Industries*, Interim report, Technical Change Centre, London.

Eyraud, F., d'Iribarne, A., Maurice, M. (1988), 'Des entrerprises face aux technologies flexibles: une analyse de la dynamique du changement', *Sociologie du Travail*, no. 1–88, pp. 55–77.

Eyraud, F., Marsden, D.W. and Silvestre, J-J. (1990), 'Occupational and internal labour markets in France and Great Britain', *International Labour Review*, forthcoming.

Finegold, D. and Soskice, D. (1988), 'The failure of training in Britain: analysis and prescription', *Oxford Review of Economic Policy*, vol. 4, no. 3, pp. 21–53.

Garonna, P. (1986), 'Youth unemployment, labour market deregulation and union strategies in Italy', *British Journal of Industrial Relations*, vol. 24, no. 1, March.

Géhin, J-P. (1989), 'L'évolution de la formation continue dans les secteurs d'activité (1973–1985)', *Formation Emploi*, no. 25, March, pp. 19–38.

Germe, J.F. (1986), 'Employment policies and the entry of young people into the labour market in France', *British Journal of Industrial Relations*, vol. 24, no. 1, March.

Guest, D. (1987), 'Human resource management and industrial relations', *Journal of Management Studies*, vol. 24, no. 5, September.

ISFOL (1982), 'L'apprendistato in Italia: problemi attuali e prospettive. ISFOL Quaderni di Formazione', no. 80, May–June.

Lorenz, E. (1987), 'Les chantiers navals en France et en Grande-Bretagne 1890–1970', *Le Mouvement Social*, no. 138, January–March, pp. 21–44.

Maase, M., Sengenberge, W. and Weltz, F. (1975), *Weiterbildung-Aktionsfeld für den Betriebsrat*, Campus Verlag, Frankfurt.

Mace, J. (1979), 'Internal labour markets for engineers in British industry', *British Journal of Industrial Relations*, vol. 17, no. 1, March, pp. 50–63.

Marsden, D.W. (1986), *The End of Economic Man? Custom and Competition in Labour Markets*, Wheatsheaf, Brighton.

Marsden, D.W. (1989), 'Institutions and labour mobility: occupational and internal labour markets in Britain, France, Italy, and West Germany', forthcoming in

Brunetta, R. and Dell'Aringa, C. (eds), *Markets, Institutions and Cooperation*, Macmillan, London.

Marsden, D.W. and Germe, J.F. (1991), 'Young people and entry paths to long term jobs and occupations in France and Great Britain', forthcoming in Ryan, P., Edward, R. and Garonna, P. (eds).

Marsden, D.W. and Ryan, P. (1988), *Youth Labour Market Structures and the Quality of Youth Employment in Major EEC Economies*, report for the Joseph Rowntree Memorial Trust, August.

Marsden, D.W. and Ryan P. (1990), 'Apprenticeship and labour market structure: UK youth employment and training in international context', *British Journal of Industrial Relations*, forthcoming.

Marshall, A. (1920), *Principles of Economics*, eighth edition, Macmillan, London.

Maurice, M., Sellier F. and Silvestre, J.J. (1978), *Production de la hiérarchie dans l'entreprise: recherche d'un effet sociétal France-Allemagne*, mimeo, Laboratoire d'Economie et de Sociologie du Travail, Aix-en-Provence.

Maurice, M., Sellier, F. and Silvestre, J.J. (1986), *The Social Foundations of Industrial Power*, MIT Press.

Maurice, M., Sorge, A. and Warner, M. (1979), 'Societal differences in organising manufacturing units. A comparison of France, West Germany and Great Britain', *International Institute of Management Working Paper IIM/79–15*, Berlin.

Maurice, M., Eyraud, F., d'Iribarne, A. and Rychner, F. (1988), *Des entreprises en mutation dans la crise: apprentissage des technologies flexibles et émergence de nouveaux acteurs*, Laboratoire d'Economie et de Sociologie du Travail, Aix-en-Provence.

Mouy, P. (1983), 'La formation professionnelle initiale des ouvriers et l'évolution du travail industriel', *Formation Emploi*, no. 1 January–March, pp. 52–70.

Noll, I., Beicht, U., Boll, G., Malcher, W. and Wiederhold, Fritz S. (1984), *Nettokosten der betrieblichen Berufsbildung*, Schriften zur Berufsbildungsforschung, Band 63, Beuth Verlag GMBH, Berlin.

OECD (1988), *New Technologies in the 1990s: A Socioeconomic Appraisal*, OECD, Paris.

Osterman, P. (1980), *Getting Started*, MIT Press, Cambridge, Mass.

Pohl, R. and Soleilhavup, J. (1981), 'Entrées des jeunes et mobilité des moins jeunes', *Economie et Statistique*, no. 134, June, pp. 85–108.

Prais, S. (1981), 'Vocational qualifications of the labour force in Britain and Germany', *National Institute Economic Review*, no. 98, pp. 47–59.

Ryan, P. (1987), 'Trade unionism and the pay of young workers', in Junankar, P.N. (ed.), *From School to Unemployment: The Labour Market for Young People*, Macmillan, London.

Ryan, P. (1988), 'Youth interventions, job substitution and trade union policy in Great Britain, 1976–1986', in Rosenberg, S. (ed.), *The State and the Labor Market: Employment Policy, Collective Bargaining and Economic Crisis*, Plenum, New York.

Ryan, P. (1990), 'Trade union responses to special employment and training measures in Great Britain: regulated inclusion or exclusion?', in Ryan P. *et al*. (eds).

Ryan, P., Garonna, P. and Edwards, R. (eds) (1991), *The Problem of Youth: The Regulation of Youth Employment in Advanced Economies*, forthcoming, Macmillan, London.

Steedman, H. and Wagner, K. (1987), 'A second look at productivity, machinery, and skills in Britain and Germany', *National Institute Economic Review*, no. 122. November, pp. 84–95.

Streeck, W. (1985), *Die reform der beruflichen Bildung in der westdeutschen Bauwirtschaft 1969–1982*, WZB, Berlin.

Venning, M., Frith, O. and Grimbley, C. (1980), *The Craftsman in Engineering*, Engineering Industry Training Board, Watford.

Conclusions

Chris Freeman and Luc Soete

I

It is rare but not unknown that an international organization sets at some stage the future research agenda in a particular area. With respect to the issues addressed in this book the OECD undertook such a leading role in initiating a programme called 'TEP' (Technology Economy Programme) in 1988. This programme, which came to an end in 1991, brought together academic researchers, policy-makers and businessmen and women from all OECD countries and sometimes from beyond the OECD area. The numerous conferences organized in different parts of the globe each addressed a different aspect of the complex technology economics interaction.[1] In doing so the OECD not only set a new policy agenda of direct relevance to member countries' policy-makers, it also initiated new theoretical debate about some of the dynamic features of the technology economics interaction and provided a forum for that debate. This book has presented a small selection of the key contributions which illustrate the new approach to technical change and economic growth. The greater realism about learning and innovation in the new theories means that a much more fruitful dialogue between policy-makers in different spheres is now possible.

The TEP policy agenda suggested by the OECD follows on from longer-term policy concerns in science and technology policy expressed, for example, by various national policy-makers that science and technology policies often appear to operate in total isolation from other policies such as education, manpower, labour market, industrial, trade, energy, environmental, etc., and often failed therefore to become integrated into a broader policy framework aimed at improving society's adjustment and capacity for long-term welfare improvement and change. From this perspective, many of the TEP concerns are not really new, but have been at the core of much policy debate within the OECD and member countries over the last decades. A number of OECD reports have been particularly influential – see among others the Brooks report (OECD, 1971), the Report on Technology Policy and Economic Performance (OECD, 1980) and the more recent Sundqvist Report (OECD, 1988). In contrast to these previous reports though, the policy agenda raised by the TEP exercise (as summarized in Soete, 1991) addresses a far greater number of issues and involves to a far greater extent the horizontal expertise of OECD and national policy-makers in different fields of activity where issues of the social and economic integration of technological change in society have become important. The new developments in economic theory greatly facilitate this dialogue and this integration.

Indeed, it is precisely at the academic theoretical level that the OECD initiative raised a number of new challenges. As the contributions in Part II especially bring to the forefront, the explicit recognition of some of the particular endogenous growth-inducing features associated with technical change, under the terminology of 'increasing returns' (Aghion and Howitt), 'learning' (Amable), 'creative destruction' (Gaffard) or 'cumulativeness' (Chiaromonte and Dosi), have led to a significant shift in economists' understanding and modelling of the dynamics of countries' growth performance. The influence of technical change on economic growth can indeed no longer be understood in terms of some long-term exogenous parameter ('manna from heaven') or some automatic catching-up process to the world's leading productivity performance, but must be understood as an endogenous, country-specific process, which in some cases could lead to a 'virtuous snowball' circle whereas in other cases it might lead to vicious low-growth performance. Growth divergence rather than growth convergence is in other words the most likely outcome of the dynamics of the wealth of nations. These new theoretical and analytical insights, referred to in the economics literature as 'new' growth theory or 'evolutionary' economics call also for policy action both at the macro- and microeconomic level.

The policy agenda of the 1980s was dominated by the view that the gains from the convergence of macroeconomic policies, orchestrated through multilateral coordination in a context of increased integration of financial markets, could not be lastingly realized and equitably distributed without the establishment of favourable structural conditions through the reform of microeconomic policies. At the end of the 1980s, the adoption by the OECD of a 'structural adjustment' strategy constituted the first step in the definition of an international consensus in the microeconomic sphere. At that time, the agreement related to the necessity and basic means of eliminating so-called rigidities in the functioning of markets blamed for the underexploitation of each country's growth potential. The great potential of revolutionary new technologies, particularly information and communication technology and biotechnology, would in this way be unleashed.

As in other areas, the focus of the suggested policy action was related to the removal of various microeconomic impediments to the realization of the potential growth offered through technological change, and again associated with 'rigidities' in the functioning of markets. The basis for such policy action at the *international* level lay in the idea of a three-fold *convergence*: convergence of objectives – all countries had to solve common problems in order to unfetter their growth potential; convergence of interests – potential growth at the international level would converge and appear more balanced; and convergence of solutions – the most effective remedies were also those that seemed most generally applicable.

At the beginning of the 1990s, the need emerged to complement this approach, which focused on the revitalization of market forces, by an approach focused far more on the assessment and international coordination of microeconomic technology policies to enhance the potential of growth itself, either through their influence on the quality of the socio-institutional environment in which the revitalized competitive mechanisms operate, or through more direct support to the so-called strategic industries. In analyzing the implications of the new characteristics of innovation processes for growth and development, TEP paved the way for this new phase of evaluation and orientation of microeconomic policies coordinated at the international level. With the new theoretical and analytical insights into the dynamics of the relationship between technology and international growth performance, it is indeed the idea of diversity and of *diverging* patterns of growth that is starting to resurface at the heart of the political debate. As reviewed in some of the chapters of Part II, the numerous recent theoretical contributions which have attempted to formalize a more endogenous and (from a technical progress perspective) more realistic theory of growth have come up with the conclusion that there may be divergences in the underlying trends accompanying national development paths unless the inequality of the initial conditions is corrected. These divergences are more fully explored in Part IV. Such theories have of course a particular bearing on the role of technological accumulation and the importance of international access to science and technology.

As a result, the previous focus within the framework of structural adjustment on improving or even restoring the system of incentives governing the use of productive resources is now in need of being broadened to include also the actual stimulation of the renewal and creation of such resources. From an international OECD perspective though, the wide scope for offensive policies in this area brings to the forefront the possible need for some general international rules of the game for the promotion of technological development, and more generally, for the 'strategic' policies regarding competitiveness that took shape in the vacuum left by the diminished importance of the old defensive adjustment policies, as the chapter by Amable tends to suggest.

Justification for such more offensive orientation of government economic policy is indeed further accentuated once international competitiveness, and with it the objective of strengthening the nation or region's development potential, is introduced. The scope of government policies is vast and will, as we discuss below in more detail, be very much dependent on the particular institutional country context. In most cases, it includes not only encouraging industrial firms to adapt to new technology by promoting R&D and encouraging technological diffusion or cooperation in the field of long-term generic research, but also the consolidation of the regional base for economic development, the strengthening of the web of small and medium-sized enterprises, and sometimes even the development of activities deemed to be of strategic importance for domestic economic growth, such as the networking activities discussed in Part III, in addition to supporting basic research, which finds new and powerful justification in the light of the analysis in Part I.

From the older, more static 'structural adjustment' perspective it is of course hard to assess the compatibility of such policies at the international level. In particular, the criterion of 'three-fold convergence' mentioned above does not apply any longer. The convergence of objectives and solutions is indeed not necessarily matched by that of interests, since it may lead to costly redundancy of efforts to develop certain activities that all parties perceive as strategically important. In contrast, the diversity of national approaches discussed in Part IV does not necessarily guarantee that they will be complementary, given the fact that they may differ in terms of both rationale and scope.

It is consequently not easy to draw general policy conclusions from these new-found insights. In line with the chapters discussed in this book we may consider the following points.

First of all, there are primarily *domestic* policy issues, which focus on the generation and/or strengthening of *national* or even *local* technological capabilities; the access to networks; the diffusion and use of technology to both large domestic and foreign firms as well as small and medium-sized firms; the training and development of national and local human resources and management capabilities. The list is far from exhaustive. The TEP exercise has brought to the forefront a number of traditionally independent, yet interconnected areas which form, so to speak, the national or local infrastructural framework within which basic science, the so-called 'transfer sciences' and innovations are being developed and diffused.

Second, and at the other end of the policy spectrum there are now an increasing number of 'global' issues which call for coordinated action and effort in the S&T area at the global level. World environmental problems are one example, in a list which includes health, famine and many other areas where national frontiers do not matter. A similar case can be made for the so-called 'big science' research activities, where no single country can any longer cover the full variety and diversity of scientific disciplines, approaches and methodologies, let alone the rapidly increasing equipment and material costs of such research. The advantages of international coordination and collaboration in these areas is obvious and many international initiatives have already been taken, even though they still fall far short of what will ultimately be required.

However, there is another set of 'globalization' issues where neither the analysis nor the policy implications are anything like so clear cut. There is a rapidly growing set of policy concerns about national policy actions on the one hand, which could have major international implications and on the other hand 'globalization' trends, e.g. in the production and technology exchanges of so-called 'multi-national' firms, which might question and undermine national policy-making. Policy conclusions take two forms here: on the one hand, those aiming at harmonizing a number of national policies in the area of competition, industrial and technology policy; on the other hand, proposals for setting up more supra-national monitoring and regulatory frameworks, within which fair and unfair actions could possibly be judged.

The desirability, feasibility and scope of such policies is neither the purpose nor the subject of this book. Rather, the chapters included here bring to the forefront and clarify the theoretical framework which defines the need for such policies. In the second section of these 'Conclusions' we therefore concentrate on these issues, because of their importance from a policy angle as well as their great interest from the standpoint of theory.

II

The issue of *globalization* is not specifically addressed except in the chapter by Lundvall. As he points out, if uncertainty, localized learning and bounded rationality are introduced as basic and more realistic assumptions about microeconomic behaviour, rather than the traditional assumptions of perfect information and hyperrationality, then it must follow as suggested above that local and national variations in circumstances could well lead to different paths of development and often to increasing diversity rather than the standardization and convergence which might be predicted using different assumptions.

At first sight the activities of multinational corporations might appear to offer a powerful countervailing force to this local variety and diversity. The largest corporations in the world, whether their original domestic base was in Europe, the United States, Japan or elsewhere, have almost all been investing in many different new locations. This investment, even though initially it may have been in distribution and service networks, has in the last few decades shifted increasingly to include production facilities. Whilst by far the greater part of the investment has been within the OECD area itself and could be more accurately described therefore as 'Triadization' rather than 'globalization', it has also flowed, even though very unevenly, to countries of the Third World and there is now a small trickle to the former socialist group of countries.

As Harry Johnson (1975) long ago pointed out, in this sense the multinationals do indeed unite the human race. Since the basic laws of physics, chemistry, biology and other sciences apply everywhere, there is an underlying unified technology which can in principle be applied anywhere with identical or very similar results. Insofar as large 'global' multinational corporations (MNCs) are able to sell their products and services world-wide and to produce them in many different locations, they can and do act as a very powerful agency tending towards the world-wide standardization of technology and output. As the model developed in Callon's chapter indicates, the diffusion process tends to enhance similarities between adopters.

Even in the case of consumer goods where it might be reasonable to suppose that there might continue to be wide variations in consumer tastes, we are all sufficiently familiar with such products as 'Coca Cola' and such services as those provided by McDonalds to recognize the reality of such global production and distribution networks, offering standardized products and services world-wide. Is it not realistic to suppose that an ever-larger proportion of world production and trade will take this form? Supporting such a view are not only the obvious examples of hotel chains, soft drinks, canned beer, tourist agencies and credit cards but theoretical economic arguments based on static and dynamic economies of scale in production, advertising, marketing, design and finance, as well as the ability of large multinationals to take advantage of surviving differences between nations in costs of capital, labour, energy and other inputs.

However, several chapters in this book suggest that it would be unwise to assume that these tendencies are the only or even necessarily the strongest tendencies within the world economy. Still less do they suggest that these tendencies are so unequivocally desirable that they should be promoted by both national and international economic policies. In fact the arguments for preserving and even encouraging diversity may sometimes outweigh the shorter-term advantages of the scale economies derived from standardization and their propagation through multinational companies, free trade and free flows of investment. In fact both processes (global standardization in some areas but increasing diversity in others) will coexist.

Whilst there are certainly some products and services, such as those already mentioned, where there is indeed a demand which is 'global' in nature and where local variations in taste, regulation, climate and other circumstances can be largely or wholly ignored, there are far more products and services where such variations certainly cannot be ignored without dire consequences. Innumerable examples leap to mind where climatic conditions affect the performance of machines, instruments, vehicles and materials and even more examples are obvious in relation to variations in national standards, specifications and regulations. Whilst it is true that international standardization is a countervailing force through the activities of the International Standards Organization (ISO) and many other bodies attempting to achieve harmonization of technical standards, it is also true that the experience of the European Community over the past 20 years demonstrates the extreme difficulties attending this process in many areas as well as the feasibility in others. And all this still does not take into account the cultural aspects of the problem which deeply affect such areas as food, clothing and personal services. Indeed, even in those products such as soft drinks and franchised restaurant chains, which as we have seen, apparently offer justification for the global standardization arguments, the evidence is by no means all in one direction. 'Kentucky Fried Chicken' has recently been one of the fastest growing companies in Japan, but it was very close to complete failure before it entered into a partnership with Mitsubishi Trading Company and modified its ingredients, its products, its service and its eating facilities to suit Japanese conditions, tastes and preferences. Even in the case of 'Coca Cola', the product mix sold in Japan is certainly not identical to that sold in the United States.

For standard products such as Nestles Instant Coffee there are big variations in productivity of plants located in different countries, even with identical technology and equipment. The same is even more true of industries such as automobiles and chemicals. Among the many sources of such variations are those identified in Part I and Part IV and especially in the chapters by Marsden and Piore – differences in skill availability, management training and traditions and labour organization.

So far we have been discussing mainly the case of established products and pointing to some factors which limit global standardization even in the simplest cases. Advocates of a strong globalization hypothesis would of course accept most of these points, although they might argue that some of them will constantly tend to diminish as the media, travel, education and international organizations all exert their long-term influence. But they would argue that such local variations can easily be dealt with inside the framework of the global strategies of the multinational corporations. Indeed, they would argue that globalization of R&D has already led to local adaptation and modification of products to meet national variations, as a normal and almost routine activity of MNCs. Companies such as Honda go one step further and claim to have a strategy of diversity in world-wide design which goes beyond the simple modification of a standard product to the idea of local variation at the design stage in several different parts of the world. However, the vast majority of Japanese-based MNCs remain essentially Japanese companies with international operations rather than truly international companies and the same is true of US and most other MNCs in relation to their home environment (Hu,

1992). Most R&D activities of MNCs are still overwhelmingly conducted in the domestic base of the company and are heavily influenced by the local national system of innovation. Moreover, ownership and control still remain overwhelmingly based on the domestic platform.

It is at this point that the arguments advanced in this book assume their greatest relevance and importance. As long as we are dealing with a static array of products and discussing only minor variations to adjust to local consumer tastes and environments, then the standardization arguments, the globalization arguments and even some of the simplifying neo-classical assumptions about perfect information are at the borderlines of credibility and usefulness. But once we leave this world and enter the more dynamic world of radical innovations, both technical and organizational, and of extremely uneven and unequal access to new developments in science and technology, then the whole picture is transformed and the neo-Schumpeterian assumptions made by the various authors in this book become essential to attain a realistic vision of what is happening in the global economy. These more realistic assumptions and this more realistic vision are of course essential if economic theory is to be of any help in policy-making.

Lundvall points out in his chapter that, even in the case of continuous incremental innovation in open economies, the drive towards standardization is inhibited by recurrent technological change. Geographical and cultural proximity to advanced users and a network of institutionalized (even if often informal) user–producer relationships are an important source of diversity and of comparative advantage, as is the local supply of managerial and technical skills and accumulated tacit knowledge. Lundvall gives several examples of such localized learning generating strong positions in the world market. Whilst he accepts that MNCs might locate in such 'national strongholds' in order to gain access to the fruits of this interactive learning process, he points out that it is not always simple to enter such markets because of the strength of the non-economic relationships involved. Competing standards for the global market may be important weapons in such situations as well as other forms of product differentiation and quality improvement.

When it comes to radical innovations the importance of institutional variety and localized learning is even greater. Posner's theory of technology gaps and imitation lags is of fundamental importance here. It may be many years before imitators are capable of assembling the mix of skills, the work organization and other institutional changes necessary to launch into the production and marketing of entirely new products and when they do so, as Amable points out in his chapter, there may be important variations in both products and processes. The chapters by Pavitt, Orsenigo, Cohendet *et al*. and Carlsson show very clearly why this is so. The simple availability of published scientific papers or patents is of some help but it is certainly not enough in itself. 'Catching up' in technology is rarely a matter of simply taking technologies off the shelf.

It is of course true that in the global diffusion of radical innovations, MNCs may have an extremely important role. They *are* in a position to transfer specialized equipment and skills to new locations if they so wish and to stimulate and organize the necessary learning processes. They are also in a position to make technology exchange agreements with rivals and to organize joint ventures in any part of the world. It is for this reason that many governments in Europe, as well as in the Third World and the ex-socialist countries have been anxious to offer incentives to attract a flow of inward investment and associated technology transfer from firms based in Japan and the United States.

The chapters in this book show conclusively that such efforts will meet with only limited success unless accompanied by a variety of institutional changes designed to strengthen autonomous technological capability within the importing countries. This is especially true of those generic technologies which have been at the centre of the world-wide diffusion process over the past two decades. Here it is essential to emphasize the interdependencies between innovations and between technical innovations and organizational innovations. A theory of technical change which ignores these interdependencies is no more helpful than a theory of economics which ignores the interdependencies of prices and quantities in the world economy.

The acronym TEP stands not only for the OECD 'Technology Economy Programme' but also for the concept 'Techno-Economic Paradigm' developed by the Venezuelan economist Carlota Perez (1983). In introducing this concept she wished to draw attention to the fact that some changes in technology are so far-reaching that they affect almost every industry and service throughout the system and change the design and management functions everywhere in the economy. Almost everyone would now accept that the process of computerization, based on order-of-magnitude changes in the costs of processing, storing and transmitting information, is indeed an extremely far-reaching set of technical changes which is associated with equally fundamental changes in the organization of production, design, distribution and administration. At the technical level, there is a cluster of still-continuing radical and incremental innovations in semi-conductors, integrated circuits, optical fibres, telecommunications, computer hardware, software and peripherals, electronic instruments and many related products and services. In some ways it is analogous to previous waves of technical change which affected numerous industries and services throughout the economy such as steam power, electric power or oil-based mass-production systems. Whether it is described as a 'technological revolution', a 'new technology system', or a 'change of techno-economic paradigm', (an expression which has the advantage of stressing the interdependence of technology and economics as well as the change in management style), is partly a matter of taste and terminology. There is no standard nomenclature in this area. But it is in any case clear that the interdependencies between clusters of technical innovations, between technical and organizational innovations and between organizational innovations and wider institutional changes cannot be ignored, least of all by economists. Arndt Sorge's introduction to Part IV highlights these connections but they emerge from every chapter in Part IV of the book.

Perez (1983) has pointed out that the social and institutional framework which is hospitable to one set of technologies will not be so suitable for a radically new technology. Whereas incremental innovations can be easily accommodated, this may not be the case with radical innovations which by definition involve an element of creative destruction. When we are talking about large clusters of radical innovations combined with rapid processes of incremental innovation then the problems of structural and social adjustment can be very great. This is quite obvious when we consider such aspects as the change in management techniques and skill-mix which are called for, but it also applies to many other types of institutional change in standards, patents, new services, new infrastructure, government policies and public organizations.

It is in this context that the concept of 'national systems of innovation' assumes such great importance and in the light of this approach it is not surprising that the recognition of the scope and depth of the computer revolution, which was accelerated by the micro-processor in the 1970s, has been followed by a growing recognition of the importance of organizational and managerial change ('multi-skilling', 'lean production systems', 'just-in-time' stock control, worker participation in technical change, quality circles, continuous learning, etc., etc.). This sequence can be clearly seen in the various OECD reports which preceded the TEP exercise and which have already been referred to. The 1980 Report on Technology Policy and Economic Performance recognized the critical importance of generic technologies and discussed various technology policies which might promote their diffusion and efficient application. But it was the 1988 Sundqvist Report which highlighted the central importance of organizational and social change and in so doing also pointed to important interconnections between employment policies, industrial policies, education policies and policies for science and technology (see also Freeman and Soete, 1987, 1991).

The TEP exercise has brought this widening of horizons one stage further. The OECD countries are now increasingly recognizing that far-reaching institutional changes, both at the national and the international level, are needed if the full potential of the new technologies is to be realized. But how does this recognition of paradigm change affect the issue of diversity and standardization and the behaviour of multinationals? Very much – because institutional and cultural diversity is very considerable. Sorge's Introduction to Part IV summarizes very clearly why it is that even when the same or a similar technology is adopted in different parts of the world, its actual mode of application will not be the same (see for example Boyer's (1992) analysis of the national variations in Fordist mass-production systems). Organizational and social innovations will vary in different parts of the world and may be more or less helpful in the effective diffusion of generic technologies. Such innovations may of course also diffuse globally. The in-house R&D department which proved very useful in the diffusion of electrical technology and new techniques in the chemical industry in the 1880s and 1890s eventually diffused in most industries and most countries.

But again, Sorge has shown why this was by no means an automatic or a rapid process. There were huge differences between countries and firms and still are. Moreover, the in-house R&D department is itself changing now to cope with the new technologies. Japanese firms in particular have learnt techniques of integrating in-house R&D activities with production and marketing ('using the factory as a laboratory' to use Baba's (1985) phrase to describe the Japanese electronic firms). Experimentation and learning from experience are just as important in organizational change as in technical change, although less codified and systematic. In both cases pluralism, diversity and competition between alternative approaches have demonstrable advantages.

Software provides an excellent example of the simultaneous value of standardization and diversity both in the technology and the social institutions. All computers need software as well as hardware but this software may either be a standard 'package' or a tailor-made customized product for one or a few users or some combination of the two. In some cases there are great advantages in buying a standard package 'off the shelf'. Russian software users have clearly suffered from the absence of an organized market for such standard packages. But in other cases a standard package may be useless or may actually lead to a fall in efficiency and big losses in down-time. Paradoxically however, the standardization of operating systems and networking packages facilitates greater flexibility and diversity at the level of users and even a good package has to be maintained and up-dated to cope both with the changing needs of the users and with the still rapidly evolving technology. No configuration stands still. This creative adaptation process is likely to be more effective if software skills reside in user organizations (such as banks or steel companies) as well as in hardware suppliers and the software 'industry'. Consequently the differences in structure between US and Japanese software activities with the United States having a much larger software 'industry' may obscure some of the relative advantages of Japanese user organizations with their greater in-house skills.

The diffusion of a new techno-economic paradigm is a trial-and-error process involving great institutional variety. The chapters in this book suggest that there are evolutionary advantages in this variety and considerable dangers in being locked in too early to a standardized technology. A technological mono-culture may be more dangerous than an ecological mono-culture. Even when a technology matures and shows clear-cut advantages and scale economies it is important to retain flexibility and to nourish alternative sources of radically new technology and work organization. As Gaffard's chapter and several others show this is not really 'wasteful' but is inherent in an evolutionary process.

III

National and international policies thus confront the need for a sophisticated approach to a complex set of problems. Policies for diffusion of standard generic technologies are certainly important and these may sometimes entail the encouragement of inward investment and technology transfer by MNCs. But also important are policies to encourage originality and diversity. As Pavitt's chapter in particular has shown, policies for fundamental research are important even for imitative learning. Japanese experience has demonstrated that import of world best-practice technology can always be accompanied by efforts to improve upon that technology based on local skills and local needs. This means that, as the most recent World Bank Report has stressed, education and training policies now move to centre stage in the promotion of world development. The far more realistic approaches to technical change and learning represented in this book have made it possible for economic theory to make a more positive contribution to world development than in the past.

These policy conclusions relate in the first instance to domestic policy concerns, aimed at strengthening the domestic technological capability of national economies. As already indicated in the first section of these conclusions there are, however, a number of features which call for more explicit international coordination and the setting of new priorities at the global level in the area of science and technology. At a general level, there is little doubt that once the increasing complexity of science and technology is taken into account in all its dimensions, the advantages of *international* interaction, networking, coordination and collaboration, not just of privately funded research but also of government-sponsored basic and long-term research, become, so to speak, self-evident. The advantages of international coordination and collaboration are obvious to most scientists and many exchange and collaboration programmes have already been set in motion. However, the policy conclusions falling under this global umbrella – and which can be drawn from the TEP experience – go well beyond the traditional call for more international collaboration in the 'big science' area.

First of all, there is now a more forceful case to be made for such international collaboration, involving proposals for international *burden-sharing* of such 'big science' research efforts. There are major differences between the OECD countries in the amount of public resources devoted to basic research. Such differences are not always closely related to the income levels of such countries, nor to their growth rate. Increasingly the question can therefore be put whether the most developed and richest OECD countries do not have a 'global' responsibility in the burden-sharing and participation in global big science research programmes.

Second, the urgency of some of the truly global problems the world is confronted with: the environment; famine; diseases; desertification; energy; etc. – the list is non-exhaustive – calls explicitly for more internationally coordinated science and technology efforts. In each of these areas it could be said that the social rate of return to the world as a whole is higher than the individual country's social rate of return. Compared to the potential waste of resources at the world level because of the duplication of effort, coordination (*not* harmonization) of research activities beyond national prestige must be an absolute priority.

Third and finally, there is the area of technology assessment where international collaboration and exchange has obvious and immediate advantages. In many OECD countries, governments have been led to take on the risk of carrying out systematic research and analysis on technology assessment. Democratic systems adjust to social change and technical innovation while preserving the checks and balances that guarantee their legitimacy. Thus, in contrast to the old so-called 'socialist' countries, democratic society's adjustment to technical progress is not just the sum of passive responses; it entails open debate, bargaining and a set of policies that come under periodical judgement at the ballot-box. Similarly, the regulation of technical change, subject to electoral scrutiny, is an unavoidable responsibility of government in our societies. The aim of such policy is not only to try to foresee the consequences of technical changes without stifling innovation, but also to regulate some of the possible negative 'externalities' of such technical change. The conditions, nature and diffusion of technical change must be in keeping with social considerations, which raises the problem of weighing risks so as not to go beyond the point where the advantages to certain groups are outweighed by the increase of threats to other groups, the population at large, regions across national frontiers or even the entire world. A more global technology assessment, seems from this perspective particularly needed whereby both the results of research and analysis carried out in one country can be exchanged and used in other countries, *and* a more internationally harmonized regulatory framework could be developed when the negative externalities have clearly international implications.

What makes these new priorities so timely, is not only their global urgency. The recent developments in East–West relations make them also suddenly more feasible and realistic. One of the greatest historical challenges in the coming decade will probably be the transformation of nationally-oriented military research and development sectors to more civilian world-oriented research aims.

Notes

1. For more detailed information on the TEP Conferences and a full list of papers and contributors see OECD (1992).

References

Baba, Y. (1985), 'Japanese colour TV firms. Decision-making from the 1950s to the 1980s', D.Phil. Dissertation, University of Sussex.

Boyer, R. (1992), 'New directions in management practices and work organisation: general principles and national trajectories', CEPREMAP, Paris, mimeo; revised draft of paper presented to OECD Conference, Helsinki, December 1989.

Freeman, C. and Soete, L. (1987), *Technical Change and Full Employment*, Pinter, London.

Freeman, C. and Soete, L. (1991), 'Synthesis report on information technology and employment', MERIT, Maastricht.

Hu, Y.S. (1992), 'Global or transnational corporations are national firms with international operations', *California Management Review*, forthcoming.

Johnson, H.G. (1975), 'Technology and economic interdependence', Macmillan, London.

OECD (1971), 'Brooks report: science growth and society', OECD, Paris.

OECD (1980), 'Technology and economic performance: science and technology in the new economic context', OECD, Paris.

OECD (1988), 'Sundqvist report: new technologies in the 1990s; a socio-economic strategy', OECD, Paris.

OECD (1992), 'TEP. Technology and the economy: the key relationships', OECD, Paris.

Perez, C. (1983), 'Structural changes and the assimilation of new technologies in the economic and social system', *Futures*, vol. 15, no. 5, pp. 357–75.

Soete, L. (1991), 'Synthesis report: technology economy and policy', OECD, Paris.

Index